Response Surface Methodology in Engineering Science

Edited by Palanikumar Kayaroganam

Published in London, United Kingdom

IntechOpen

Supporting open minds since 2005

Response Surface Methodology in Engineering Science
http://dx.doi.org/10.5772/intechopen.90965
Edited by Palanikumar Kayaroganam

Contributors
solomon Worku Kidane, Irina Ioannou, Morad Chadni, Valentin Reungoat, Yang Zhang, Yue Wu, Nam-Ky Nguyen, Mai Phuong Vuong, Tung-Dinh Pham, Jian J. James Zhang, Dr. Amit Kumar Dey, Abhijit Dey, Nefise Gönül Şengöz, José Manuel Pais-Chanfrau, Jimmy Núñez Pérez, Marcos Vinicio Lara-Fiallos, Luis Enrique Trujillo-Toledo, Rosario del Carmen Espín-Valladares, Ichiro Hagiwara, Sankha Bhattacharya, Seyed Reza Omranian, Aliyu Usman, Muslich Hartadi Sutanto, Madzlan Bin Napiah, Nura Shehu Aliyu Yaro, Palanikumar Kayaroganam

First published in London, United Kingdom, 2021 by IntechOpen
IntechOpen is the global imprint of INTECHOPEN LIMITED, registered in England and Wales, registration number: 11086078, 5 Princes Gate Court, London, SW7 2QJ, United Kingdom
Printed in Croatia

British Library Cataloguing-in-Publication Data
A catalogue record for this book is available from the British Library

Additional hard and PDF copies can be obtained from orders@intechopen.com

Response Surface Methodology in Engineering Science
Edited by Palanikumar Kayaroganam
p. cm.
Print ISBN 978-1-83968-917-8
Online ISBN 978-1-83968-918-5
eBook (PDF) ISBN 978-1-83968-919-2

Meet the editor

Prof. Dr. Palanikumar Kayaroganamis currently a professor and principal at Sri Sai Ram Institute of Technology, Chennai, India. He is involved in teaching, research, development, and innovation in the field of higher technical education. He is among the world's top 2% of scientists on a list compiled by Stanford University. He has numerous journal publications to his credit as well as several awards, including the AICTE-Visvesvaraya Best Teacher award and a Global Peer Review Award from Publons Web of Science. His areas of interest include statistical techniques, modeling and optimization, total quality management, composite materials, machining, and additive manufacturing.

Contents

Preface

Response Surface Methodology (RSM) is one of the major methodologies for analysts but not so much for engineers and scientists. However, there has been a paradigm shift and, as such, this book explores the use of RSM in engineering science.

The book includes thirteen chapters with case studies that discuss the basics of RSM and its potential uses and applications in engineering science. Chapter 1 provides an introduction to RSM in engineering science and Chapter 2 discusses machine learning models of RSM.

Chapter 3 highlights the global optimization and surface approximation methodologies along with the applications to eliminate the industrial problems. Chapter 4 centers on response surface design robust against nuisance factors. Chapter 5 deals with the central composite design (CCD) for RSM for possible application in pharmacy. Chapter 6 discusses RSM optimization in asphalt mixtures. Chapter 7 examines the application of RSM for analyzing pavement performance. Chapter 8 emphasizes the optimal condition selection for removal of methylene blue dye implementing a desirability-based RSM approach. Chapter 9 talks about how RSM can be used to optimize agro-industrial processes. Chapter 10 covers optimal laser settings for lithotripsy by numerical response surfaces of ablation and retropulsion. Chapter 11 calls attention to RSM applied to optimize phenolic compound extraction from Brassica. Chapter 12 discusses response surface designs in textile engineering, and Chapter 13 concludes with an overview of the use of RSM in food process modeling and optimization.

This volume is designed to counter the notion that RSM is only useful for analysts. It is a useful resource for readers interested in this methodology and those who wish to use it in new and innovative research.

Palanikumar Kayaroganam
Department of Mechanical Engineering,
Sri Sai Ram Institute of Technology,
Chennai, India

Section 1

Introduction

Chapter 1

Introductory Chapter: Response Surface Methodology in Engineering Science

Kayaroganam Palanikumar

1. Introduction

RSM (response surface methodology) is a factorial/experimental design for optimising a strong bond between one or more target variables. This pragmatic method has been recommended by Box and Wilson [1]. Moreover, it is the key component for gaining the finest results through an array of exclusive experiments. To achieve it, normally, the second-degree polynomial model is used. Even though RSM is a hypothetical model, the scientists and researchers practice it for estimations.

RSM helps to examine the connection between the input variables and the responses (output) of any process or system. Its main intention is to advance the response time or to reach the scope of betterment in the work.

2. Response surface methodology - A view

The response surface methodology (RSM) is used to analyse the rapport when the input factors are quantitative. Additionally, the variables (x_1) and (x_2) maximise the yield of a process (Y). To put it briefly, the variables influence the process yield, as mentioned below:

$$y = f(x_1, x_2) + \varepsilon \tag{1}$$

Indeed, RSM is expedient in creating and analysing the system, as the ultimate aim is to optimise the impact response by various factors.

As RSM is very important in the formulation, creation, and implementation of new engineering research and products, it is more predominant in the field of industrial, manufacturing, clinical sciences, material development, social science, and physical, biological, food engineering, and engineering sciences. It is a known fact that it has many applications; therefore, the researchers explore its origin. In addition, many researchers have used RSM for manufacturing and engineering applications. For example, surface roughness and cutting force components are modelled analytically by using the response surface methodology (RSM). Its outcome determines the workpiece's feed rate and its hardness impacts the cutting force components. Nevertheless, the surface roughness is affected by both the feed rate and the hardness of the workpiece [2].

RMS's efficiency depends on the accuracy of y at various points throughout the response surface; therefore, the researcher determines its optimal or improved system's reaction. From the very beginning of the research, the researcher studies the components or variables of the response surface. Correspondingly, non-significant independent variables are isolated from the critical ones with a view to conducting a good experiment. Also, it is necessary to focus on the essential components before the examination study. In addition, Hill and Hunter divide the response surface analysis into four phases [3]. The specifics are as follows:

1. The experiments can be planned as a statistically valid experiment.

2. Compute the coefficients for the response surface equation.

3. Test the equation by using the lack-of-fit test to see if this is correct.

4. Test the response surface regions target by examination.

RSM manipulates the precise design of experiments (DoE), which has recently acquired favour for formulation. Along with it, its statistical approach is used to assess the interaction effect between the process factors than the traditional approach.

RSM (response surface methodology) is a computational and scientific technique for modelling and analysing the situations that consider various factors that impact the desired response and attempt to maximise the result [4]. The connection between the output and the independent variables is unknown in most of the RSM issues [5]. In a similar fashion, RSM estimates the outcome variable 'y' and the set of independent variables 'x' first. In several portions of the response variable, a low-order polynomial is commonly used. The functional approximation in the first-order model is well represented by the independent parameters and it is given below [4]:

$$y = \beta_0 + \beta_1 x_1 + \beta_2 x_2 + \ldots + \beta_k x_k + \varepsilon \tag{2}$$

When the structure is having curvature in the second-order form, a higher-grade polynomial is used [4].

$$y = \beta_0 + \sum_{i=1}^{k} \beta_i x_i + \sum_{i=1}^{k} \beta_{ii} x_i^2 + \sum\sum_{i<j} \beta_{ij} x_i x_j + \varepsilon \tag{3}$$

Approximation polynomials' parameters are estimated by using the least-square approach. So, a surface response analysis is performed on the linked surface. Besides that, the analysis of the installed surface is compared with the analysis of the actual system, when the installed surface is a reasonable approximation to the genuine response function [6].

Consecutively, the model variables envisage the most suitable experimental designs successfully for data collection. Eventually, the surface design is considered as the suitable responsive surface. Undoubtedly, RSM is a method for identifying the system's optimal process parameters or a factor space area that successively meets [7]. Likewise, simultaneous analysis of multiple responses commences with appropriate response surface models for each result. Later, it pursues to optimise a set of operating conditions and keeps all responses within the necessary range at the lower limit [8].

DOI: http://dx.doi.org/10.5772/intechopen.100484

3. Conclusions

From the literature and other details as on date, RSM is extensively used in engineering science and it has found numerous applications. By using it properly, the process output can be improved to many folds and RSM can be utilised as an important technique in engineering and science applications.

Author details

Kayaroganam Palanikumar
Department of Mechanical Engineering, Sri Sai Ram Institute of Technology, Chennai, India

*Address all correspondence to: palanikumar@sairamit.edu.in; palanikumar_k@yahoo.com

References

[1] Box GE, Wilson KB. On the experimental attainment of optimum conditions. Journal of the Royal Statistical Society, Series B (Methodological). 1951;**13**(1):1-45

[2] NandanSit AKJ. Comparison of response surface methodology (RSM) and artificial neural network (ANN) modelling for supercritical fluid extraction of phytochemicals from Terminalia chebula pulp and optimization using RSM coupled with desirability function (DF) and genetic algorithm (GA) and ANN with GA. Industrial Crops and Products. 2021;**170**. DOI: 10.1016/j.indcrop.2021.113769. (In Press)

[3] Myers R, Khuri A, Carter W. Response Surface Methodology: 1966-1988. Technometrics. 1989;**31**(2):137-157. DOI: 10.2307/1268813

[4] Montgomery C. Design and Analysis of Experiments. 8th ed. Wiley: New Delhi; M66 2013;519:478-544. ISBN 978-1118-14692-7.

[5] Noordin MY, Venkatesh VC, Sharif S, Elting S, Abdullah A. Application of response surface methodology in describing the performance of coated carbide tools when turning AISI 1045 steel. Journal of Materials Processing Technology. 2004;**145**:1

[6] Palanikumar K, Karunamoorthy L, Karthikeyan R. Assessment of factors influencing surface roughness on the machining of glass fiber-reinforced polymer composites. Materials & Design. 2006;**27**:10

[7] Alimohammadi HR, Naseh H, Ommi F. A novel framework for liquid propellant engine's cooling system design by sensitivity analysis based on RSM and multi-objective optimization using PSO. Advances in Space Research;**67**(5):202

[8] Yadav RN. A hybrid approach of Taguchi-response surface methodology for modeling and optimization of duplex turning process. Measurement. 2017;**100**:131-138

Section 2

Methods for Response Surface Methodology

Chapter 2

Introducing Machine Learning Models to Response Surface Methodologies

Yang Zhang and Yue Wu

Abstract

Traditional response surface methodology (RSM) has utilized the ordinary least squared (OLS) technique to numerically estimate the coefficients for multiple influence factors to achieve the values of the responsive factor while considering the intersection and quadratic terms of the influencers if any. With the emergence and popularization of machine learning (ML), more competitive methods has been developed which can be adopted to complement or replace the tradition RSM method, i.e. the OLS with or without the polynomial terms. In this chapter, several commonly used regression models in the ML including the improved linear models (the least absolute shrinkage and selection operator model and the generalized linear model), the decision trees family (decision trees, random forests and gradient boosting trees), the model of the neural nets, (the multi-layer perceptrons) and the support vector machine will be introduced. Those ML models will provide a more flexible way to estimate the response surface function that is difficult to be represented by a polynomial as deployed in the traditional RSM. The advantage of the ML models in predicting precise response factor values is then demonstrated by implementation on an engineering case study. The case study has shown that the various choices of the ML models can reach a more satisfactory estimation for the responsive surface function in comparison to the RSM. The GDBT has exhibited to outperform the RSM with an accuracy improvement for 50% on unseen experimental data.

Keywords: response surface methodology (RSM), machine leaning (ML), regression, the least absolute shrinkage and selection operator (LASSO), generalized linear model (GLM), decision trees, random forests, gradient boosting decision trees (GBDT), multiple-layer perceptrons (MLP), support vector regression (SVR)

1. Introduction

Response surface methodology (RSM) is an intersection of the experimental design, the objective optimization and the statistical modeling. RSM aims to identify the adequate mathematical model with the optimal selection of the influence factors to predict the responsive factor and to obtain the extremum of the model under the constrained numerical intervals or categorical levels of the influence factors. The objectives are achieved through designing and conducting a series of experiments to

 IntechOpen

collect the necessary amount of experimental data to approximate the mathematical model.

RSM was originally proposed and described in [1] and some other papers are subsequently published to contribute to the development of the RSM [2, 3]. Review papers began to appear starting from [4] that summarized the utilization of RSM in the chemical and processing fields. It is followed by a few more recent review papers in the application of the physical and science area [5, 6] and the biometric area [7] and some books in the related subjects as well [8, 9]. RSM has quickly gained popularity in empirical modeling in physical experiments to replace or complement the traditional presumed approach [10] where the theoretical knowledge of the experimental systems are available since its appearing. It is as well utilized in modeling the numerical experiments together with the simulation-based methods such as the finite element analysis in the application of the design optimization [11]. RSM contains three skeletal concepts including the estimation of the mathematical model or function, the design of experiments (DoE) and the validation and representation of the postulated mathematical function. Those three steps are likely to be insufficient to conduct only once and will require iteration in practice to achieve a satisfactory result [4].

The most commonly used postulation for the mathematical model is first-order or higher-order polynomials [12, 13]. Despite the wide implementation in chemistry and chemical engineering, it is inevitable that under certain circumstances, the polynomials are inadequate to approximate the underlying RMS functions (e.g. non-linear systems). Especially with the rapid development of information technology in the recent few decades, the utilization of RMS has been spreading to cover many other fields such as civil [14], advanced manufacturing [15, 16], and biomedical engineering [17, 18] and agricultural and food science [19, 20]. Experimental data has become much easier to collect, process and cache, parallel to which is the emergence of machine learning.

Machine learning (ML) is a process of the model building using experimental data or past experience in order to solve a given problem [21, 22]. It enhances the RMS by fitting a rather wide range of approximation models to achieve the responsive surface function. The whole process of ML model construction innately cooperates with the model estimation and model validation stages of the RSM [22]. Additionally, many ML models intrinsically select the most significant influence factors during the model construction process (e.g. the LASSO [23], the GLM and decision trees based models [24, 25]). This nature of the ML models will help to reduce the chance for attempting different DoE to identify the most appropriate influence factor combination and therefore contribute to diminishing the repetitiveness of the three-step cycle and reducing the total experimental runs and cost. Indeed, in cases where the experimental data is extremely expensive or difficult to obtain such as those in the biochemistry field [18, 19], the polynomial approximation and the corresponding DoE methods such as the full or partial factorial design [26], central composite design [27] and the Taguchi's experimental designs [26, 28] are still of utmost importance.

While the ML methods may not be suitable in certain scenarios where expensive time and financial cost are associated with the physical experiments, some techniques are still worth to be explored and utilized to replace or complement the traditional polynomial approach. The objective of the chapter is to introduce some of the linear and non-linear ML models to estimate the responsive surface function under the fair assumption of a reasonable cost and easiness in obtaining enough amount of experimental data. The book chapter is divided into 4 sections, the following section describing the frequently used ML models in detail, Section 3 implementing an engineering example to demonstrate the advantages of those ML models in comparison to the traditional polynomial postulation and Section 4 summarizing the content in the chapter and discussing further research direction.

2. The machine learning approach

2.1 The model construction, validation and testing

Before diving into the various ML models to estimate the responsive surface functions, it is worthy to comprehend the overall model construction and assessment process of the ML approach. Similar to the cyclic three-concept of the RSM, ML also contains an iterative process to reach a satisfactory estimation result.

ML requires the pre-acquisition of a good amount of experimental data (i.e., the number of samples should be more than the number of coefficients to be estimated. The more samples, the better). A model is constructed using part of the data (training set) and assessed on the remaining data (test set). This technique will help to eliminate the risk of the overfitting problem (the model predicts superior on the current data, yet inferior on unseen data), and therefore to ensure the reliability of the constructed model even when new experimental data becoming available.

In addition to the train-test split technique, ML also employs the cross-validation technique to further assist in preventing the overfitting issue and simultaneously help to select the hyper-parameters (parameters that requires pre-definition by the researcher). When deploying the cross-validation technique, the training set is divided into multiple folds of smaller sets. One of the fold is held as the validation set to assess the prediction power of the constructed model and the rest is utilized to construct the model. The process of model building and validation repeats until each sub-fold having been used as a validation set. The final goodness-of-fitness of the model going through the cross-validation technique is computed as the average of each performance values in the loop.

The following part of the section will introduce a few commonly used regression ML models (learning a function mapping from the influence factors to the responsive factor based on available influence-response pair data) as the responsive surface function:

- the advanced linear regression models (the least absolute shrinkage and selection operator model and the generalized linear model).
- the tree-based models (decision trees, random forest and the gradient boosting decision trees).
- a basic type of the Neutral Nets, i.e. the multiple layer perceptrons and
- support vector regression.

2.2 Linear regression methods

2.2.1 Traditional RSM method/the ordinary Least Square (OLS)

RSM favors the low-order polynomial as the postulation of the mathematical function where the coefficients of the polynomial are estimated by finding the optimal solution to minimize the sum of squared error of the observed response values and the predicted response values. In ML term, the traditional RSM approach to approximate the responsive surface function is referred to as the ordinary least squared model (OLS).

Take the first-order polynomial as an example [29], i.e.

$$\hat{y} = \boldsymbol{\beta}\mathbf{X}, \tag{1}$$

where $\boldsymbol{\beta} = (\beta_1, \beta_2, \cdots, \beta_m)$ stands for the m coefficients to be estimated, $\hat{\boldsymbol{y}}$ is the approximated values of the responsive factor and $\mathbf{X}$ is the matrix of the influence factors. The optimization problem is to find the $\boldsymbol{\beta}$ that minimizes the sum of the squared error [29], i.e.

$$arg\,min_{\beta} \|\boldsymbol{\beta}\mathbf{X} - \mathbf{y}\|_2^2, \tag{2}$$

where the l^p-norm of a vector $\boldsymbol{u} = (u_1, u_2, \cdots, u_N)$ is defined as: $\|\boldsymbol{u}\|_p = \left(\sum_{i=1}^{N} |u_i|^p\right)^{1/p}$ [30].

2.2.2 The least absolute shrinkage and selection operator model (LASSO)

LASSO is an upgraded version of the OLS as it allows influence factors selection during the coefficients estimation process and generalization (a technique to avoid overfitting) [23]. LASSO is also a linear regression model yet differs from the OLS by adding an extra regularization term to the loss function to realize the two additional functions [31], i.e. the optimization problem then becomes:

$$argmin_{\beta} \overbrace{\frac{1}{2n} \|\boldsymbol{\beta}\mathbf{X} - \mathbf{y}\|_2^2 + \underbrace{\alpha\|\boldsymbol{\beta}\|_1}_{regularization\ term}}^{loss\ function}, \tag{3}$$

where α is a constant of the l^1-norm regularization term. Other ML models offering similar functions include the bridge model (where the loss function contains an extra l^2-norm term other than a l^1-norm term) and the Elastic-Net model (where the loss function contains both the l1-norm and l^2-norm terms).

2.2.3 The generalized linear model

Another useful linear ML model is the generalized linear model (GLM) which allows estimating the response factor when the residuals of the responses do not follow a Gaussian distribution, e.g. the response factor is always positive, or constant value changes in the influence factors leads to exponential value varying other than constant varying of the response factor. In this case, GLM can be utilized to approximate the responsive surface function. It elevates the OLS by differing in two aspects, the predicted value of the response factor is linked to an inverse function of the linear combination of influence factors, i.e. $\hat{\boldsymbol{y}} = h(\boldsymbol{\beta}\mathbf{X})$ and the residual term in the loss function is replaced by the unit deviance of a reproductive exponential dispersion model (EDM) [32], i.e.

$$arg\,min_{\beta} \frac{1}{2n} \sum_{i=1}^{n} d(y_i, \hat{y}_i) + \frac{\alpha}{2} \|\boldsymbol{\beta}\|_2, \tag{4}$$

where n is the number of samples in the training set. Since the loss function also contains a regularization term, the GLM provides feature selection and generalization during model construction as LASSO does.

Distribution	Response domain	Unit deviance $d(y,\hat{y})$
Gaussian	$y \in (-\infty, \infty)$	$(y,\hat{y})^2$
Poisson	$y \in [0, \infty)$	$2\left(y\ln\frac{y}{\hat{y}} - y + \hat{y}\right)$
Gamma	$y \in (0, \infty)$	$2\left(\ln\frac{\hat{y}}{y} + \frac{\hat{y}}{y} - 1\right)$
Inverse Gaussian	$y \in (0, \infty)$	$\frac{(y,\hat{y})^2}{y\hat{y}^2}$

Table 1.
The unit deviance functions for response factors following various distributions.

Examples of the unit deviance of the EDM are given in **Table 1** [33].

The 1st-order polynomial assumption is made for purpose of simplifying the concept to introduce the above linear ML models. The intersection and quadratic terms of a higher-order polynomial can be easily computed and added to by performing a transformation on the 1st-order polynomial. The linear regression models described above are still suitable to estimate the coefficients of the transformed influence factors to form a higher-order polynomial responsive surface function.

As mentioned, non-linearity may exist between the influence factors and as such, the polynomials will become inadequate to approximate the mapping from the influence factors to the responsive factor. In this case, the non-linear ML models will become distinctive.

2.3 Tree-based methods

2.3.1 Decision trees

Decision trees (DT), instead of the linear regression models described in Section 2.2, is a non-parametric ML model (models defined without coefficients). The problem is to build a model to predict the values of the responsive factor by means of defining a series of decision principles deduced from the training data [29].

A DT model is built in a top-down manner with each split node partitioning the influence factors into a subgroup and the process eventually reaches a value of the responsive factor [24]. The more important the split node, the higher the node in the tree. The common criteria to minimize as to decide the orders for future split nodes are mean squared error, Poisson deviance and the mean absolute error [34]. Assuming to use the Poisson deviance, suppose the data at the node m is represented by Q_m with N_m samples, the loss function at the split node is defined as:

$$H(Q_m) = \frac{1}{N_m} \sum_{y \in Q_m} y\ln\frac{y}{\overline{y}_m} - y + \overline{y}_m, \qquad (5)$$

where $\overline{y}_m$ is the mean of the responsive values at the node m and the optimization problem is to identify the node m that minimizes $H(Q_m)$ [25].

In comparison to the linear regression models, DT can fit non-linear systems due to the nature of how it is constructed. Besides, since the importance of each influence factor is computed during the tree construction process, it also helps to select the most significant influencers as a procedure of the feature selection and thus to improve the prediction accuracy of the resulted model. However, there are hidden drawbacks when implementing the DT [34]:

- Overfitting problem, i.e. an over-complex DT is built, resulting in good prediction in existing data but poor prediction in unseen data.

- Unstable model, i.e. slight variation in the data can lead to a completely different DT.

2.3.2 Random forests and gradient boosting decision trees (GBDT)

In order to address the disadvantages of the DT and to estimate a satisfactory model, ensemble techniques are explored. The ensemble technique combines the prediction of multiple base estimators to achieve reduced variance and enhanced robustness over a single meta-estimator. DT is a common model used as a type of base estimator to form the final meta-estimator. The DT used in an ensemble model is usually simple-structured by limiting the maximum depth of the trees or the maximum leaf nodes of the trees and as a consequence to ease the overfitting issue of using a single DT model.

Two types of ensemble trees, the random forests and the GBDT are introduced here.

Random forest constructs multiple DT with each DT built from a subset drawn with replacement from the training dataset and uses the averaged prediction of the individual DT as the final prediction for the meta-estimator [29]. GBDT builds a sequence of DT with each preceding DT attempting to eliminate the error of the current sequential DT model and uses a weighted sum of the predictions generated by the sequentially built DT to produce the final prediction [29]. More details regarding the two models can be found in [35, 36].

2.4 The neural-nets method

Neutral-nets models are a group of models originally inspired by the biological neural networks and are able to learn a complex function mapping from the influence factors to the responsive factor. Neural-nets models have popularized since their development due to their accuracy in predicting without knowing the underlying relationship between the influencers and the responders and therefore massive descendent models have been published recently. In this section, the first generation and the most fundamental neural-nest model, the multiple layer perceptrons (MLP) is presented [37].

MLP builds a non-linear function approximation to map the set of influence factors to the responsive factor using the training data. Between the influence factors and the final response factor, there may exist one or multiple non-linear layers, as illustrated in **Figure 1** [29].

Each circle in **Figure 1** is a neuron. The leftmost layer is the input layer that consists of the neurons representing the influence factors. Each neuron in the hidden layer is a weighted linear summation of the neurons in the previous layer followed by a non-linear transformation by applying an activation function. The value of the response factor is given by the neuron in the output layer after receiving the values from the last hidden layer and transforming the values using the linear summation and an appropriate activation function. For the MLP regression model, the activation function in the last step is an identity function, i.e. no activation function is applied in the last step. Similar to the advanced linear models, MLP employs the sum of the squared error loss and an additional l^2-norm regularization term as the loss function [38], the optimization problem is thus to identify the $\boldsymbol{\beta}$ that minimizes:

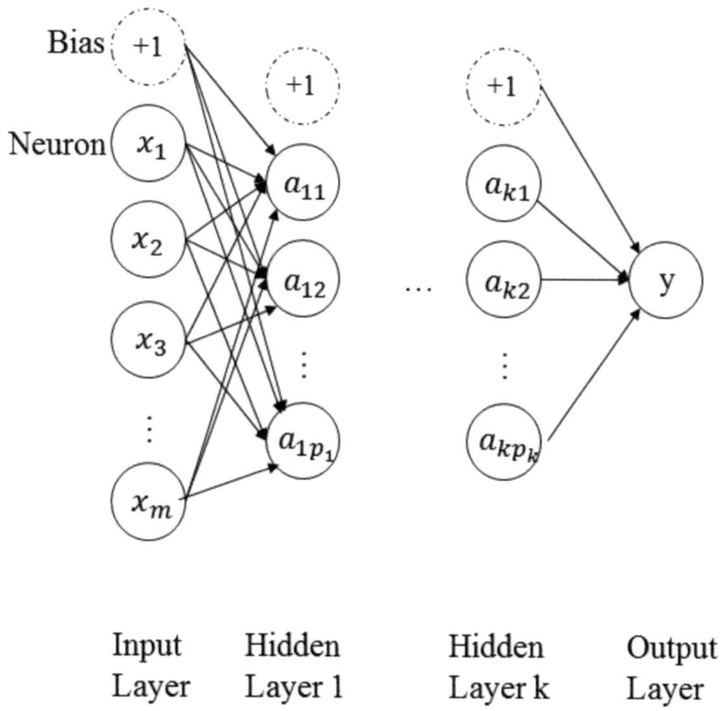

Figure 1.
Structure of a multiple layer perceptrons (MLP) model.

$$argmin_{\beta} \overbrace{\frac{1}{2}\|\beta\mathbf{X} - \mathbf{y}\|_2^2 + \underbrace{\frac{\alpha}{2}\|\beta\|_2^2}_{\text{regulization term}}}^{\text{loss function}}, \tag{6}$$

MLP utilizes the stochastic gradient descent (SGD) to update the coefficients based on the gradient of the regularization term and the loss function at each iteration in order to obtain the optimal values of the coefficients [38], i.e.

$$\boldsymbol{\beta} \leftarrow \boldsymbol{\beta} - \gamma\left(\alpha\frac{\partial R(\boldsymbol{\beta})}{\partial\beta} + \frac{\partial cost}{\partial\beta}\right), \tag{7}$$

where γ is a pre-defined learning rate that controls the step-size in the iteration to reach a local extreme minimum, $R(\boldsymbol{\beta})$ is the regularization term and *loss* is the loss function in Eq. (7) respectively.

2.5 Support vector regression

Support Vector Regression (SVR) was developed in the 1990s [39], a decade late than the surge of the neutral-nets [40]. Unlike the Neutral-nets, which are result-oriented 'black-box' models, the SVR has well-defined underpinned theoretical properties.

Support vectors are the data points from the training set that have a direct determining impact on the optimum location of the decision surface (a hyperplane separating one class of data points from anther) [41]. The SVR outstands when there is a limited number of experimental data, in particular, when the number of samples in the training set is less than the number of influence factors, as the SVR fit a mathematical model by utilizing a subset of the training samples to decide the decision surface. It enhances the linear approximation models by means of fitting non-linear properties in the data as during the model construction process, the influence factors can go through a pre-define non-linear kernel function and as such to realize a non-linear transformation to obtain the response [29].

The procedure to obtain the predicted response values using SVR is similar to the MLP and is illustrated in **Figure 2**. The support vectors are first transformed using and a map φ and then conduct the dot product to evaluate the kernel function $k(\boldsymbol{x}\ \boldsymbol{x}_i)$ (for examples, the Gaussian kernel function is given $k(\boldsymbol{x}, \boldsymbol{y}) = e^{-\frac{\|x-y\|^2}{2\sigma^2}}$). The values of the kernel functions are then added up using certain weights to achieve the predicted value of the response factor.

Assuming for the 1st-order polynomial approximation (as represented in Eq. (1)), SVR searches for the coefficients by minimizing the inner product of the coefficients, i.e.

$$\frac{1}{2}\|\boldsymbol{\beta}\|_2 = <\boldsymbol{\beta}\bullet\boldsymbol{\beta}>, \tag{8}$$

subjected to the condition of limiting the prediction error into a certain threshold, i.e. $|\boldsymbol{\beta}\mathbf{X} - \mathbf{y}| < \varepsilon$, where ε is a pre-defined threshold value. However, it is inevitable that not all data points will fall into the threshold and as such, the equation needs to consider the possibility of errors larger than ε [41]. Therefore, the equation is completed with an inclusion of the slack variable ξ as the deviation from the error threshold ε. The optimization problem then becomes:

$$argmin_{\beta}\left(\frac{1}{2}\|\boldsymbol{\beta}\|_2 + C\sum_{i=1}^{n}\xi_i\right), \tag{9}$$

with constrains $|y_i - \beta x_i| \leq \varepsilon + \xi_i$ for each $i = 1, 2, \cdots, n$, and $\xi_i \geq 0$ [41].

This minimization problem can be resolved by finding the solution of an equivalent Langrangian Dual problem, i.e. finding the ξ that maximize:

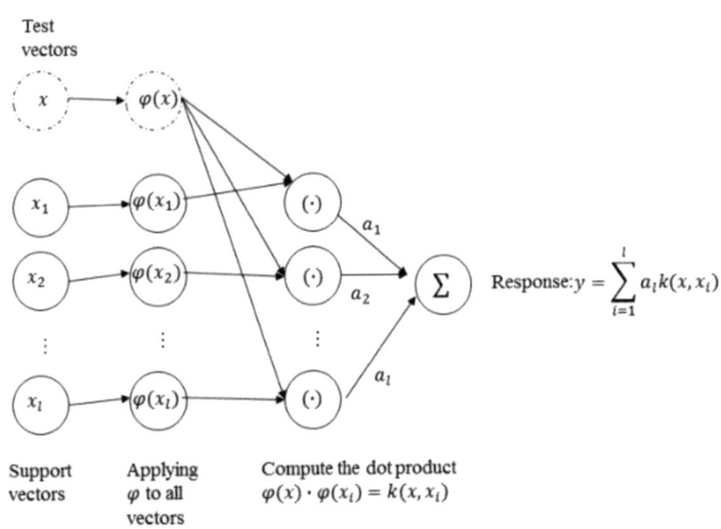

Figure 2.
The architecture of the support vector regression (SVR) to obtain a prediction.

$$argmax_{\xi}\left(\sum_{i=1}^{n}\xi_i - \frac{1}{2}\sum_{i=1}^{n}\xi_i\xi_j y_i y_j k(\boldsymbol{x}_i \boldsymbol{x}_j)\right), \tag{10}$$

subjected to the conditions $\sum_{i=1}^{n}\xi_i y_i = 0$ and $\varepsilon \geq \xi_i \geq 0$. [41]

3. A mechanical engineering case study

The traditional RSM method and the introduced ML models above will be employed to approximate the underlying mathematical model for a set of mechanical engineering data used in [42].

3.1 Description of the experiment data

In the manufacturing processes process, the machine vibration severity level is a critical index to indicate the status of the machine tool and the finish of the cutting material. If a high severity level occurs, the manufacturing process is likely going wrong such as the occurrence of chatter or breakage of a machine tool. However, it is not always loss-effective to have the corresponding devices installed and human technicians in-place to monitor the vibration severity continuously. Instead, a numerical approximation can be explored by collecting machining data, which is easier and less expensive to access.

The experimental dataset contains 56519 samples and 74 variables collected from the sensors installed on the shop floor machine and the central controller computer [42]. To simplify the case to a single response factor problem, the severity variables along different directions measured on various ranges are compacted together using their standardized l^2-norm values, i.e., the 8 severity columns are first standardized by removing their column mean and scaling to the unit variance. The unified columns are then used to compute the l^2-norm in the horizontal direction to get the target response variable column. After removing the two time-related columns, the repeated program number column, finally, the processed data provides 63 influence factors and one response factor.

3.2 Implementation of the traditional RSM and the ML methods

A piece of program code is written in Python language and utilizes a package named Scikit-learn [29] to realize the discussed ML models. The main objective here is to demonstrate the advantages of ML in estimating the response surface function to predict the values of the response factor with higher accuracy compared with the traditional RSM polynomial approximation.

First of all, the dataset is standardized to its unit variance form in the same manner as to compute the single severity column in Section 3.1, i.e., using the formula: $z = \frac{x-\text{mean}(x)}{std(x)}$. This is due to that many models to be implemented later require the standardization of the data to avoid the multiple issues (obscuring the conclusion of the statistical significance of the model terms, producing imprecise coefficients or incorrect model structure) caused by the collinearity.

The standardized dataset (containing the 56512 samples data) is divided into a training set and a held-out test set with a 7/3 ratio. The training dataset is further split into three subfolders with each subfolder utilized as a validation set to estimate the prediction power of the model constructed using the remainder two subfolders to adopt the cross-validation technique described in Section 2.1. Employment of the cross-validation technique will further prevent the over-fitting problem and as such to reassure the robustness of the estimated model and better prediction on unseen data. The training set is used to estimate and select the optimal responsive function and the held-out test set is used to exam the prediction accuracy and the reliability of the fitted RSM function on unseen data. The goodness-of-fitness of the estimated function on the training set and the prediction accuracy of the function on unseen data is measured using three different measuring metrics and the outcomes are shown in the cross-validation results and the held-out test set results sections respectively in **Table 2**.

The three measuring metrics evaluate the model performance from different statistical perspectives to ensure a solid conclusion to be reached. The statistical meanings and the computation equations for the three measuring metrics are given below [43]:

- The explained variance (EV) measures the proportion to which a mathematical model accounts for the variance of a given data set and is computed as

$$EV(y, \hat{y}) = 1 - \frac{var(y - \hat{y})}{var(y)}, \tag{11}$$

where *var* stands for the variance. The optimal possible of an EV value is 1. The lower the value, the worse the model performs.

- The mean absolute percentage error (MAPE) measures the percentage difference between the actual and the predicted response factor values and is defined as:

$$MAPE(y, \hat{y}) = \frac{1}{n}\sum_{i=1}^{n} \left| \frac{y_i - \hat{y}_i}{y_i} \right|, \tag{12}$$

The best possible MAPE value is 0% when every single predicted value matching the actual one. The greater the score, the worse the model performs.

Metric name/model type	**Linear models**						**Non-linear models**			
Model name	**OLS/RSM**		**LASSO**		**GLM**		**RF**	**GBDT**	**MLP**	**SVR**
Polynomial order	1st	2nd	1st	2nd	1st	2nd	N/A			
Cross-validation Results										
EV	0.697	0.841	0.700	0.877	0.700	0.865	0.923	0.924	0.905	0.901
MAPE	31.4%	88.4%	31.4%	19.7%	30.2%	17.8%	13.5%	13.0%	15.3%	14.9%
RMSE	0.660	1.62	0.656	0.630	0.656	0.660	0.333	0.331	0.372	0.352
Held-out Test Set Results										
EV	0.699	−1.08	0.699	0.878	0.699	0.865	0.925	0.926	0.908	0.900
MAPE	31.4%	160%	31.5%	19.7%	30.3%	18.0%	13.2%	12.9%	14.6%	14.3%
RMSE	0.657	7.74	0.658	0.609	0.657	0.639	0.328	0.325	0.367	0.348

Table 2.
Experimental results of applying the traditional RSM method and the ML methods on a set of engineering data.

- The rooted mean squared error (RMSE) measures the actual differences between the actual value and the predicted value from the model and is defined by:

$$\mathrm{RMSE}(y, \hat{y}) = \left(\frac{1}{n} \sum_{i=1}^{n} (y_i - \hat{y}_i)^2 \right)^{1/2} \tag{13}$$

The smaller the value, the better the model performs.

The functions that realize the traditional RSM linear model (i.e. OLS), and the six ML models, including the LASSO, the GLM, the random forests (RF), the GBDT, the MLP and the SVR) in the Sciket-Learn package [29] are implemented and utilized to estimate the underlying mathematical function that maps the 63 influence factors to the target responsive factor for the case study dataset. The technique that generates the 2nd order polynomial terms is also deployed to obtain the polynomial estimation of the traditional RSM function. The goodness-of-fitness and the prediction accuracy of the estimated responsive function using the RSM model or each of the ML models are displayed in the corresponding column labeled with the name of the model in **Table 2**. Particularly, for linear models, both 1st-order and 2nd-order polynomial terms have been attempted.

Furthermore, in order to investigate whether the existing experimental data is enough to achieve a robust and accurate responsive surface function, the learning curve, which determines the cross-validation training and validation accuracy scores under different training sample sizes, is also drawn for each of the models and shown in **Figure 3**. More specifically, using a proportion of the training data to perform the cross-validation technique described in Section 2.1. The mean, the minimum and the maximum of the cross-validation results will be shown on the learning curve for each of the subset used.

3.3 Experimental results and discussion

3.3.1 Accuracy of the estimated responsive surface function

The scores in **Table 2** display the cross-validation accuracy and the prediction accuracy on a held-out test set of the RSM and the ML models assessed under three different measuring metrics, the explained variance (EV), the mean absolute percentage error (MAPE) and the rooted mean squared error (RMSE).

Results in **Table 2** has demonstrated that assuming for 1st-order polynomial, the performance of the tradition RSM method, i.e. OLS, is equally mediocre with the other two linear ML models, LASSO and GLM, as all of them produce similar testing results under each of the measuring metrics. This indicates that a 1st-order polynomial assumption may be not enough to include all information between and within the influence factors, which has been validated by the improved model performance for 2nd-order polynomial approximation using a corresponding ML linear model and using the non-linear approximation models.

Under the 2nd-order polynomial postulation, the two linear ML learning models, LASSO and GLM greatly surpass the RSM method as the two models allow influencer selection during model construction and as such to eliminate the influence factors that interferes with estimating precise coefficients. Therefore, more accurate polynomial coefficients are estimated; Besides, both LASSO and GLM perform better under the 2nd-order polynomial assumption than under the 1st-order assumption. The two points imply that some intersection terms of the 2nd-polynomials are redundant and intrusive while the others are necessary to be taken into account. Estimating the

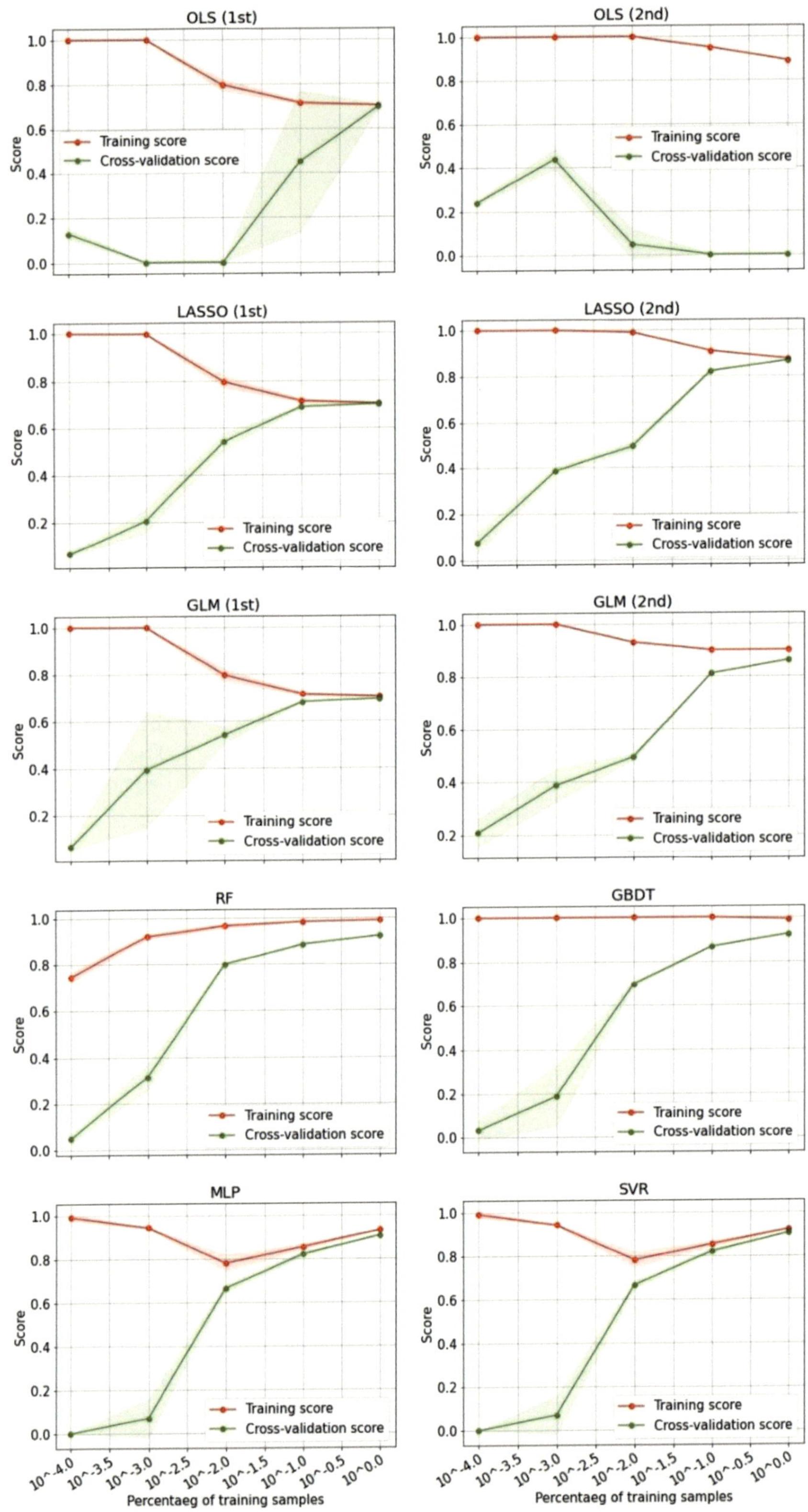

Figure 3.
The learning curves obtained using the traditional RSM method (OLS) and the ML models (LASSO, GLM, RF, GBDT, MLP, and SVR).

coefficients of the intrusive influencers can be a bottleneck for the traditional RSM approximation method as the 2nd-order polynomial approximation using OLS has seen obvious chaos with unreasonably large prediction error.

The table has also demonstrated that the non-linear ML models (RF, GBDT, MLP and SVR) outperform the linear models with or without the intersection and quadratic terms as the measuring metrics have leapt when switching from the linear models to the non-linear ones. Within all the non-linear ML models, the GBDT has exceeded all the others though the performance advantage is relatively small (<10%). In comparison with the traditional RSM method (1st-order polynomial), GBDT has seen a significant improvement (about 50%) on prediction accuracy measured using each of the metrics.

The scores obtained on the held-out test set are almost equal with those achieved through the cross-validation stage for all experimented ML models. The consistency in the metric values reassures the reliability of the constructed model and the prediction accuracy of the model on future unseen data.

3.3.2 Size of experimental data

Figure 3 depicted the learning curve for each of the model described above trained using a proportion of 10^{-4}, 10^{-3}, 10^{-2}, 10^{-1} and 10^{0} of the original training set (which contains 39563 samples).

Diving into the detail of an individual subplot in **Figure 3**, the red points represent the mean of the training scores while the green ones represent the mean of the validation scores of the cross-validation stage under the training sets with 5 different sizes. The shadowed interval shows the distance between the minimum and maximum of either the training scores (red shadow) or the validation scores (green shadow) under each subset. In cases where the scores are close, the interval can be invisible. The x-axis represents the proportion of the original training set used and the y–axis represents the prediction scores of the trained model. Here, the explained variance is used. The model used to produce the scores is shown on the top of each subplot.

Looking at OLS approximation with the 1st order polynomial terms (the 1st plot on the left), the model is apparently over-fitted when the training set is small. With the increase of the training data size, the validation scores improve but the training scores decrease and the two come to parallel with each other when all training data is used. For the OLS approximation with the 2nd order polynomial terms (the 1st plot on the right), the model is always over-fitted, increasing the training samples does not seem to improve the problem.

For the LASSO approximation with the 1st order polynomial terms (the 2nd plot on the left) and with the 2nd order polynomial terms (the 2nd plot on the right), both of the models improve on predicting the validation set when the overall training set is up-scaled. However, the training scores decrease at the same time.

The same trend applies to the GLM approximations (as shown in the two plots on the 3rd line).

The RF model and the GBDT model have displayed a more satisfying pattern. As shown in the two plots on the 4th line, the validation scores increase or remain as a constant together with the increase of the validation scores when the training samples grow. Even when all training samples are used, the validation score is still lower than the training score. This indicates that there is a chance for a better-fitted model if more experimental data is to be used.

For the MLP model and the SVR model (as shown in two plots on the last line), the training scores are climbing after experienced an inflection point when 0.01 proportion of the training data is used and the validation scores continue to rise with more data involved. Though the training and the validation scores are neck and neck when the whole training set is used, both of training and the validation scores still get space for improvement.

Considering the general law that the testing scores will not exceed the training scores and the actual scores shown in the plots, we can conclude that all of the RF, GBDT, MLP and SVR models have the potential to train a more accurate responsive surface function if more experimental data becoming available in the future. The other models have already reached their limitations using the current training set and will see little improvement with larger training data.

3.3.3 Embedding the simulation technique

A finite element method such as the Monte Carlo simulation can be applied in two ways to complement the RSM study, either to assist in obtaining a better responsive surface function estimation or to achieve the extremums of the existing function.

Instead of collecting more experimental data, synthetic data can be generated using the known knowledge of value intervals, categorical levels or distributions of the influence factors obtained from the existing data. These synthetic data can be populated in pairs of inputs and responses and then be used as a supplement of the current training set aiming to train a more accurate responsive surface function (i.e., to train new RF, GBDT, MLP and SVR models using the up-scaled data). Alternatively, generating the inputs data solely to feed into the obtained mathematical model to attain a corresponding response. Then, picking the pair of influence factors and responsive factor values leading to the global minimum or maximum as the extremums. The simulation technique will not be further discussed here in this book chapter.

4. Conclusion

In summary, this book chapter has introduced and discussed

1. two linear ML models (LASSO, GLM) and four non-linear ML models (random forests, GBDT, MLP, and SVR) as alternatives to estimate the responsive surface function.

2. The two linear models use the same optimization function as the traditional RSM method (i.e., OLS in the ML term) to estimate the optimal coefficients of the assumed polynomial yet exceed the OLS by adding an extra regularization term to help to eliminate the redundant and intrusive influencer factors. The improvement can greatly save the efforts on attempting different combinations of influence factors (especially with the higher-order polynomial terms) and solving the optimization function repetitively.

3. The non-linear ML models can pick up the non-linearity across the influence factors that cannot be modeled by the linear models and therefore lead to more precise prediction accuracy.

The advantages of using the ML approached models have been demonstrated by the mechanical engineering case study in Section 3.

1. Results in **Table 2** has shown that all the ML models outperform the traditional RSM polynomial approach. The non-linear models have produced dramatically

improved prediction accuracy. In particular, the GDBT model has shown to exceed the OLS for about 50% on prediction accuracy under each of the measuring metrics.

2. The investigation on the size of the experimental data, i.e. the learning curves in **Figure 3**, has shown that the four non-linear ML models are capable to produce a model with higher accuracy if more training data becoming available.

3. Last yet not least, the simulation technique has been introduced. This technique is essential in either the physical-based or the computer-based experiments to assist in further improving the estimation of the responsive surface function or obtaining the extremums of the current function.

Despite the case study is to predict the vibration severity of the manufacturing machine, the utilization of the ML methods in the RSM can be extended to solve many other engineering problems such as (but not limited) to predict the machine tool life, to estimate the reliability of a structural material or to optimize a bioengineering process where appropriate.

Author details

Yang Zhang[1*] and Yue Wu[2]

1 Data Scientist, London, UK

2 Mathematical Institute, University of Oxford, Oxford, UK

*Address all correspondence to: y.zhang1205@gmail.com

References

[1] Box GEP, Wilson KB. On the experimental attainment of optimum conditions. J R Stat Soc Ser B. 1951;

[2] Box GEP, Draper NR. A basis for the selection of a response surface design. J Am Stat Assoc. 1959;54(287):622–654.

[3] Box GEP, Hunter JS. Multi-Factor Experimental Designs for Exploring Response Surfaces. Ann Math Stat [Internet]. 1957 Mar 1 [cited 2021 Apr 11];28(1):195–241. Available from: http://projecteuclid.org/euclid.aoms/1177707047

[4] Hill WJ, Hunter WG. A Review of Response Surface Methodology: A Literature Survey. 1966;8(4):571–590.

[5] Myers RH. Response surface methodology - Current status and future directions. J Qual Technol [Internet]. 1999 [cited 2021 Apr 11];31(1):30–44. Available from: https://www.tandfonline.com/doi/abs/10.1080/00224065.1999.11979891

[6] Myers RH, Montgomery DC, Geoffrey Vining G, Borror CM, Kowalski SM. Response Surface Methodology: A Retrospective and Literature Survey [Internet]. Vol. 36, Journal of Quality Technology. American Society for Quality; 2004 [cited 2021 Apr 11]. p. 53–78. Available from: https://asu.pure.elsevier.com/en/publications/response-surface-methodology-a-retrospective-and-literature-surve

[7] Mead R, Pike DJ. A Biometrics Invited Paper. A Review of Response Surface Methodology from a Biometric Viewpoint. Biometrics [Internet]. 1975 Dec 1 [cited 2021 Apr 11];31(4):803. Available from: https://www.jstor.org/stable/2529809?origin=crossref

[8] Box GEP, Draper NR. Response Surfaces, Mixtures, and Ridge Analyses [Internet]. 2nd ed. 2007 [cited 2021 Apr 11]. Available from: https://www.wiley.com/en-sg/Response+Surfaces%2C+Mixtures%2C+and+Ridge+Analyses%2C+2nd+Edition-p-9780470053577

[9] Myers RH, Montgomery DC, Anderson-Cook C. Response Surface Methodology. 4th ed. New Jersey: Wiley Online Library; 2016. 856 p.

[10] Pike DJ, Box GEP, Draper NR. Empirical Model-Building and Response Surfaces. J R Stat Soc Ser A (Statistics Soc. 1988;

[11] Venter G, Haftka RT, Starnes JH. Construction of response surfaces for design optimization applications. In: 6th Symposium on Multidisciplinary Analysis and Optimization. 1996.

[12] Box GEP, Tidwell PW. Transformation of the Independent Variables. Technometrics. 1962;

[13] Box GEP, Cox DR. An analysis of transformations. J R Stat Soc Ser B. 1964;

[14] Long T, Wu D, Guo X, Wang GG, Liu L. Efficient adaptive response surface method using intelligent space exploration strategy. Struct Multidiscip Optim. 2015;

[15] Kumar A, Kumar V, Kumar J. Multi-response optimization of process parameters based on response surface methodology for pure titanium using WEDM process. Int J Adv Manuf Technol. 2013;

[16] de Oliveira LG, de Paiva AP, Balestrassi PP, Ferreira JR, da Costa SC, da Silva Campos PH. Response surface methodology for advanced manufacturing technology optimization: Theoretical fundamentals, practical guidelines, and survey literature review. Int J Adv Manuf Technol. 2019;

[17] Mandenius CF, Brundin A. Bioprocess optimization using design-of-experiments methodology. Biotechnology Progress. 2008.

[18] Gilmour SG. Response surface designs for experiments in bioprocessing. Biometrics. 2006;

[19] Edmondson RN. Agricultural response surface experiments based on four-level factorial designs. Biometrics. 1991 Dec;47(4):1435.

[20] Khuri AI. Response Surface Methodology and Its Applications In Agricultural and Food Sciences. Biometrics Biostat Int J [Internet]. 2017 Apr 11 [cited 2021 Apr 11];5(5). Available from: http://medcraveonline.com

[21] Mitchell TM. Machine learning and data mining. Commun ACM. 1999;

[22] Alpaydin E. Introduction to Machine Learning. 2014.

[23] Friedman J, Hastie T, Tibshirani R. Regularization paths for generalized linear models via coordinate descent. J Stat Softw. 2010;

[24] Breiman L, Friedman JH, Olshen RA, Stone CJ. Classification and regression trees. Belmont; 2017.

[25] Hastie T, Tibshirani R, Friedman JH (Jerome H. The elements of statistical learning : data mining, inference, and prediction [Internet]. 2nd ed. New York: Springer; 2009 [cited 2018 Jun 1]. 745 p. Available from: http://www-stat.stanford.edu/~tibs/ElemStatLearn/

[26] Montgomery DC. Design and Analysis of Experiments. 8th Editio. John Wiley & Sons; 2012.

[27] Gunst RF, Myers RH, Montgomery DC. Response Surface Methodology: Process and Product Optimization Using Designed Experiments. Technometrics. 1996;

[28] What are Taguchi designs. In: e-Handbook of Statistical Methods [Internet]. 2012. Available from: https://www.itl.nist.gov/div898/handbook/pri/section5/pri56.htm

[29] Pedregosa, F. Varoquaux, G. Gramfort A, Michel V, Thirion B, Grisel O, Blondel M, Prettenhofer P, et al. Scikit-learn: Machine learning in Python. J Mach Learn Res [Internet]. 2011 [cited 2021 Mar 4]; Available from: https://scikit-learn.org/0.23/

[30] Vector Norm [Internet]. Wolfram Mathworld. [cited 2021 Apr 11]. Available from: https://mathworld.wolfram.com/VectorNorm.html

[31] Kim S-J, Koh K, Lustig M, Boyd S, Gorinevsky D. An Interior-Point Method for Large-ScalèScalè 1-Regularized Least Squares. IEEE J Sel Top Signal Process. 2007;1(4).

[32] McCullagh P, Nelder J. Generalized Linear Models [Internet]. Chapman and Hall; 1983. Available from: http://www.utstat.toronto.edu/~brunner/oldclass/2201s11/readings/glmbook.pdf

[33] Generalized Linear Regression [Internet]. scikit learn. 2021 [cited 2021 Feb 24]. Available from: https://scikit-learn.org/stable/modules/linear_model.html#generalized-linear-regression

[34] Hastie T, Tibshirani R, Friedman J. Elements of Statistical Learning. 2nd ed. Springer; 2009.

[35] Breiman LEO. Random Forests. Mach Learn [Internet]. 2001;5–32. Available from: https://link.springer.com/content/pdf/10.1023/A:1010933404324.pdf

[36] Chen T, He T, Michael B, Vadim K, Yuan T, Hyunsu C, et al. Extreme Gradient Boosting [Internet]. CRAN.

[cited 2020 Sep 2]. Available from: https://cran.r-project.org/web/packages/xgboost/xgboost.pdf

[37] Rumelhart E. David, Geoffrey HE, J. WR. Learning representations by back-propagating errors. Lett to Nat [Internet]. 1986;323(9). Available from: https://www.iro.umontreal.ca/~pift6266/A06/refs/backprop_old.pdf

[38] Kingma DP, Lei Ba J. Adam: a method for stochastic optimization. In: The 3rd International Conference on Learning Representations [Internet]. San Diego; 2015. p. 1–15. Available from: https://arxiv.org/pdf/1412.6980v8.pdf

[39] Boser BE, Guyon IM, Vapnik VN. A training algorithm for optimal margin classifiers. In: Proceedings of the fifth annual workshop on Computational learning theory [Internet]. 1992. Available from: https://dl.acm.org/doi/10.1145/130385.130401

[40] Berwick R. An Idiot's guide to Support vector machines [Internet]. p. 1–28. Available from: http://web.mit.edu/6.034/wwwbob/svm-notes-long-08.pdf

[41] Smola AJ, Schölkopf B. A tutorial on support vector regression. Stat Comput. 2004;

[42] Zhang Y, Beudaert X, Argandona J, Ratchev S, Munoa J. A CPPS based on GBDT for predicting failure events in milling. Int J Adv Manuf Technol. 2020;

[43] Hyndman RJ, Athanasopoulos G. Forecasting: Principles and Practice [Internet]. 2018 [cited 2018 Dec 6]. Available from: https://otexts.org/fpp2/

Chapter 3

Global Optimization Method to Multiple Local Optimals with the Surface Approximation Methodology and Its Application for Industry Problems

Ichiro Hagiwara

Abstract

Although generally speaking, a great number of functional evaluations may be required until convergence, it can be solved by using neural network effectively. Here, techniques to search the region of interest containing the global optimal design selected by random seeds is investigated. Also techniques for finding more accurate approximation using Holographic Neural Network (HNN) improved by using penalty function for generalized inverse matrix is investigated. Furthermore, the mapping method of extrapolation is proposed to make the technique available to general application in structural optimization. Application examples show that HNN may be expected as potential activate and feasible surface functions in response surface methodology than the polynomials in function approximations. Finally, the real design examples of a vehicle performance such as idling vibration, booming noise, vehicle component crash worthiness and combination problem between vehicle crashworthiness and restraint device performance at the head-on collision are used to show the effectiveness of the proposed method.

Keywords: nonlinear problem, holographic neural network, function approximation, response surface methodology, structural analysis, vehicle idling vibration, vehicle booming noise, vehicle crashworthiness, vehicle occupant restraints

1. Introduction

Due to the recent developments in computer science and software technology, fluid analysis, nonlinear structural analysis and their combined analysis have become possible for large-scale models and will eventually be employed in manufacturing design. The importance of being able to make rational and correct decisions during the early stages of design is currently being emphasized. However, non-linear problems such as crashworthiness analysis require considerable time for one functional evaluation even using supercomputers. Optimization techniques which cover several design disciplines (multiple discipline optimization-MDO [1]) have been hot topics recently. Sensitivity-based optimization approaches [2–15], which sequentially construct local approximations, are widely employed and provide valuable contributions to linear optimization problems. Unfortunately, sensitivity approaches are not available for nonlinear optimization problems. For this purpose, response surface methodology (RSM) works effectively to achieve these design optimizations. The most important research work on RSM is to considerably reduce the extremely numerous number of function evaluations for computational cost problem and to get the robust approximation design model. That means instead of the time-consuming finite element method (FEM) analysis, robust approximation models can be constructed to evaluate responses rapidly in the design space of interest by RSM, when the response surface of the structure is infiltrated with computational noise. Early examples are such that White Jr. K.P. used RSM in their study of passenger-car crash worthiness [16] and Giunta, A.A. applied RSM to high speed civil transportation design [17]. Several key issues must be considered to develop the response surface modeling. One issue for RSM may be the selection of the activation function. Most response surfaces are expressed in terms of polynomials and among them *quadratic polynomial* are the most popular [18–25]. Unfortunately, this is not always accepted as a proper activation function for response surfaces and may not provide a satisfactory solution, particularly for wide ranges of design space with heavy non linearity. This limits the validity of RSM to only cover a part of the design space. And so in the past, feed forward multiple layers back-propagation neural network seemed to be a potential activate function in response surface function methodology [26]. Carpenter, W. made the comparison between polynomials and multiple layers neural network [27]. There are some problems when using back-propagation neural network: (1) the learned result is influenced by the initial parameters of the network due to the nonlinear gradient descent algorithm. (2) Training of neural network costs computational time, etc. In contrast, Hagiwara et al. firstly employed the Holographic neural network (HNN) [28] for the activation function of the response surface and compared it to the multilayer feed-forward neural network and demonstrated that the former type of neural networks had a much higher approximation accuracy and the training cost of HNN was about 1/1000 the computational time compared to that of a multilayer feed-forward neural network in the case on the assumption that the designer roughly knows the location of local optimal [29, 30].

Another problem in using RSM is the difficulty in predicting the error incurred using response surface function values rather than the "true" values from the analysis code. Therefore, the extensive RSM research is focused on the selection of design trial points in design space, such as Box–Behnken Designs (BBD [20, 21, 25]) and Central Composite Design (CCD [19, 22–24]) which are two typical choices. And Central Composite Rotatable Design (CCRD [31]) and Face Central Composite Design (FCCD [18]) have also been applied to optimization studies in recent years.

By the way, now they say 3rd AI boom has come due to the born of deep learning [32]. And so recently the use of artificial neural network models to derive response surface plot has found increasing use in optimization modeling. In most of them, CCD is used such as in [33–35]. These researches mainly consist of 2 types. One is neural network itself works as RSM step [33]. The other is artificial neural network function is used as an activation function of RSM [34, 35]. Deep learning must have a lot of data. It also needs a lot of data to search the global optimization for multiple local optimal problem shown here by RSM.

Since the exact form of response surface remains unclear in general application and the accuracy of the function approximation near the global optimal design is very important, techniques to search the design space containing the global optimal design are examined. For the purpose of determining the global optimal design, techniques for searching the most-likely global optimal design (MPOD) over the entire design space and techniques for determining a more accurate approximation near the global optimal designs using the extrapolation ability of the HNN are proposed [36–38].

The MPOD is as mentioned above a sequential procedure composed of two steps. The first step is employed to determine the most likely global optimal design over the entire design space by conditioned random seeds which control the mutual distances between every two designs, to scan the wide design space rapidly. The second step is to verify and improve the accuracy of the approximation of the objective function and constraint functions at optimal design. The trade-off between the approximation accuracy and computational cost can be well balanced by the preset threshold of the mutual distance of designs. Some design examples for vehicle structure such as idling vibration, interior noise, crashworthiness and combination problem between vehicle crashworthiness and restraint device performance at the head-on collision demonstrate the validity and utility of this method. The method developed by us can omit many useless data and very unique from both of the 2 important issues for RSM such as activation function and selected design points. This approach is also useful for the execution of the deep learning. Let us get started to show this splendid method.

2. Theory of holographic neural network (HNN)

2.1 Basic theory of HNN

Here, $x = \{x_1, x_2, \ldots, x_k\}^T$ is the stimulus data, $y = \{y_1, y_2, \ldots, y_m\}^T$ is the training data. If there is learning data for n, it is presented in Eq. (1).

$$X = \begin{pmatrix} X_1^T \\ \vdots \\ X_n^T \end{pmatrix} = \begin{pmatrix} x_{11} & x_{12} & \cdots & x_{1k} \\ x_{21} & x_{22} & \cdots & x_{2k} \\ \vdots & \vdots & \ddots & \vdots \\ x_{n1} & x_{n2} & \cdots & x_{nk} \end{pmatrix} \quad Y = \begin{pmatrix} Y_1^T \\ \vdots \\ Y_n^T \end{pmatrix} = \begin{pmatrix} y_{11} & y_{12} & \cdots & y_{1m} \\ y_{21} & y_{22} & \cdots & y_{2m} \\ \vdots & \vdots & \ddots & \vdots \\ y_{n1} & y_{n2} & \cdots & y_{nm} \end{pmatrix} \tag{1}$$

Here, X, Y are input and output matrices. Each element of X and Y is converted to angles $\theta_{ai}(a = 1, \cdots, n; i = 1, \cdots, k)$ and $\phi_{aj}(a = 1, \cdots, n; j = 1, \cdots, m)$.

$$\theta_{ai} = f_x(x_{ai})\phi_{aj} = f_y\left(y_{aj}\right) \tag{2}$$

Linear, Sigmoid and Arc Tangent functions are applied for f_x and f_y. Especially, for the stimulus data, Sigmoid function is used as follows.

$$\theta_{ai} = \frac{2\pi}{\left(1 + e^{-c_{ai}(x_{ai} - \mu_{ai})/\sigma_{ai}}\right)} \tag{3}$$

Here, μ_{ai} and σ_{ai} are the mean value and standard deviation of x_{ai} respectively and λ_{ai} defines a vector magnitude bound within the unit circle (0 to 1), expressing a weighting or dominance for each element of the stimulus field. If no design variable dominates, $\lambda_{ai} = 1$. c_{ai} is the coefficient to adjust the transformation slope shown in **Figure 1**.

Next, the above angles are mapped on complex representation by using exponential function as follows.

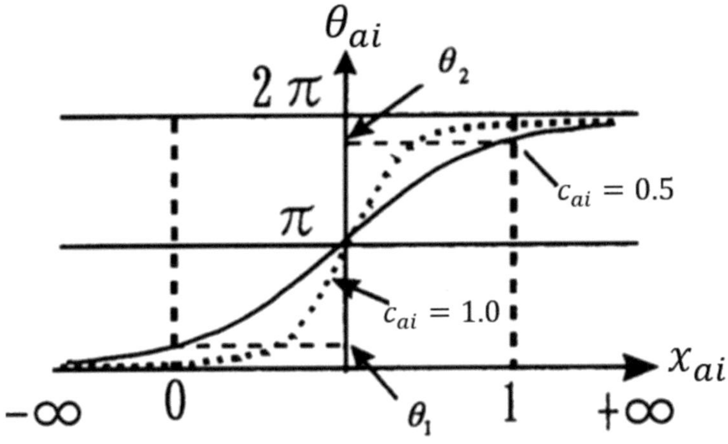

Figure 1.
Transform function of phase angle: If the untrained designs are outside the trained design space, the predicted values are same values as those at trained design edges 0 or 1 for the case of $c_{ai} = 1.0$. In contrast, the predicted function values are extended outside the trained design edges 0 or 1 for the case of $c_{ai} = 0.5$. In this way, the extrapolation can be done and the trained designs are not necessary on the boundary.

$$s_{ai} = \lambda_{ai} e^{\hat{i}\theta_{ai}} \; r_{aj} = \gamma_{aj} e^{\hat{i}\phi_{aj}} \tag{4}$$

From the above, X and Y are converted to stimulus S and response R.

$$S = \begin{pmatrix} s_{11} & s_{12} & \cdots & s_{1k} \\ s_{21} & s_{22} & \cdots & s_{2k} \\ \vdots & \vdots & \ddots & \vdots \\ s_{n1} & s_{n2} & \cdots & s_{nk} \end{pmatrix} R = \begin{pmatrix} r_{11} & r_{12} & \cdots & r_{1m} \\ r_{21} & r_{22} & \cdots & r_{2m} \\ \vdots & \vdots & \ddots & \vdots \\ r_{n1} & r_{n2} & \cdots & r_{nm} \end{pmatrix} \tag{5}$$

$H = [h_1, \dots, h_m]$, the transfer function of HNN is presented by minimizing the difference between teaching data R and $S \cdot \mathrm{H}$ as follows;

$$\mathrm{H} = (S^* S)^{-1} S^* R \tag{6}$$

Here, symbol * presents conjugate transpose. Output V can be predicted to be $V = U \cdot H$ for new input U. H is the transfer function of HNN. It is not efficient to treat Eq. (6) because it has the inverse matrix operating. When the number of activation functions is large, it will lead to considerable calculation time and may result in calculation error. The following iteration process is used to save calculation time.

$$H_0 = S^* \mathrm{Y} \tag{7}$$

The enhancement of parameters will be performed by

$$\Delta H_{k+1} = S^* (Y - SH_k) \tag{8}$$

$$H_{k+1} = H_k + \Delta H_{k+1} \tag{9}$$

The iteration may be terminated when the error between the approximating response and the true response is orthogonal to every column vector of S in a complex domain. The iteration is terminated when the error becomes smaller than the preset threshold value, eps.

$$J(H_{k+1}) = (Y - S\,H_{k+1})^* (Y - S\,H_{k+1}) \le \text{eps} \tag{10}$$

By the way, within the HNN, limitations of mapping accuracy can be largely overcome through a pre-processing operation involving the generation of higher-order product terms. For example, for two design variables ($n = 2$), $s = \left[\lambda_1 e^{i\theta_1}, \lambda_2 e^{i\theta_2}\right]$ can be increased to 12 terms (n = 12) if performed to the second order.

$$\begin{aligned}\overline{S} = \Big[&\lambda_1 e^{i\theta_1}, \lambda_2 e^{i\theta_2}, \lambda_1 e^{-i\theta_1}, \lambda_2 e^{-i\theta_2}, \\ &\lambda^2{}_1 e^{i2\theta_1}, \lambda^2{}_2 e^{i2\theta_2}, \lambda^2{}_1 e^{-i2\theta_1}, \lambda^2{}_2 e^{-i2\theta_2}, \lambda_1\lambda_2 e^{i(\theta_1+\theta_2)}, \\ &\lambda_1\lambda_2 e^{i(-\theta_1+\theta_2)}, \lambda_1\lambda_2 e^{-i(\theta_1+\theta_2)}, \lambda_1\lambda_2 e^{i(\theta_1-\theta_2)}\Big]\end{aligned} \tag{11}$$

Furthermore, the extrapolation of the approximated response function may be easily performed with trained HNN by presetting the coefficient c_k in Eq. (3) to a small value to adjust the slope to be a smaller. **Figure 1** shows the effect of the coefficient. In **Figure 1**, the experimental designs used for training the neural network are located within the range [0–1], the dotted line expresses the transform when $c_k = 1.0$ and the solid line expresses the transform when $c_k = 0.5$. We can observe from the figure that if the predicted designs are outside the trained design space, then the predicted function values outside the trained design space may be considered to be the same values as those at trained design edges 0 or 1 for the case of $c_k = 1.0$. In contrast, the predicted function values outside the trained design space may keep the trend beyond the trained design edges 0 or 1 for the case of $c_k = 0.5$. **Figure 2** shows the effect of extrapolation. In this figure, space "I" expresses the design space inside which designs are used for approximation by training and space "D" expresses the entire feasible design space. Approximation using designs in space "I" may extrapolate the values of space "D". This is an important issue for reducing functional evaluations and some advantages of this are: 1) design spaces with irregular shapes can be considered, 2) any number of experimental designs can be considered, and 3) searching for reasonable optimal designs over a wide space becomes possible.

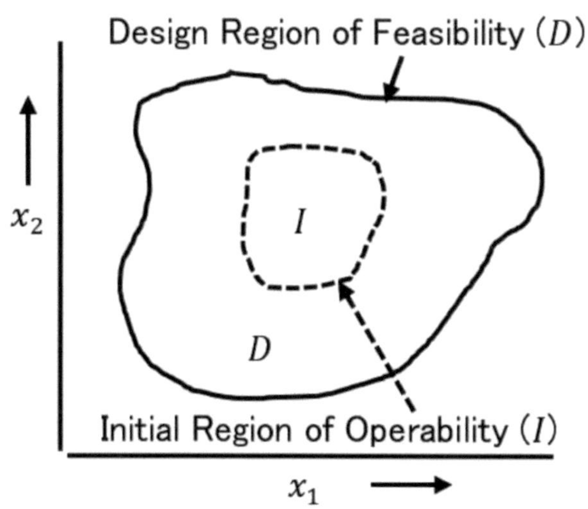

Figure 2.
Explanation of extrapolation: Space "I" expresses the design space inside which designs are used for approximation by training and space "D" expresses the entire feasible design space. Approximation using designs in space "I" may extrapolate the values of space "D".

2.2 Examination for getting the inverse matrix of transfer matrix

Although for calculation inverse matrix $(\overline{S}^*\overline{S})$, we can use the procedure from Eqs. (7) to (10), for not so big problem, the inverse of $(\overline{S}^*\overline{S})$ is calculated directly. If $(\overline{S}^*\overline{S})$ has not the inverse matrix in Eq. (6), we can use Moore Penrose generalized inverse matrix $\hat{H}_{MP}$ which is generated by operating $\|h_k\|^2$ (k~1, … .,m) to be the minimum. Here h_k is the column vector of H.

$$\hat{H}_{MP} = \overline{S}^*\overline{S}\left(\overline{S}^*\overline{SS}^*\overline{S}\right)_G^{-1}\overline{S}^*R \tag{12}$$

Here $()_G^{-1}$ is the generalized inverse matrix.

But following is the example where it has not satisfied predicted value for **Figure 3** with $\hat{H}_{MP}$.

It is shown the approximation function in **Figure 4** which is generated with Moore Penrose *pseudo-inverse matrix* by using 23 sampling points as shown ● in **Figure 5**. Here the base function of HNN is extended to the 3rd order where the total number of the base function is 30.

It is recognized that the surface shape in **Figure 4** is quite different from the one in **Figure 3**. This is because over fitting phenomena occurs which is unique to neural network. This phenomena occurs in such way that the precision of the predicted values of the points other than sampling points become worth because the values of the sampling points are forced fit to the given HNN function. In this case we cannot use HNN for the prediction.

Instead, the following penalty function was developed [39].

$$f(h_k) = (r_k - Sh_k)^*(r_k - Sh_k) + \rho_1 h_k^* O h_k + \rho_2 h_k^* h_k (k = 1, \dots m) \tag{13}$$

Here O is the square matrix whose all elements are 1. The right side paragraph 1 means approximation error to output, the paragraph 2 means the average of regression factor h_k and the paragraph 3 means variation. The paragraphs 2 and 3 are weighted with ρ_1, ρ_2. In Eq. (13), it cannot be predicted uniquely without some

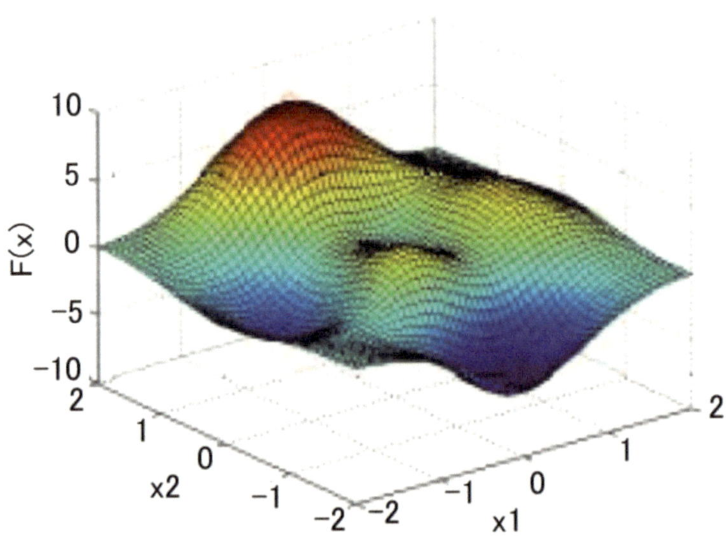

Figure 3.
Visualization of function used for test.

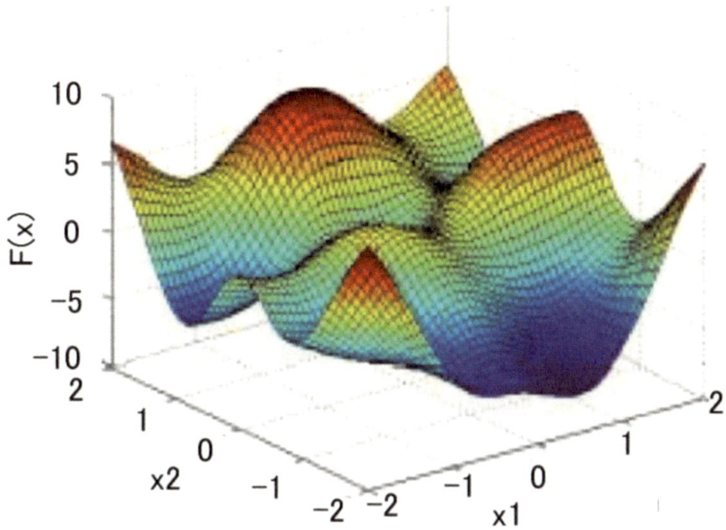

Figure 4.
Approximated function using Moore-Penrose generalized inverse: The base function of HNN is extended to the 3rd order. The surface shape is quite different from the original one because over fitting phenomena occurs.

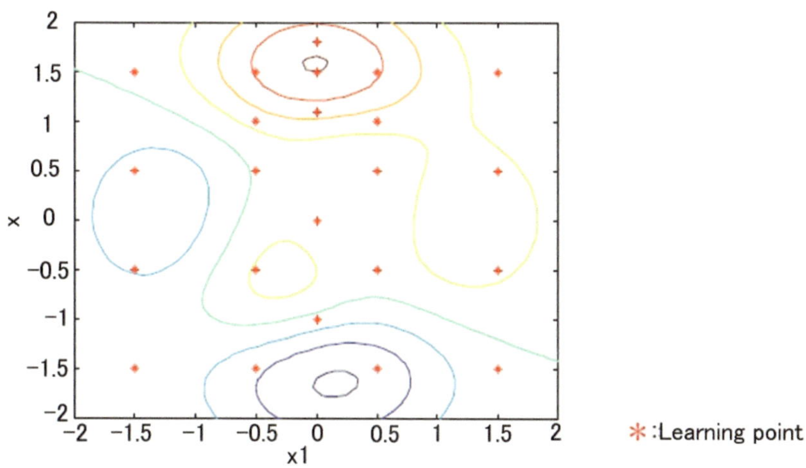

Figure 5.
Layout of sampling point on design space.

constraints and so such conditions are added. Although theoretically, H is predicted uniquely without the paragraph 2, it is more stable with this paragraph. H is as follows by minimizing the penalty function (13);

$$H_P = (S^* \cdot S + \rho_1 O + \rho_2 I)^{-1} \cdot S^* \cdot R \tag{14}$$

Here, I is identity matrix. The approximation function by using this H_P is depicted in **Figure 6** where 23 sampling data are used as above, ρ_1, ρ_2, are 1.0. It can be said it does not occur the over fitting phenomena in **Figure 6**. The approximation functions by both methods are estimated by mean square error.

$$E = \sqrt{\frac{1}{N}\sum_{k=1}^{N}(y_k - \hat{y}_k)^2} \tag{15}$$

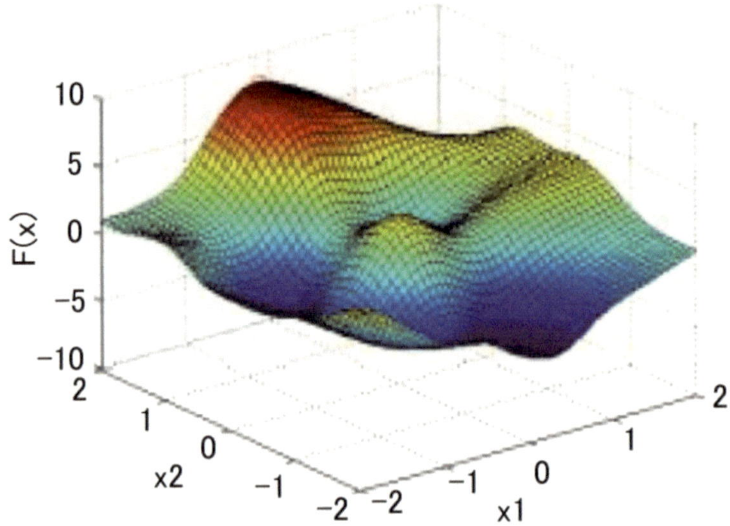

Figure 6.
Approximated function using penalty coefficient method: The over fitting phenomena does not occur.

Here N is number of estimation points, y_k is the real function value and $\hat{y}_k$ is the value of approximation function. Here it is called "approximation error" which is estimated by using 23 sampling points and called "prediction error" which is estimated by 2601 points, generated by dividing design region into each of the two axes 50 equally separated. From **Table 1**, it can be estimated in the case of Moore Penrose, the approximation error is very small but the prediction error gets worth. On the other hand, in the case of proposed method, the approximation error is not smaller than in the case of Moore Penrose but the prediction error is moderately suppressed. In the case of proposed one, it is important how much ρ_1, ρ_2 to be set. If they are very small, it approaches to the approximation function by Moore Penrose because the part of right side paragraph 1 of Eq. (14) is larger. On the other hand, if they are too large, the part of right side paragraph 1 of Eq. (14) is too small to be used effectively the information of sampling data. And so when the number of sampling data is large and the sampling data are selected equally over the design region, ρ_1, ρ_2 are small so that the information of sampling data can be used effectively. On the other hand, if it is difficult to have enough sampling data, it is better ρ_1, ρ_2 are made reasonably large so that design region is roughly approximated by small information in order not to have over fitting phenomena.

In the case of Moore Penrose, the approximation error is very small but the prediction error gets worth. In the case of Penalty Coefficient Method, the approximation error is not smaller than in the case of Moore Penrose but the prediction error is moderately suppressed.

	$\hat{H}_{MP}$	$\hat{H}_P$
Error in learning	0.0447	0.4993
Error in prediction	1.8681	0.9373

Table 1.
Comparison between two methods by square-mean-error.

3. Optimization method with holographic neural network

3.1 Procedure for MPOD

A two-step sequential optimization is proposed to determine the optimal design as shown in **Figure 7**. The first step is to scan the entire design space to determine where the global optimum design exists, most probably by designs selected using conditioned random seeds. The second step is to increase the accuracy of the most-likely global optimum design by imputing additional information where required. The details regarding these two steps are described below. The optimum problem is defined as.

$$\text{Max} f(x) \tag{16}$$

$$\text{Subject to}\, g_j(x) \leq 0, (j = 1, 2, 3, \ldots, \text{m}) \tag{17}$$

Here m is the number of constraints.

Step 1: To search the most-likely global optimum design over the entire design space. Selecting initial trial designs x_k (k = 1, 2, ..., N) where N is the initial number of design points. Although one should understand that this is not the only choice, it is generally recommended $N = (n+1)(n+2)/2$, n is the number of design variables. Actually the determination of the N value depends on the scale of the problem such as the number of design variables and the computation time of each functional evaluation. For instance, as far as the formulation of CCD (Center Composite Design), points $N=2n+1$ is recommended for problems with a large number of designs. It should be emphasized that the designs produced by conditioned random seeds are employed to search the entire design space. The uniformly distributed conditioned random seeds are used to provide a good valance of search in the design space. Design conditions are required to satisfy Eq. (18).

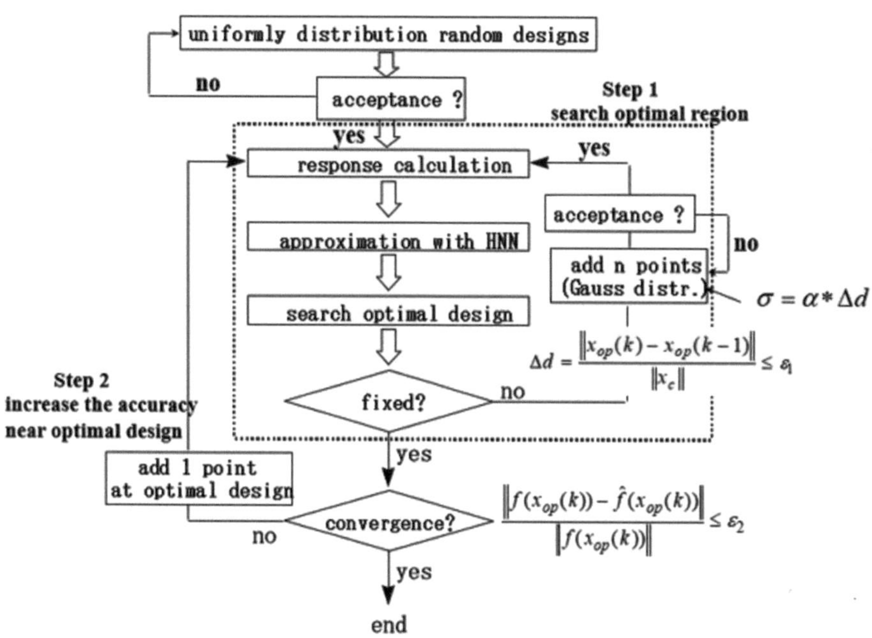

Figure 7.
Flowchart of MPOD: The first step is to scan the entire design space to determine where the global optimum design exists, most probably by designs selected using conditioned random seeds. The second step is to increase the accuracy of the most-likely global optimum design by imputing additional information where required.

$$d_m = \min\left\{\frac{\|x_i - x_j\|}{\|x_{CG}\|}\right\} \ (1 \le i, j \le N, i \ne j, m = 1, 2, \ldots \ldots, p) \tag{18}$$

Where x_{CG} is the normalization parameter which can be selected to be the geometrical center of design $x_k = \{x_{1,k}, x_{2,k}, \ldots ., x_{n,k}\}$, $(k=1, \ldots \ldots, N)$, $d_m \ge d_{min}$ $(m=1,2, \ldots ., p)$, d_{min} is the threshold of distance, p is the total number of design combinations $p = {}_NC_2$. If no new design is obtained after 5000 random trials within the threshold distance, the new threshold distance must be decreased by $d_{min} = \beta d_{min}$ $(0 < \beta < 1)$. Designs satisfying Eq. (18) and their objective function $f(x_i)$ $(i = 1, \ldots \ldots, N)$. and constraint $g_j(x_i)$ $(i=1, \ldots \ldots, N, j=1, \ldots \ldots, m)$ values are trained using neural networks to approximate the response surface $\bar{f}$ (x) and $\bar{g}_j(x)$. A global optimization design $x_{op}(0)$ with an approximation function can be solved by optimization codes. Then, more S designs are created with the same conditions $d_m \ge d_{min}$ $(m=1,2, \ldots ., p)$,which are used to improve the accuracy of the approximation. Here, S is generally set to n. But it is also not the only choice. The random S designs are selected as the Gaussian normal distribution with the mean value of $x_{op}(0)$ and the deviation must be carefully chosen. The approximation accuracy can be judged by the change in the movement of approximation optimal design location. Therefore, the deviation should change with the movement distance $\sigma = \alpha \Delta d$. Here, Δd is the distance of designs calculated by Eq. (19) and α is the coefficient which controls the deviation value and affects the approximation accuracy and the number of functional evaluations, as discussed below. The kth sequential improved approximation functions by the neural network is to be used to obtain new optimal design $x_{op}(k)$. The convergent conformation for the kth sequential approximation is taken to terminate the random seeds operation

$$\Delta d = \frac{\|x_{op}(\mathrm{k}) - x_{op}(\mathrm{k}-1)\|}{\|x_{CG}\|} \le \varepsilon_1 \tag{19}$$

Thus far, the above sequential approximation is used to determine the approximation global optimal location.

Step 2: To verify and improve the accuracy of optimal design and its performance. The accuracy check is based on

$$\frac{\| f(x_{OP}(\mathrm{k}) - \bar{f}(x_{OP}(\mathrm{k})\|}{\| f(x_{OP}(\mathrm{k}))\|} \le \varepsilon_2 \tag{20}$$

where $x_{OP}(k)$ is the optimal design in Step 1, ε_2 is the threshold of the termination and $f(x_{OP}(\mathrm{k}))$ and $\bar{f}(x_{OP}(\mathrm{k}))$ are the corresponding exact and approximation function values. If Eq. (20) is not satisfied, n learning points are added by normal distribution random number of which center is $x_{OP}(\mathrm{k})$ until Eq. (20) is satisfied. Here n is number of design values and variance σ_i is decided in Eq. (21).

$$\sigma_i = \alpha_i(x_{OP}(k) - x_{OP}(k-1)) \ (i = 1, 2, \ldots \ldots \ldots, n) \tag{21}$$

3.2 Necessary designs used for optimal design by MPOD in step 1

The convergence and efficiency of the proposed method are based on the deviation of the Gaussian random seeds. The probability of determining the global optimal design is increased when a larger deviation is used which can search a wider feasible design space resulting in an increase of the functional evaluations. In contrast, if the deviation is small, the number of functional evaluations will decrease but the possibility of searching the smaller cluster will reduce. As illustrated in **Figure 8**, the scatter of design points is adjusted by the deviation value. This method is sufficiently flexible to satisfy any user's requirements through the adjustment of coefficient α. The necessary designs used for solving the global optimal can definitely be estimated. For sake of convergence and without a loss of generality, a one-dimensional example is employed to express the estimator as shown in **Figure 9**. Assume the design space $\mathcal{R}$ is d_0 and the size of the entire feasible design space is D.

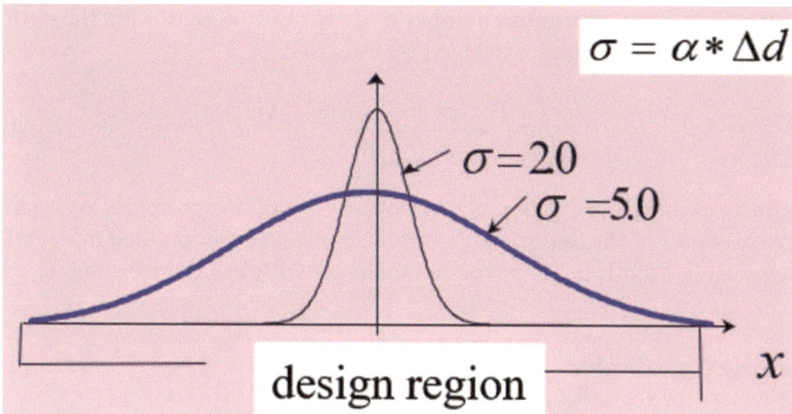

Figure 8.
Random data distribution according to standard deviation: The scatter of design points is adjusted by the deviation value.

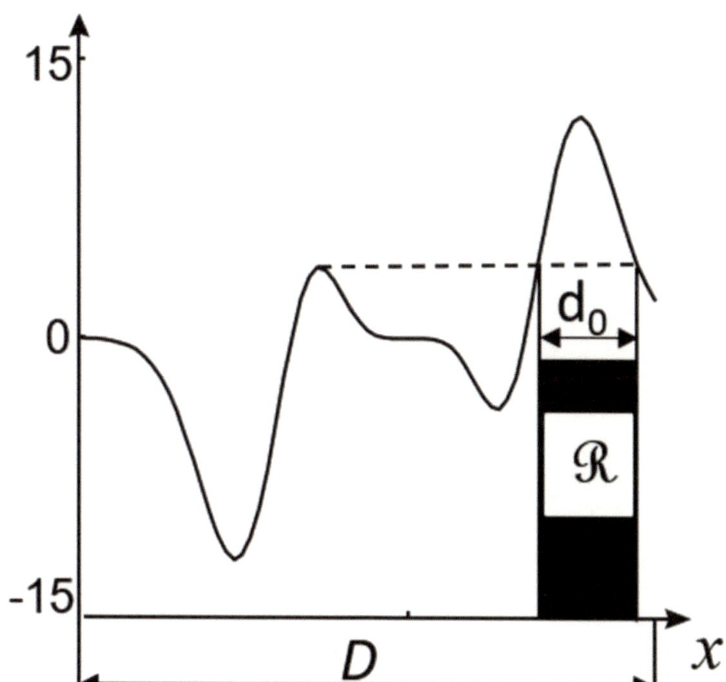

Figure 9.
Explanation of necessary designs to find global optimal: The design space $\mathcal{R}$ *is* d_0 *and the size of the entire feasible design space is* D.

$$d_0 = \frac{2min\,\|x_e - x_c\|}{\|x_c\|} \tag{22}$$

$$f(x_i) > \mathrm{f}(x_j); x_i \in \mathcal{R}; x_j \notin \mathcal{R} \tag{23}$$

We define the space $\mathcal{R}$ as an optimal cluster, which is mathematically defined by Eq. (22).which indicates that the function values within the optimal cluster $\mathcal{R}$ as an optimal cluster are longer than those outside the optimal cluster. x_e is the edge point of the optimal cluster. x_c is the geometrical center of optimal cluster and d_0 expresses the size of the optimal cluster. If the size of the optimal cluster is known, it is estimated that the maximum number of designs for determining the optimum cluster with probability unity is decided by Eq. (24).

$$P = \mathrm{int}\left(\frac{2max\,\|x_E - x_{CG}\|/\|x_{CG}\|}{d_0}\right) \tag{24}$$

Where x_E expresses the group points on the edge of design space, x_{CG} is the geometrical center of the design space and int. expresses the rounded integer. In this case, if $\varepsilon_1 = d_0$ in Step 1, it can be moved to Step 2 with less than P designs.

4. Numerical examples

4.1 Comparison of approximation by holographic neural network and polynomials

The comparison of the HNN with polynomials for the function approximation is performed by the following multiple peaks function, which is shown in **Figure 9**.

$$f\ (\mathrm{x}) = 10\,x^5\,\left(x^2 - 2\right)\,\mathrm{e}^{-x^2},\ -4.0 \leq \mathrm{x} \leq 3.0 \tag{25}$$

For the above numerical example, the number of training designs is 15 and they are evenly spaced over the design space. The approximation accuracy of the HNN and polynomials is shown in **Figure 10**. In this figure, the horizontal axis expresses the orders of polynomials (a) or orders of the HNN (b), the vertical axis expresses

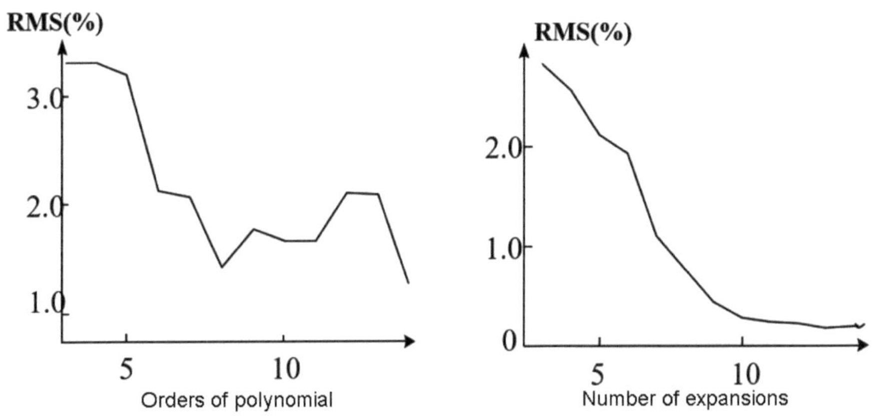

Figure 10.
Performance comparison of polynomial and holographic neural networks: The horizontal axis is the orders of polynomials (left) or number of expansions for the HNN (right), the vertical axis is the standard variance calculated with the 200 designs evenly spaced in the design space [−4.0,3.0].

the standard variance calculated with the 200 designs evenly spaced in the design space $[-4.0,3.0]$. We can see from the results that the HNN is much more accurate than the polynomials and maintains good accuracy after the ninth order of expansion. The standard deviation of the HNN at the ninth order is 0.3, while those of polynomials at the 14th order is higher than 1.0. In other words, a larger number of coefficients must be selected for the polynomials and thus, more experimental designs must be used compared to that of the HNN. Therefore, the HNN is a more suitable activation response function for this problem than polynomials.

4.2 Numerical example with holographic neural network for extrapolation application

To demonstrate the efficiency of this method, a numerical example of multiple local maximums with constraints is used to search the global optimal design in the feasible design space. The optimization problem is defined as

$$\max f(x_1, x_2)$$

$$f\ (x_1, x_2) = 3(1 - x_1)^2\ e^{-x_1{}^2-(x_2+1)^2} - 10\left(\frac{x_1}{5} - x_1{}^3 - x_2{}^5\right)\ e^{-\left(x_1{}^2+x_2{}^2\right)} - \frac{1}{3}\ e^{-(1+x_1)^2-x_2^2}$$

$$\textit{Submit to}\ \{-2,-2\} \le \{x_1, x_2\} \le \{2, 2\}$$

$$g_1\ (\mathrm{x}) = \frac{15}{80}x_1^2 - \frac{5}{8}x_1 + x_2 - 2 \le 0$$

$$g_2\ (\mathrm{x}) = \frac{1}{2}x_1 - x_2 - 1 \le 0$$

Three local optimal designs, marked ①, ②, ③, exist within the feasible design space depicted in **Figure 11**. It is a contour plot, and the feasible design space is enclosed within the shadowed area. The following two cases are performed to investigate the performances of convergent accuracy and the number of functional evaluations.

1. Case of large deviation (σ=5Δd): a sequential approximation is performed using the proposed two steps as shown in **Figure** 7. Value of β is set to 0.6. The method of training the HNN is: to start with expansion terms of 2, and training accuracy of 0.02,which if not satisfied after 1000 epochs, is gradually increased to a higher-order expansion by two orders. The thresholds of Eqs. (19) and (20) are ε_1 =0.15, ε_2 =0.01. Step 1 is terminated after nine iterations (corresponding to 22 functional evaluations) of random design searches. Then, Eq. (20) is satisfied after two iterations and the total number of functional evaluations until convergence is 24. The approximation optimal design is $x_{OP}(\mathrm{k})$ = (−0.013, 1.583), $\bar{f}(x_{OP}(\mathrm{k}))$ =8.106. The history of each iteration is illustrated in **Figure 12**, in which the horizontal axis stands for the distance (ds) between the original position (0,0) and the approximation optimal design.

2. Case of small deviation (σ=2Δd): The training procedure and stop thresholds are similar to the case of large deviation. Step 1 is terminated after six iterations (corresponding to 16 functional evaluations) of random design searches. Then, Eq. (20) is satisfied after two iterations and the total number

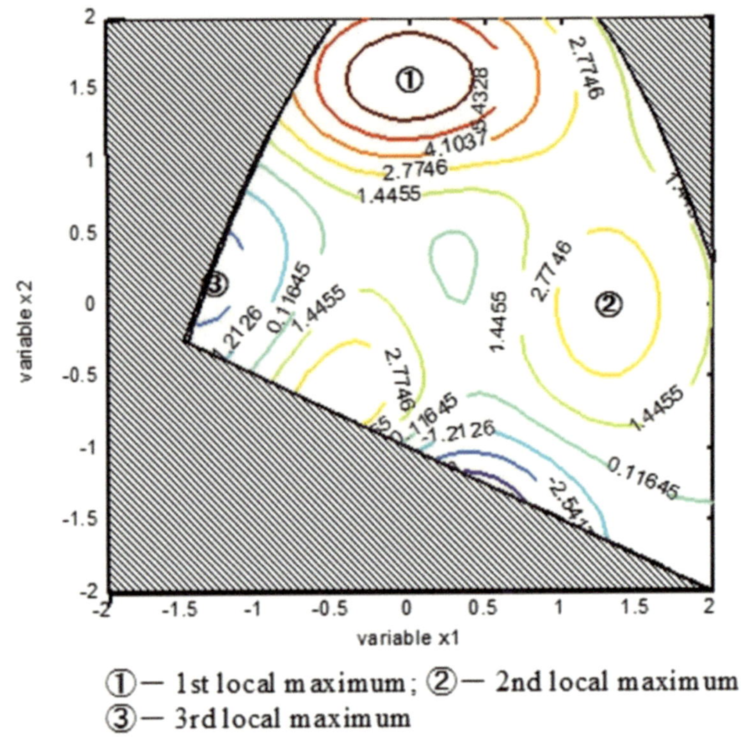

Figure 11.
Contour plot of objective function and constraints: It is a contour plot, and the feasible design space is enclosed within the shadowed area. Three local optimal designs, marked ①, ②, ③ exist within the feasible design space.

of functional evaluations until convergence is 18. The approximation optimal design is =8.105. The history of each iteration is illustrated in **Figure 13**, in which the horizontal and vertical axes have the same meaning as those of **Figure 12**. Further inspection of the global approximation between the above two cases is performed using 1600 (40×40) lattice design points covering the entire design space. The deviations of cases (1) and (2) are 1.12 and 1.32,

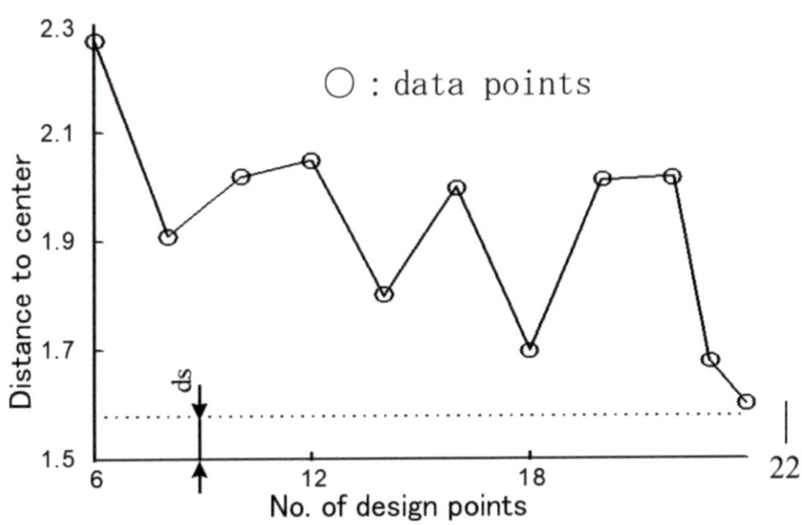

Figure 12.
Convergence of large standard deviation: The horizontal axis stands for the distance (ds) between the original position (0,0) and the approximation optimal design.

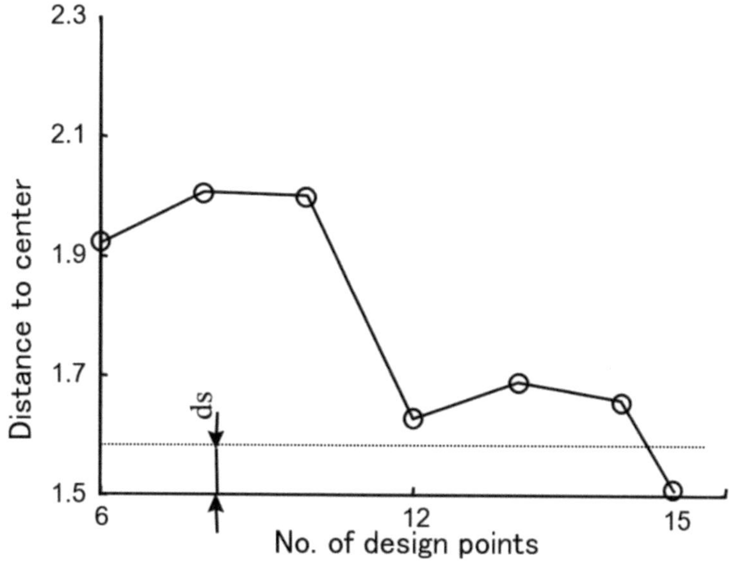

Figure 13.
Convergence of small standard deviation: The horizontal axis stands for the distance (ds) between the original position (0,0) and the approximation optimal design.

respectively. That is, a larger deviation of Gauss random designs gives slower convergence but higher accuracy of approximation over the entire design space. On the other hand, a smaller deviation of Gauss random designs gives rapid convergence but lower accuracy of approximation over the entire design space. Let us confirm the effect of Eqs. (22) and (23). If the threshold termination in Step 1 is set to ε_1=0.15, the number of designs is 12 and 22 in case α=2 and α=5, respectively. If the threshold termination in Step 1 is set sufficiently small, say, $\varepsilon_1=1.0\times10^{-3}$, then the history of convergence is depicted in **Figure 14** and the number of functional evaluations in Step 1 becomes 36. However, on the basis of the result shown in **Figure 14** we know

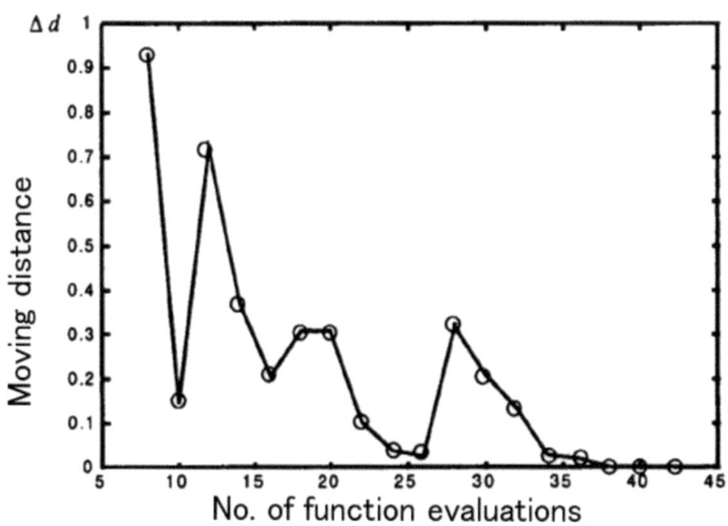

Figure 14.
Change of distance in large standard deviation: The number of functional evaluations in step 1 becomes 36.

that there is no further improvement in determining the global optimal design after 36 functional evaluations, except to increase the probability. On the other hand, d_0=0.9 can be determined for this problem, and Step 1 is terminated when ε_1=0.9 at eight designs. In practice, however, this is not known; therefore, if one attempts to search for the global optimal design including small clusters, one is expected to select a stricter termination threshold in Step 1 and a larger deviation of Gauss random designs.

5. Application to vehicle problems

5.1 Vehicle idling vibration problem

5.1.1 Modeling for optimal problem

NVH such as noise, vibration and harshness is very important for *merchantability of vehicle. Here idling vibration is treated as one of* NVH. *Many studies have been done related to idling vibration where two types of modeling exist. One modeling is that body is separated from powertrain and engine mount [40]. The other one consists of powertrain, engine mount and body together where body is modeled very simply such as solid or as only 1st bending and twist modes in most of studies [41, 42]. Here powertrain and engine mount are modeled together with rather precise body model.*

Here powertrain and engine mount are modeled as shown in **Figure 15** *where weight of powertrain is supported by 3 main mounts. Each mount is modeled with 3 springs in x, y, z directions and damping. The weight of each mount is so small compared with body and powertrain that its weight is ignored.*

Powertrain is so stiff compared with body that it is modeled as 6 rigid body modes. Body is presented with 44 modes under 50 Hz which are calculated by eigenvalue analysis for the FEM model in **Figure 16**. Total model is generated by component mode synthesis [43] by using spring parameters, *d*amping for engine

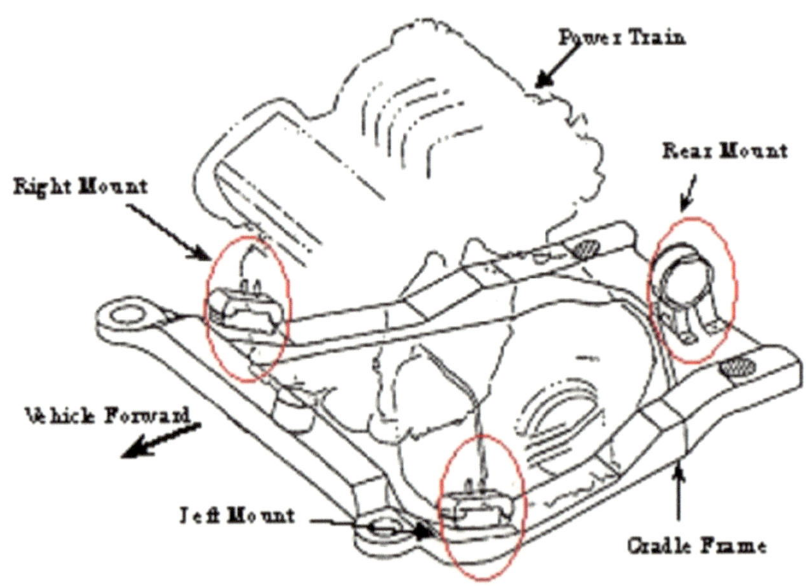

Figure 15.
Mounts model: Weight of powertrain is supported by 3 main mounts. Each mount is modeled with 3 springs in x, y, z directions and damping.

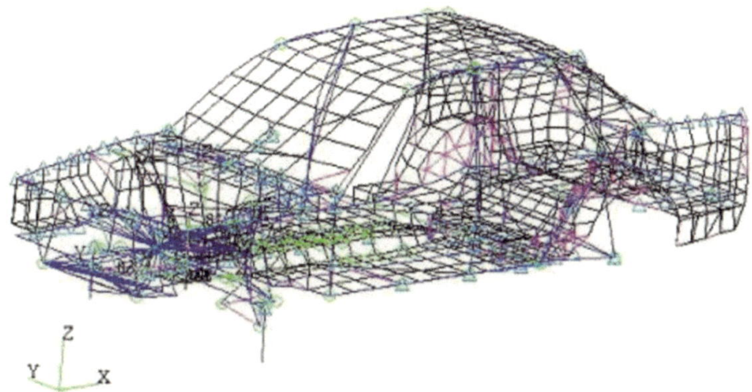

Figure 16.
Vehicle FEM model: Body is presented with 44 modes under 50 Hz.

mount and by vectors, eigenvalues and eigen damping for body and powertrain. This total model has 50 numbers of freedom which is much lower than that of physical coordinates. Only engine mount is changed in repeated optimal calculations.

5.1.2 Setting up optimization problems

5.1.2.1 Position arrangement of engine mount

It is shown 34 candidates for position arrangement of engine mount in **Figure 17**. They are arranged so that powertrain is enclosed. 3 points are selected among these 34 points. Among the 3 selected points p_{t1}, p_{t2}, p_{t3}, the next relationship is held:

$$1 \le p_{t1} < p_{t2} < p_{t3} \le 34 \tag{26}$$

When these 3 mounts are placed closely, support for powertrain becomes unstable.

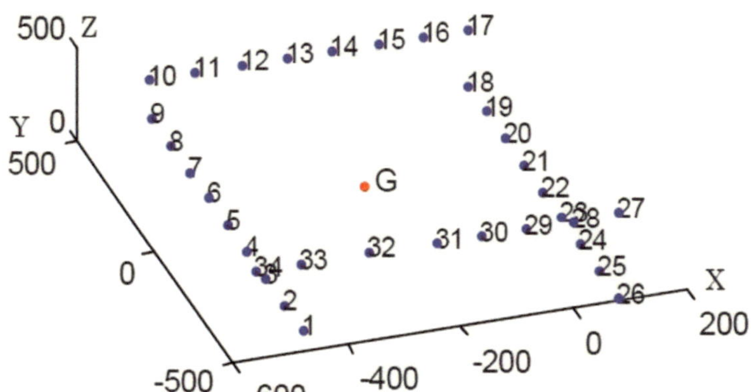

Figure 17.
Candidates for location of mounts: 34 candidates for position arrangement of engine mount are arranged so that powertrain is enclosed. 3 points are selected among these 34 points.

To deal with this, the following restrictions are added:

$$\begin{aligned} p_{t2} - p_{t1} &\geq 5 \\ p_{t3} - p_{t2} &\geq 5 \\ p_{t1} + 34 - p_{t3} &\geq 5 \end{aligned} \tag{27}$$

Spring and damping parameters are continuous variables. Following constraints are given to guarantee other characteristics than idling vibration.

- Front and rear stiffness for dynamic vibration and acceleration shock when the engine starts:

$$\sum_{i=1}^{3} \text{kxi} = 35.0\ [\text{kgf/mm}] \tag{28}$$

- Left and right stiffness for handling and stability

$$\sum_{i=1}^{3} \text{kyi} = 45.0[\text{kgf/mm}] \tag{29}$$

- Up and down stiffness for engine shake

$$\sum_{i=1}^{3} \text{kzi} = 45.0[\text{kgf/mm}] \tag{30}$$

Here k_{di} (d=x, y, z, i=1,2,3) are stiffness parameters of number i, direction d. All parameters have upper and lower limits as follows:

$$\begin{aligned} 7.0 \leq k_{xi} &\leq 21.0 \\ 9.0 \leq k_{yi} &\leq 27.0 \\ 9.0 \leq k_{zi} &\leq 27.0 \end{aligned} \tag{31}$$

$$0.02 \leq c_{di} \leq 0.1\ (d = x, y, z, i = 1, 2, 3)$$

Damping parameter is also constrained as follows:

$$\sum_{i=1}^{3} c_{di} = 0.1\ (d = x, y, z) \tag{32}$$

Total number of design values is 21 such as mount location (3×3), stiffness parameters of 3 directions (3×3) and damping parameter (3).

From Eqs. (27) and (30), follows are induced:

$$\begin{aligned} K_{d3} &= 35.0 - (k_{d1} + k_{d2}) \\ 14.0 \leq k_{d1} + k_{d2} &\leq 28.0\ (d = x, y, z) \end{aligned} \tag{33}$$

This means that 21 design values and 6 linear equation constraints induce 15 design values (as shown in **Table 2**) and 6 boundary values (as shown in **Table 3**).

	Mount1	Mount2	Mount3
Location	p_{t1}	p_{t2}	p_{t3}
Stiffness (x)	kx1	kx2	kx3
Stiffness (y)	ky1	ky2	ky3
Stiffness (z)	kz1	kz2	kz3
Damper (x)	Cx1	Cx2	Cx3
Damper (y)	Cy1	Cy2	Cy3
Damper (z)	Cz1	Cz2	Cz3

Table 2.
21 design variable candidates. Since there are 6 boundary conditions as shown in ***Table*** *3, 15 of them are design variables.*

Lower		Upper
14.0	$k_{x1} + k_{x2}$	28.0
18.0	$K_{y1} + k_{y2}$	36.0
18.0	$K_{z1}+ k_{z2}$	36.0
0.04	$C_{d1} + C_{d2(d=x,y,z)}$	0.08

Table 3.
Boundary conditions.

5.1.2.2 Settings of objective function

Idling vibration is estimated at 2 observer points where the occupant seat is installed on the floor. These points are intersection points between left and right side sills and cross member where the seat is installed. The frequency vertical responses of these 2 points are shown in **Figure 18**. Idling vibration is the phenomena during engine rpm 600–900 and so it is estimated by frequency response from 20 Hz to 30 Hz. As a result, the objective function is the sum of 2 integral functions with frequency in this range. Trapezoidal rule is used for integration with integral interval 0.2 Hz.

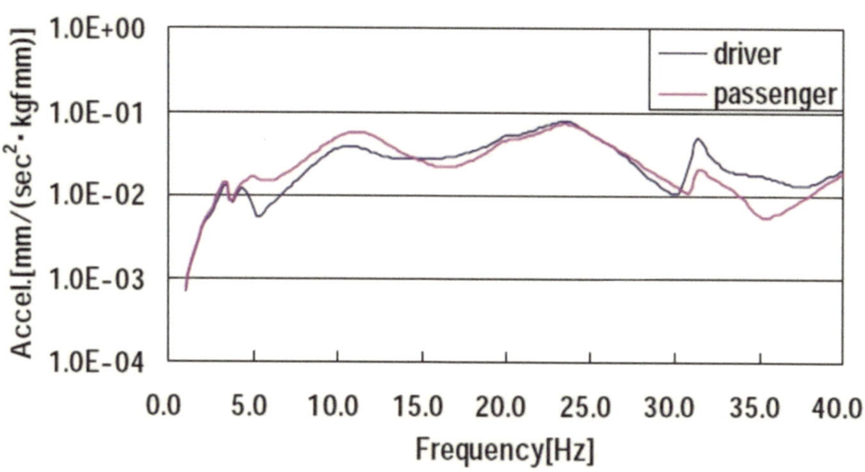

Figure 18.
Acceleration response for frequency at observation point: Idling vibration is the phenomena during engine rpm 600–900 and so it is estimated by frequency response from 20 Hz to 30 Hz.

5.1.3 Optimization results

First, with N initial samplings, response values are calculated. N means at least necessary number for the approximation with secondary polynomials;

N= (n+1)(n+2)/2 (n is the number of design variables, here, n=15. So N=136). Each sampling point is selected among feasible region which is given from Eqs. (25), (26), (30) and **Tables 2** and **3**. Both of ε_1, ε_2 are set up 0.05 and α_i is set up 1.5. Both of GA and SQP (Sequential Quadratic Programming) are used to explore the optimal value on the approximated response surfaces.

Three types of calculation are performed as far as initial point for optimal analysis by SQP. In 1st one, it is selected the point where it is the optimal point using GA. In 2nd one, last optimal point is selected as the initial point. In the last one, the best solution point among sample points is selected as the initial point. In each step, these 3 calculations are performed. However, because the solution by SQP is continuous value, at last stage discrete value is searched around the solution point. It is the optimal point which has the least value among three types calculation in each step.

5.1.3.1 Comparison with GA

To certify the effectiveness of MPOD, it is compared with GA. Here general GA is selected although there are many types of GA. Real-coded Genetic Algorithms is adopted with elite preservation strategy where population size is 50 and number of generations is 100. It is shown in **Table 4** the comparison between GA and MPOD.

As far as the problem where it takes much time to calculate objective function, it is very effective to use MPOD.

In term A, it means the precision of optimal value is almost same. In term B, it is compared the number of calculation of objective function. It is fewer in MPOD than in GA. But in term C, total time is shorter in GA than in MPOD. The calculation time of objective function for one selected point is 0.23 sec. In term D, all calculation times for object functions are filled out. In term E, in MPOD, the calculation time of objective function accounts for about 13% in the total calculation time although in GA it accounts about 90%. In other words, the total calculation time in GA is more influenced by the calculation time of object function than in MPOD. On the other hand, the most time-spent process in MPOD is to search the optimal solution on the approximation function updated every time. In GA, this process is finished within few seconds although in SQP, it takes sometimes few minutes with some initial values. And so in MPOD, it influences very much to the calculation time how the optimal value can be searched on the approximation function. As far as the problem where it takes much time to calculate objective function, it is very effective to use MPOD.

		MPOD	GA
A	Response of optimum design	0.244	0.240
B	Number of sampling points	189	5000
C	Total calculation time [sec]	1571.8	1256.8
D	Calculation time of response for all sampling points [sec]	43.7	1150.0
E	Rate of D for C [%]	2.78	91.50

Table 4.
Comparison between result of MPOD and one of GA.

5.2 Booming noise problem for simplified vehicle model

5.2.1 Modeling for optimal problem

Here booming noise for wagon-type automobile is treated. This noise is from 40 to 60 hertz when a vehicle gets over a small protrusion.

This is the coupled acoustic-structural vibration problem [44–48].

It is shown the simplified model simulated vehicle system in **Figure 19** which has average dimensions such as width 1.5 m, length 3 m, height 1.2 m. It is supposed that the nodal shape box is body and inside the box is interior space. The FEM model is generated such that the body has 1980 rectangular shell elements with length of one side 0.1 m, thickness 1 mm and inside the box has 5400 cube acoustic elements with length of one side 0.1 m. To treat the coupled forward and rear direction acoustic-upper panel structural phenomena, the 5 faces except upper panel are high stiffness by 100,000 times Young modulus of these 5 faces such that the 1st eigen-frequency of acoustic system and second panel bending one are near. The mode attenuation ratio is 3.5% both for the structural system and acoustic system.

5.2.2 Setting up optimization problems

The problem is set as follows:

The objective function: the average of 8 frequency integral values of each sound pressure response absolute value at observation points shown in **Figure 19**. The eight coordinate positions are 1 m from the front or the back panel in Z direction, 0.3 m from the right or the left panel in Y direction and 0.3 m from the upper or the bottom panel in X direction symmetrically. Integral values of each sound pressure response absolute value are calculated by trapezoid official over from 40 to 50 hertz divided into 200 pieces:

Design values: X_1,X_2.

X_1: the position coordinate of the roof bow on the roof panel

0.5 $\leq X_1 \leq$2.5, 0.1 m increments, 21 point discrete values

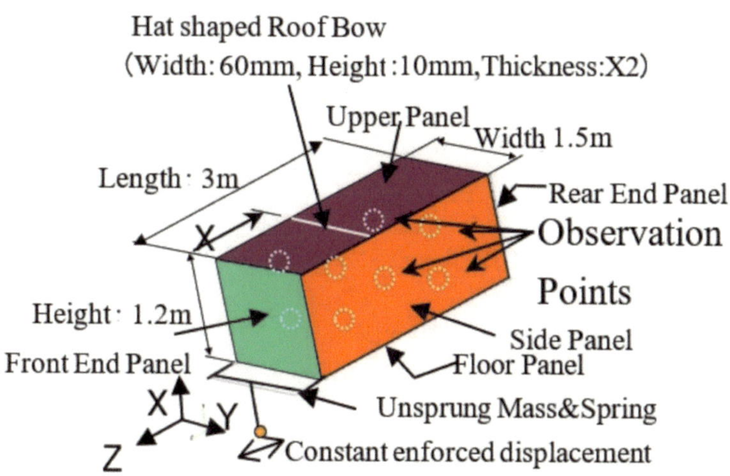

Figure 19.
Simplified vehicle FE model: To treat the coupled forward and rear direction acoustic-upper panel structural phenomena, the 5 faces except upper panel are high stiffness by 100,000 times young modulus of these 5 faces such that the 1st eigen-frequency of acoustic system and second panel bending one are near.

X_2: panel thickness $0.6 \le X_2 \le 1.8$, 0.3 mm increments, 5 point discrete values.
Input force: Sinusoid forced displacement in forward and rear direction with amplitude 0.2 mm which is equivalent to the force which is generated when the vehicle gets over a small protrusion.

5.2.3 Optimization results

Because the number of design values is 2, 6 initial learning points are selected through uniform random number by keeping longer than d_{min}. Here d_{min} is 0.33, ε_1 in Eq. (19) is 0.05, ε_2 in Eq. (20) is 0.01 and $\alpha_1 = \alpha_2$ in Eq. (21) are 2.0. The distribution of learning points is shown in **Figure 20**. In step 1, 6 points marked 1st and 3 points marked 7th–9th are set over entire design space. On the other hand, 3 points marked 10th–12th in step 2 are concentrated on one part of the space. In step 1, although after 3 repetition, Eq. (19) is satisfied with X_1 =2.5 and X_2 =0.6, Eq. (20) is not satisfied. And so in step 2 it is repeated. The first optimal point in step 2 is X_1 =1.3 and X_2 =0.6 apart from X_1 =2.5 and X_2 =0.6. And finally, Eq. (20) is satisfied with X_1 =1.5 and X_2 =0.9. This is because there are 2 design points which have almost same minimum values, one design point is X_1 =1.5, the other is X_1 =2.5. And so the optimal point moves between 2 minimum points on the way to be improved accuracy on response surface. As a result, it converges to the optimal solution X_1 =1.5, X_2 =0.9 and Y=95.59 with 12 learning points. In such way, optimal value is searched by repeating step 1where global solution is searched and by repeating step 2 where approximation precision is improved on the response surface. The approximation response surface is shown in **Figure 21**. Although referring to **Figure 19**, originally, the response surface must be symmetric with respect to X_1 =1.5, it is asymmetric because the learning points have biased distribution. This shows we can abbreviate unnecessary learning points for getting minimum value by taking many learning points in neighborhood of optimal point.

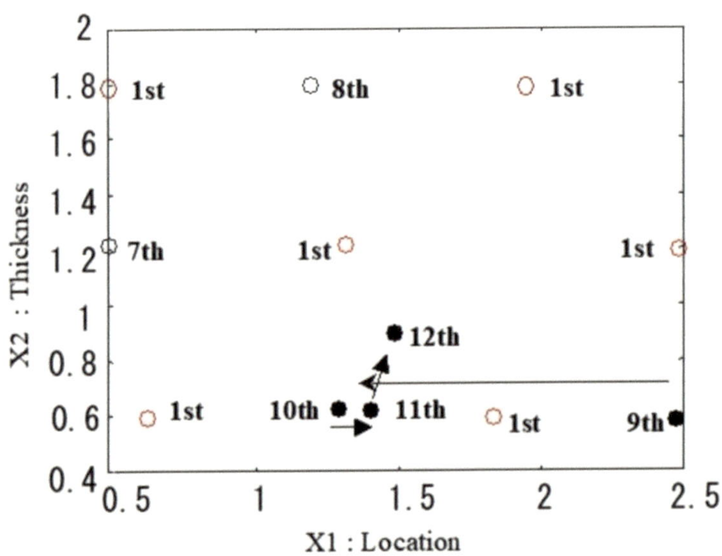

Figure 20.
Learning points distribution in MPOD: In step 1, 6 points marked 1st and 3 points marked 7th–9th are set over entire design space. On the other hand, 3 points marked 10th–12th in step 2 are concentrated on one part of the space.

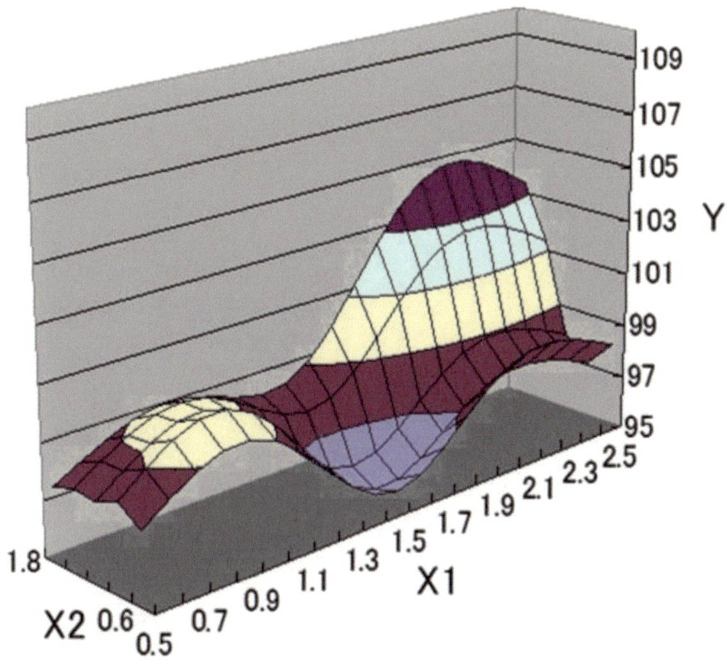

Figure 21.
Approximate response surface by MPOD: The response surface must be symmetric with respect to X_1 =1.5, it is asymmetric because the learning points have biased distribution. This shows we can abbreviate unnecessary learning points for getting minimum value by taking many learning points in neighborhood of optimal point.

5.2.4 Comparison between MPOD and experimental planning method using orthogonal table (EPO)

As shown in **Figures 22**, 25 points at 5 levels are selected by quartering upper and lower design limits. **Figure 23** shows the response surface approximated by 4th order direct polynomial. It is high in the symmetric level of response surface with respect to X_1 =1.5 because the design points are uniformly selected. It has 3 valleys

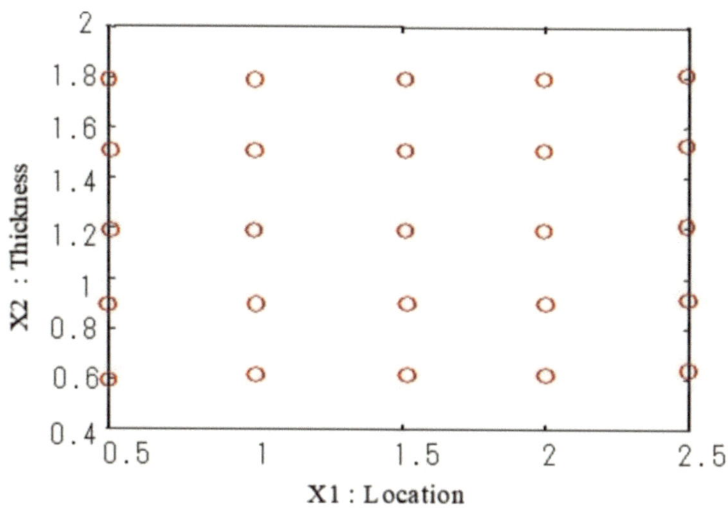

Figure 22.
Designing point distribution in EPO: 25 points at 5 levels are selected by quartering upper and lower design limits.

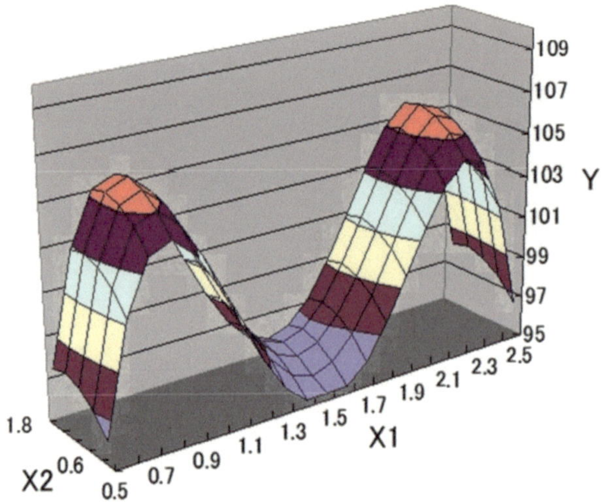

Figure 23.
Approximate response surface by EPO: This response surface is approximated by 4th order direct polynomial. It is high in the symmetric level of response surface with respect to X_1 =1.5 because the design points are uniformly selected.

in the neighborhood of X_1 =0.5, 1.5, 2.5 and 2 mountains in the neighborhood of X_1 =1.0, 2.0. Gradient of response surface for X_2 is extremely smaller than the one for X_1. The optimal solution by this response surface is X_1 =1.5, X_2 =0.6 and Y=94.36 which are slightly different from X_1 =1.5, X_2 =0.9 and Y=95.59 by the MPOD. The reason for this difference is that although in EPO the trial points must be placed uniformly, in MPOD they are placed adaptably such as the distance between any 2 points is longer than d_{min} The minimum point by EPO exits in the region within d_{min} =0.33 where it is out of this problem.

5.3 Booming noise problem for real vehicle model

5.3.1 Modeling for optimal problem

Following the previous section, it is applied the MPOD also to the booming coupled noise and structural problem for a real wagon-type vehicle where the dynamic damper should be set. It is shown the overview of vehicle FEM model in **Figure 24** where the structural system is modeled with shell elements, the acoustic

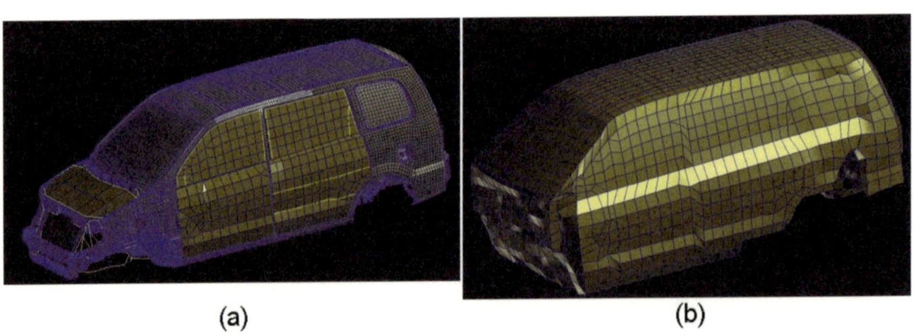

Figure 24.
(a) Structural FE model for trimmed body. (b) Acoustic FE model of cabin. Overview of vehicle FEM model: The structural system is modeled with shell elements, the acoustic system is modeled with solid elements and the dynamic damper is modeled with lumped mass and one degree of freedom scalar spring.

system is modeled with solid elements and the dynamic damper is modeled with lumped mass and one degree of freedom scalar spring.

5.3.2 Setting up optimization problems

Let us set up the objective function Y and the design variables X1, X2 as follows:

Y: Minimize the maximum SPL (Sound Pressure Level) during 30 and 60 Hertz at the observation point ● which is set ear height position of the center of the front seat under the condition such that forward and rear unit input is loaded at the connected point of the arm to the front suspension for giving the same-phase vibration on the left and right. This load is generated when the vehicle gets over a small protrusion.

X1: the position of the damper which is selected among 22 candidate from ① to ㉒ in **Figure 25.**

X2: Spring constant which is selected among 5 candidates: 7.0,9.0,11.0,13.0,15.0 kgf/mm. The weight and the structural damping coefficient are constant such as is 1.5 kg and 0.1 each other.

5.3.3 Optimization results

As far as the convergence judgment for the optimal, ε_1 is 0.05, ε_2 is 0.01 and also as far as the control coefficient, α_1 and α_2 are 2.0. From the approximation response surface with the initial 6 points, the 1st optimal is X1=16, X2=11.0, Y=0.659. The 2nd optimal is X1=15, X2=7.0, Y=0.567 by adding 7th, 8th 2 points. The step 1 is terminated because the 3rd optimal is X1=15, X2=7.0 that is same as last time by adding 9th, 10th 2 points. It is finished because the optimal solution X1=14, X2=7.0, Y=0.556 are gained by adding the 11th point and this satisfies with Eq. (20). It is shown the learning distribution points in **Figure 26** and the approximated response surface using 11 learning points in **Figure 27**. This means the minimal solution over the total design region is gained with very few learning points compared with total 110 design points. As shown in **Figure 28**. The objective function Y is reduced 6 dB around 45Herz by the optimal analysis. As shown in **Figure 29**, the coupled acoustic structural mode at 44.3 Hz gives the entire car body elastic deformation induced by the deformation of the upper panel and rear floor. Moreover the acoustic system gives 1st mode of which node is a little rear from the center of passenger cabin and of which anti-node is front to back end of passenger cabin. It is shown the difference of the panel vibration acceleration along the center line of vehicle body in **Figure 30** with the optimally optimum arranged damper and without damper.

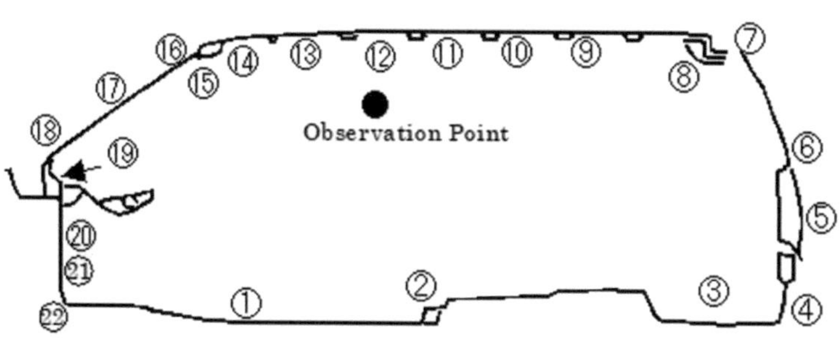

Figure 25.
Candidates of dynamic damper location: The position of the damper which is selected among 22 candidate from ① to ㉒.

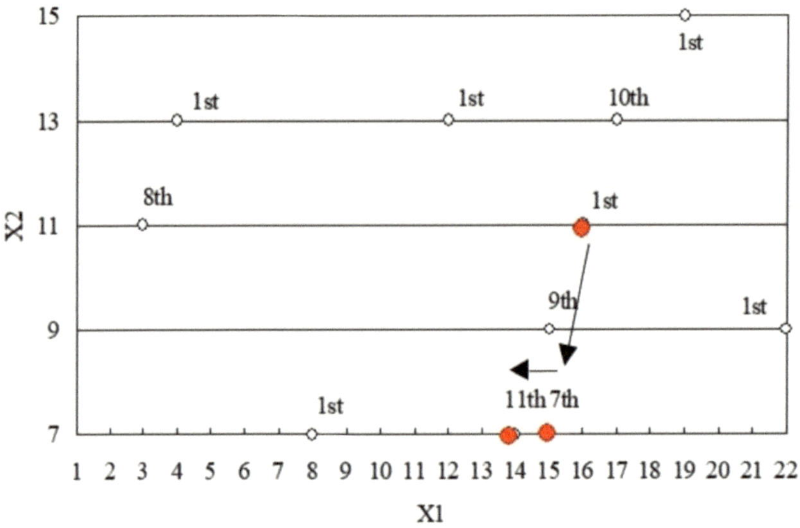

Figure 26.
Learning points distribution: The minimal solution over the total design region is gained with very few learning points compared with total 110 design points.

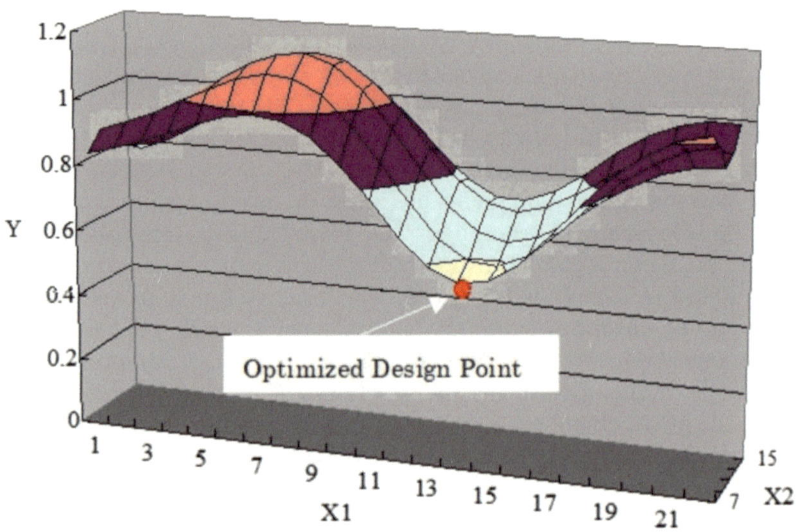

Figure 27.
Response surface by 11 learning points:

From **Figure 30**, the effect of optimal arrangement of damper can be interpreted to be the vibration-absorbing effect for the global coupled acoustic-structural mode at 44.3 Hz rather than the local vibration reduction around the front roof rail.

To certify the effect of the optimal damper arrangement, bench test using smooth drum is performed with the optimal arrangement damper and without damper. It is shown the noise level–frequency relationship in **Figure 31**. Maximum approximately 5 dB reduction of SPL is realized among 45~50 Hz. Although it needs many man-hours to find out the optimal arrangement of dynamic damper by experiment trial and error, it has been certified the method for booming noise reduction can be realized efficiently by MPOD.

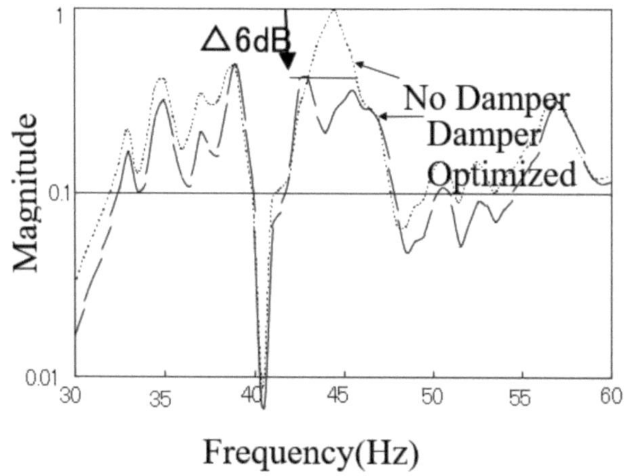

Figure 28.
Estimation of optimized dynamic damper effect: The objective function Y is reduced 6 dB around 45Herz by the optimal analysis.

5.4 Vehicle crash problem

5.4.1 Setting up optimization problems

The vehicle side member (component parallel to the central axis of vehicle) plays a big role in energy absorption while the frontal collision and energy absorption of the vehicle is determined by its size, shape and welding [49–53]. In this example, the uniform hollow square-cross section, a perfectly straight side member with uniform thickness is investigated and reinforcement of the component is considered as a way to increase the energy absorption. The size and material property of the member are ρ = 7.85E-3 g/mm^3, E = 210GPa, E_t = 2.5GPa, σ_y = 220 MPa, v = 0.3, and the form of reinforcement is depicted in **Figure 32**.

All degrees of freedom at the bottom end are rigidly fixed, and at the top end, a rigid mass of 500 kg, and velocity of 54 km/h are used as a load to simulate a crush. The load–displacement behavior of the member while crushing is calculated by FEM solver LS-DYNA3D [54], in which 2700 shell elements and 2754 nodes are used. The design variables include the thickness of the base plate (t_1), the upper plate (t_2) and the location of reinforcement (z). The total weight of the component is constant (w=780 g), therefore only two design variables are independent. The objective function is the energy absorption of the component, 10 ms after crushing. The mathematical definition of the problem is as follows;

$$\text{Max} f\ (t_1, t_2, z)$$
$$\text{Subject to}: 0.5 \le t_1, t_2 \le 1.0, 1.5 \le z \le 250$$
$$t_2 = (1.5 + (1 - t_1))/2.0$$

5.4.2 Optimization results

The training procedure and end thresholds are identical to the case of large deviation, that is, α=5, β=0.6 and ε_1=0.15, ε_2=0.01. Step 1 is terminated after four iterations (corresponding to 12 functional evaluations) of random seed searches. Then, Eq. (20) is satisfied after two iterations and the total number of functional evaluations until convergence, is 14. The approximation optimal is (1.21, 1.39,160), f $(x_{op}(k))$=8484.11 kJ, which is 11.6% higher than that of the original design

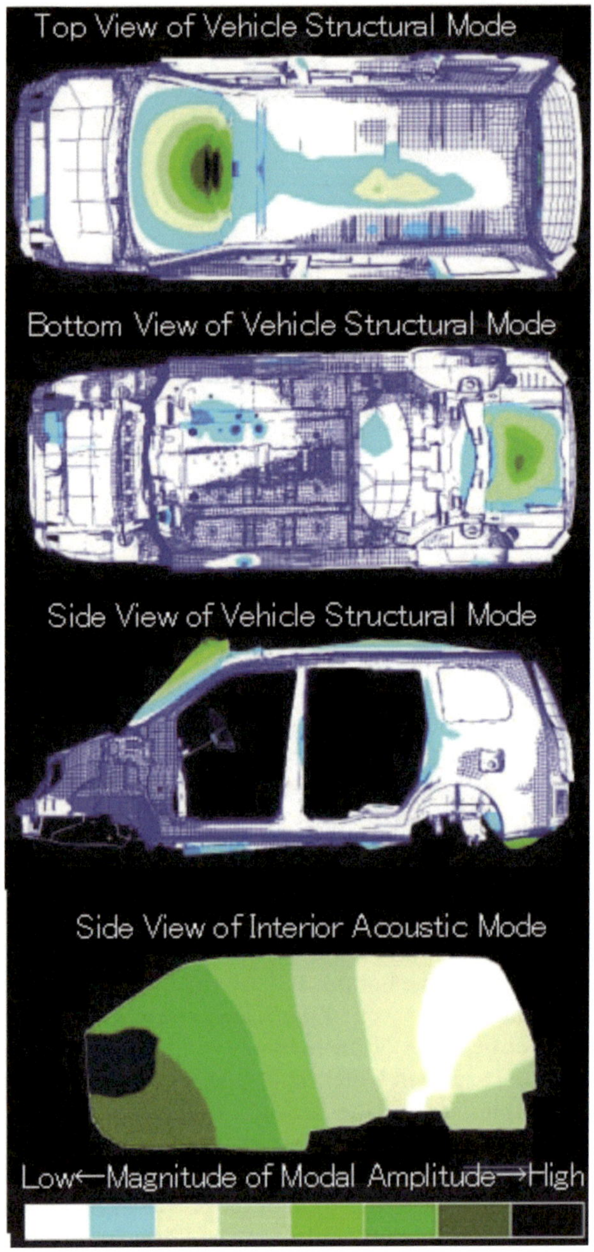

Figure 29.
Fluid–structure coupled eigen mode (44.3 Hz): The coupled acoustic structural mode at 44.3 Hz gives the entirecar body elastic deformation induced by the deformation of the upper panel and rear floor.

(1.00,1.50,0.0). The contour plot of the approximation function is depicted in **Figure 33**. In the figure, the open circle ○ indicates the optimal design using the proposed MPOD method and the plus sign+ expresses the designs used for approximation. The comparison between the approximation function and FEM values with t_1=1.21 and t_2=1.39 are fixed while the changing reinforcement location z is plotted in **Figure 34**. It can be seen from the figure, and also mentioned in Refs. [49–53], that the FEM value varies, because of the heavy nonlinearity of the crashworthiness problems [49–53]. A smooth and robust approximation function is obtained by the HNN approximation and an optimal design is realized successfully. The MPOD

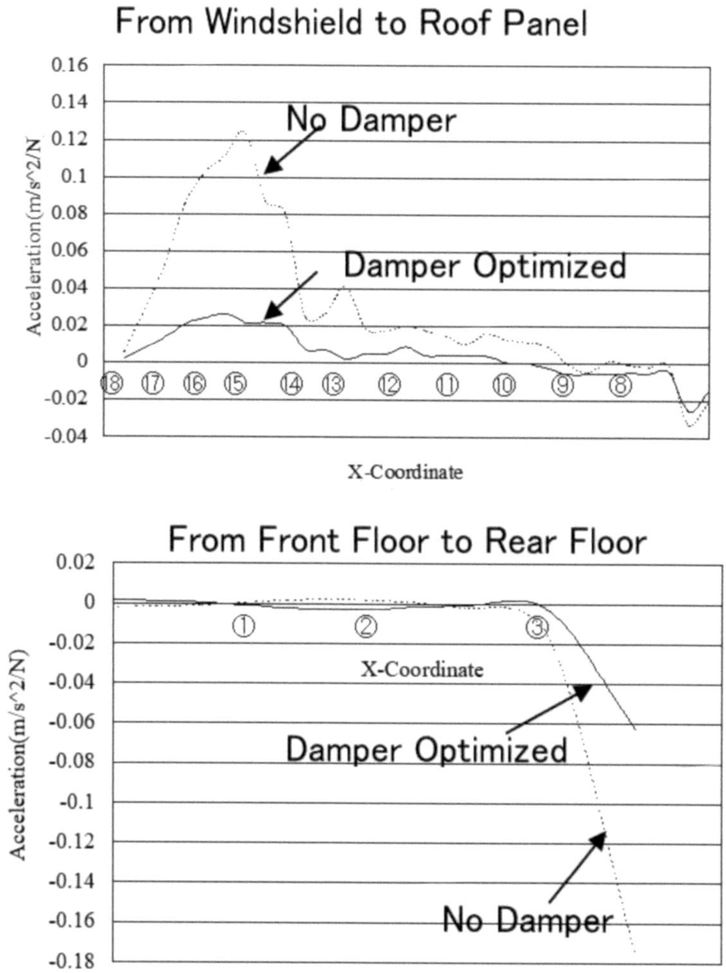

Figure 30.
Acceleration of vehicle body panel (44.3 Hz): Optimal arrangement of damper can give the vibration-absorbing effect for the global coupled acoustic-structural mode at 44.3 Hz rather than the local vibration reduction around the front roof rail.

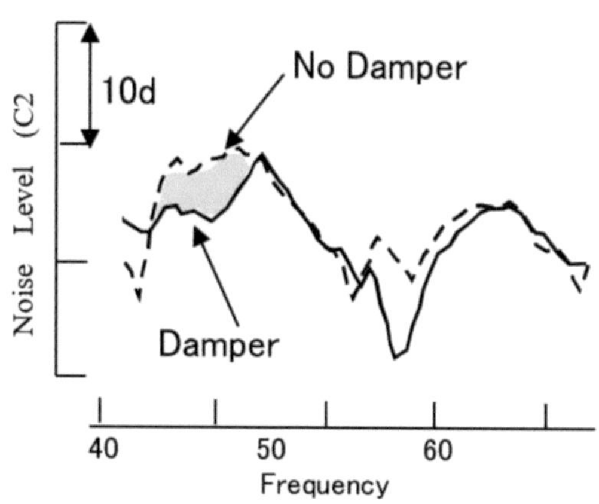

Figure 31.
Validation of optimized dynamic damper effect: Maximum approximately 5 dB reduction of SPL is realized among 45~50 Hz.

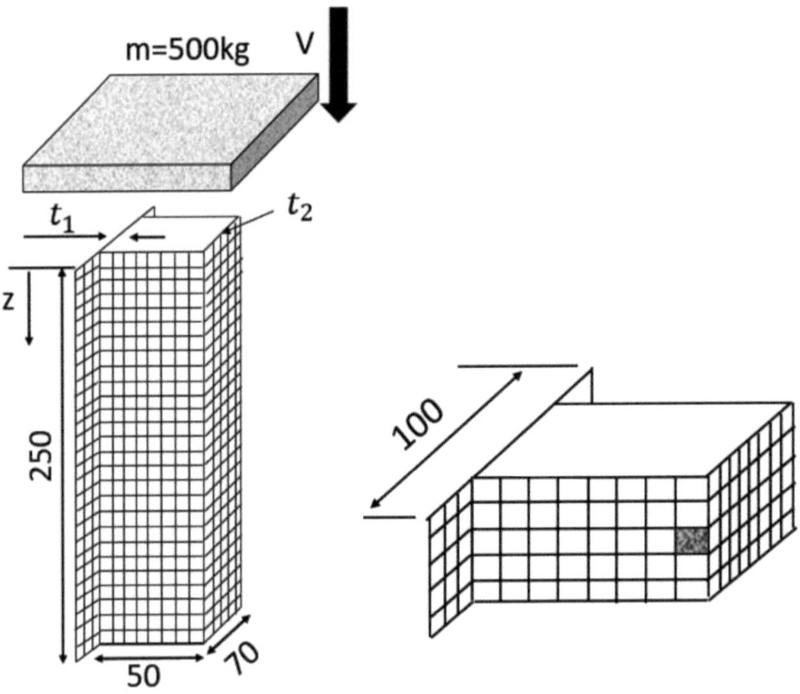

Figure 32.
Analysis model of crashworthiness optimization: The member is perfectly straight with uniform hollow square-cross section and uniform thickness.

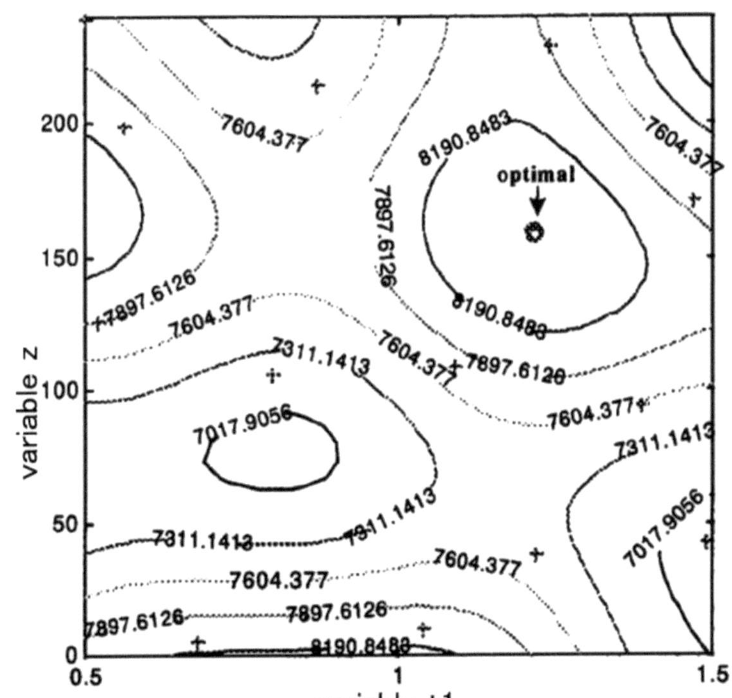

Figure 33.
Approximation function of the present method: The open circle ◯ indicates the optimal design using the proposed MPOD method and the plus sign+ expresses the designs used for approximation.

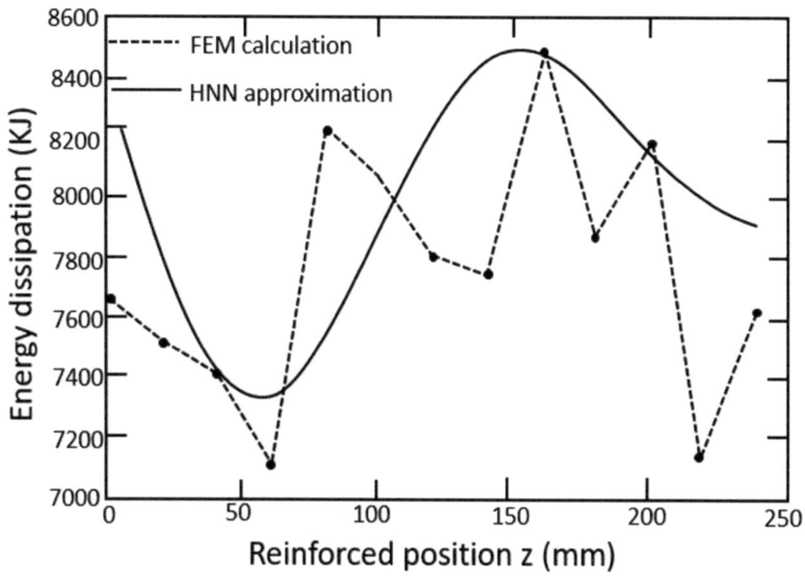

Figure 34.
Comparison of FEM and HNN when design variable z is changed: The FEM value varies, because of the heavy nonlinearity of the crashworthiness problems. A smooth and robust approximation function is obtained by the HNN approximation and an optimal design is realized successfully.

approach based on the HNN has a robust property against calculation noise. However, the deviation of the coefficient is dependent on the design distribution. This topic will be discussed further in future works.

5.5 Combination problem between vehicle crashworthiness and restraint device performance at the head-on collision

5.5.1 Setting up optimization problems

Here the problem is the head-on collision where the vehicle collides perpendicular to a concrete wall. This phenomena is that the vehicle finishes the moving after the crush energy is absorbed completely by the vehicle structure [55]. During this phenomena, the occupant moves length L+I in the axial direction where L is the amount of crush and I is the amount of movement of the occupant in the cabin. The longer L+I is the smaller the average deceleration of the occupant is. And so it is necessary to raise L+I. On the other hand, if the amount of crush of the vehicle front-end is too large, the possible amount of occupant moving becomes smaller. Also if the movement amount of the occupant to the vehicle front side, it arises the possibilities of the crush between the occupant and the steering wheel, etc. In such way, the most important issue is to get the optimal valance between the vehicle crashworthiness and occupant restraints performances. So let us start the feasibility study concerned with this. Ever before, representative 4 patterns have been considered as the best vehicle performance for the occupant. As shown in **Figure 35**, these 4 patterns of vehicle crush performance have been referred to the standard numbers which are generated when the small passenger cars collide perpendicular to a concrete wall with 35mph based on the criterion of NHTSA (National Highway Traffic Safety Administration). Here it is selected among these 4 vehicle crush performances as the design of vehicle. That means the design vehicle value becomes integer for the vehicle performance. As far as the occupant restraint, it is supposed

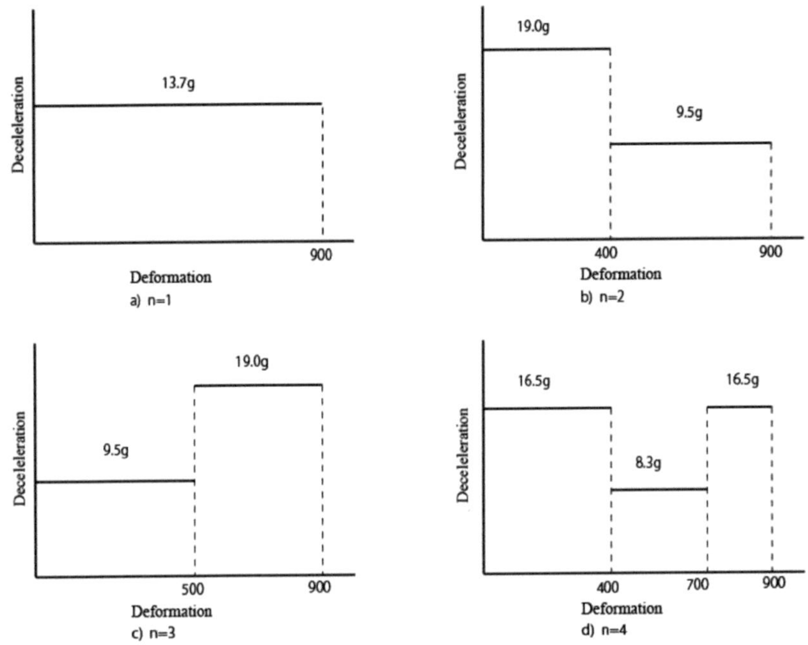

Figure 35.
Type of deceleration curve of vehicle: These 4 patterns of vehicle crush performance have been referred to the standard numbers which are generated when the small passenger cars collide perpendicular to a concrete wall with 35mph based on the criterion of NHTSA.

that there is not the contact between the occupant and the steering wheel for simplicity so that only seat belt is considered. For the design values of seat belt, the magnification x_1 of standard value 2.5GPa of seat belt Young modulus and the magnification x_2 of standard value 3000 N of working load of the load limiter are selected. Here the load limiter is the device which relieves the impact on the chest by sending seat belt with the load remained constant when the load applied to the seat belt winder reaches to the working load as shown in **Figure 36**. Here the send-off amount of load limiter is 100 mm. Next it is shown in **Figure 37** the analysis model of dummy system which consists of dummy, seat belt and sled. The dummy is Hybrid III whose size is the average of adult male. The number of total nodal points is 3700. The number of shell elements is 3203 for dummy and 116 for sled. Seat belt is basically modeled with 58 beams. But the parts of the seat belt for the chest and the waist are modeled with 192 shell elements so that the contact between dummy and seat belt can be considered precisely. The dummy and the sled are solid and seat belt is elastic. Contact by penalty method is defined for the friction between dummy and seat. The coefficient of friction is 0.2. The model for seat is comparatively coarse because it is only for the in-plane force to get friction as shown in this figure. The objective function is Eq. (35) by using HIC in Eq. (34) and chest G which are generated by head-on collision with crush velocity 35mph referred to the criterion of NHTSA.

$$\mathrm{HIC} = \left[\frac{1}{t_{2-t_1}} \int_{t_1}^{t_2} \alpha \, dt\right]^{2.5} (t_2 - t_1) \tag{34}$$

$$\mathrm{Min}\ \mathrm{f}(\mathrm{n}, x_1, x_2) = \sqrt{\left(\frac{HIC - 200}{200}\right)^2 + \left(\frac{ChestG - 20}{10}\right)^2} \tag{35}$$

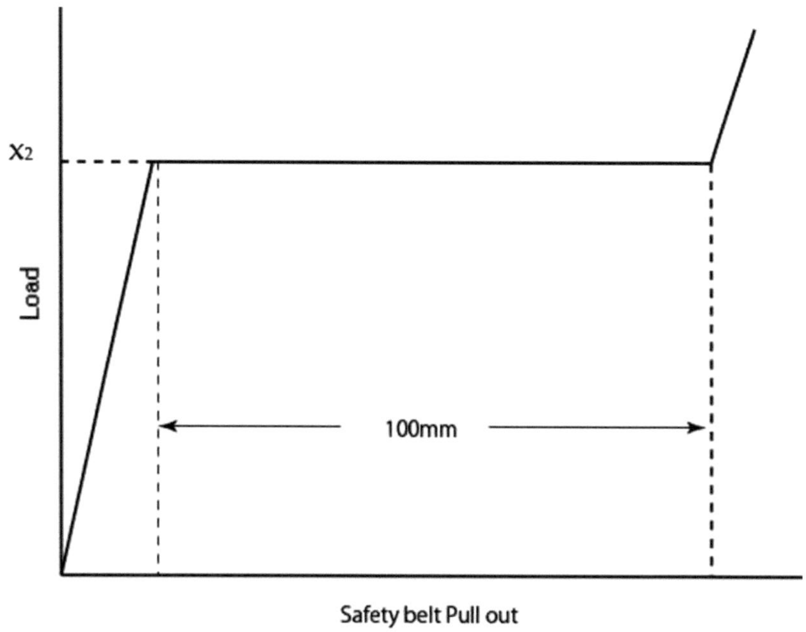

Figure 36.
Safety belt pull out vs. load of load limiter: The load limiter is the device which relieves the impact on the chest by sending seat belt with the load remained constant when the load applied to the seat belt winder reaches to the working load.

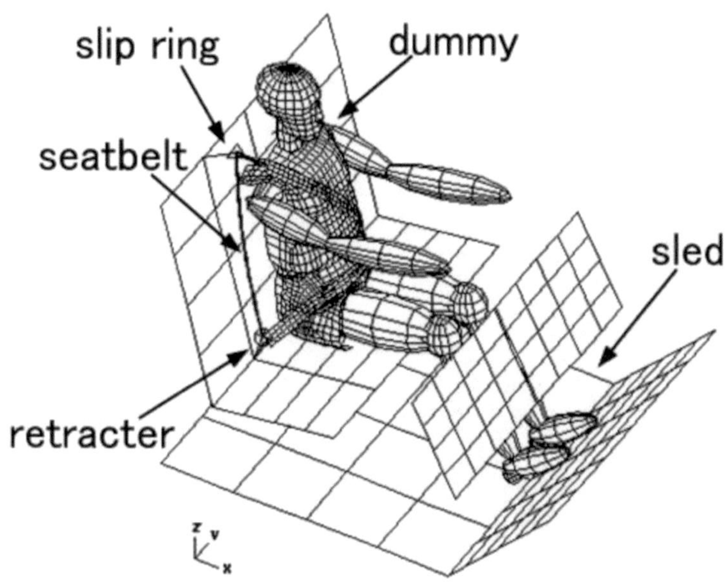

Figure 37.
Hybrid III dummy and sled FEM model: The analysis model of dummy system which consists of dummy, seat belt and sled.

$$\begin{gathered}\text{Subject to } 1 \leq \mathrm{n} \leq 4 \\ 0.7 \leq x_1 \leq 1.3 \\ 0.5 \leq x_2 \leq 1.5 \\ H_{disp}\left(\mathrm{n}, x_1, x_2\right) \leq 420\mathrm{mm}\end{gathered} \tag{36}$$

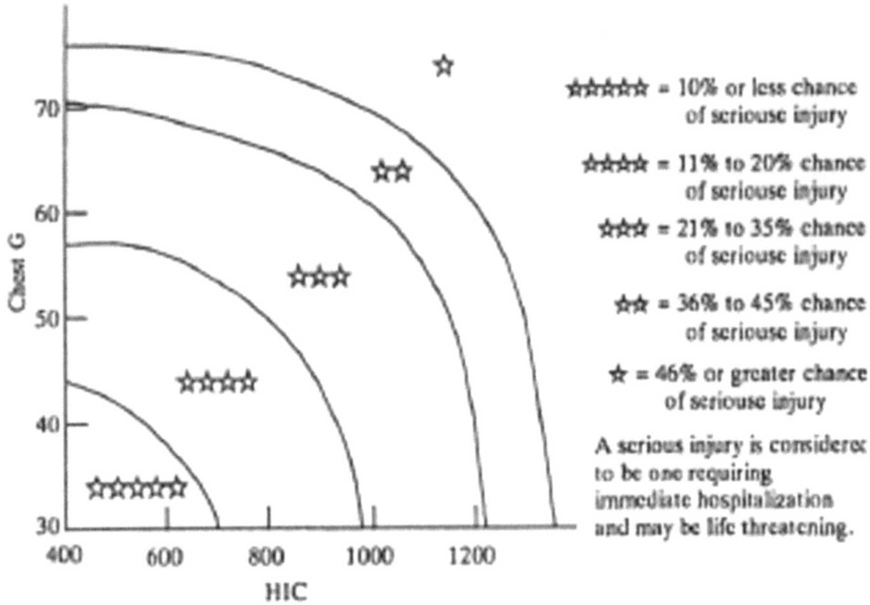

Figure 38.
Safety criteria using NHTSA: The optimal design point is n=3, $x_1 = 0.70, x_2 = 1.20$. The value of the objective function is 1.34 (HIC=355.8、 chest G=30.9) estimated as 5☆. The other local optimal design point is n=3, $x_1 = 1.30, x_2 = 1.15$. The objective function is 3.12 (HIC=620.2, chest G=43.1) estimated as 4☆.

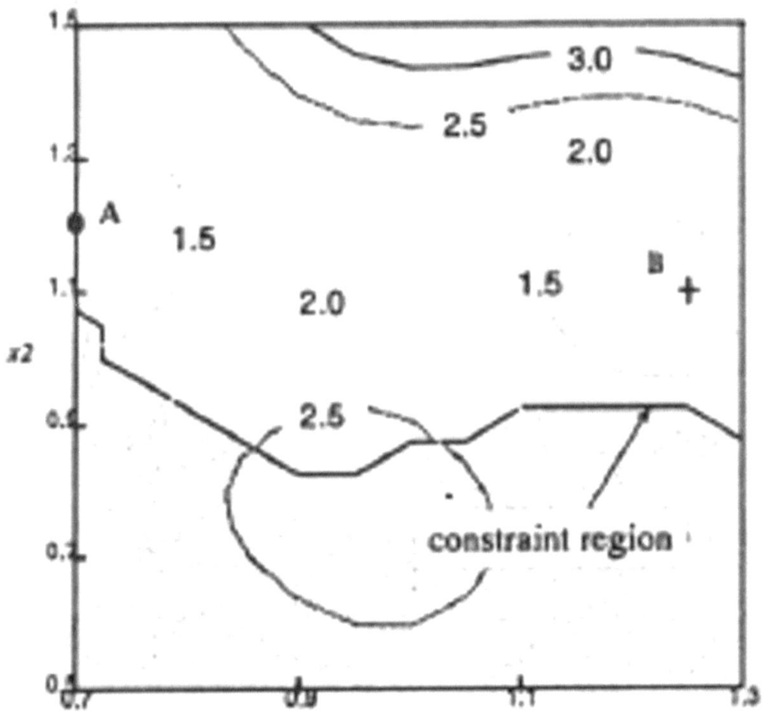

Figure 39.
Contour of approximation response surface at n=3: A is the given optimal point. Although all trial points are given within the region, a is on the boundary. This means the optimal point a is never found without the MPOD's intrinsic extrapolation ability.

Here, α in Eq. (34) is deceleration of head. The unit of deceleration for HIC and chest G is used g(1 g=9.8 m/ s^2). Chest G is the maximum deceleration which continuously generates longer than 3 ms.

The objective function in Eq. (35) is rated on a 5-point stage. Here H_{disp} in Eq. (36) means the moving amount of head to the front and it is shorter than 420 mm so that the occupant does not crush to the steering.

5.5.2 Optimization results

6 points are selected for the initial learning points. $\varepsilon_1 = 0.02$ in Eq. (19), $\varepsilon_2 = 0.05$ in Eq. (20), $\alpha_1 = \alpha_2 = \alpha_3 = 1.0$ in Eq. (21). Firstly step 1 is terminated with 5 iterations which use total 22 points. In step 2, it is terminated with 1 iteration which uses 2 points. So 24 points are used. The optimal design point is n=3, $x_1 = 0.70, x_2 = 1.20$. The value of the objective function is 1.34 (HIC=355.8、Chest G=30.9) of which estimation is 5☆ in **Figure 38**. The other local optimal design point is n=3, $x_1 = 1.30, x_2 = 1.15$. of which objective function is 3.12 (HIC=620.2, Chest G=43.1). This corresponds to 4☆ in **Figure 38**. It is shown the contour of the objective function with n=3 in **Figure 39**. In this figure, A is the given optimal point, B is the local optimal point and the shadow part means the one where it cannot be designed. Although all trial points are given within the region, A is on the boundary.

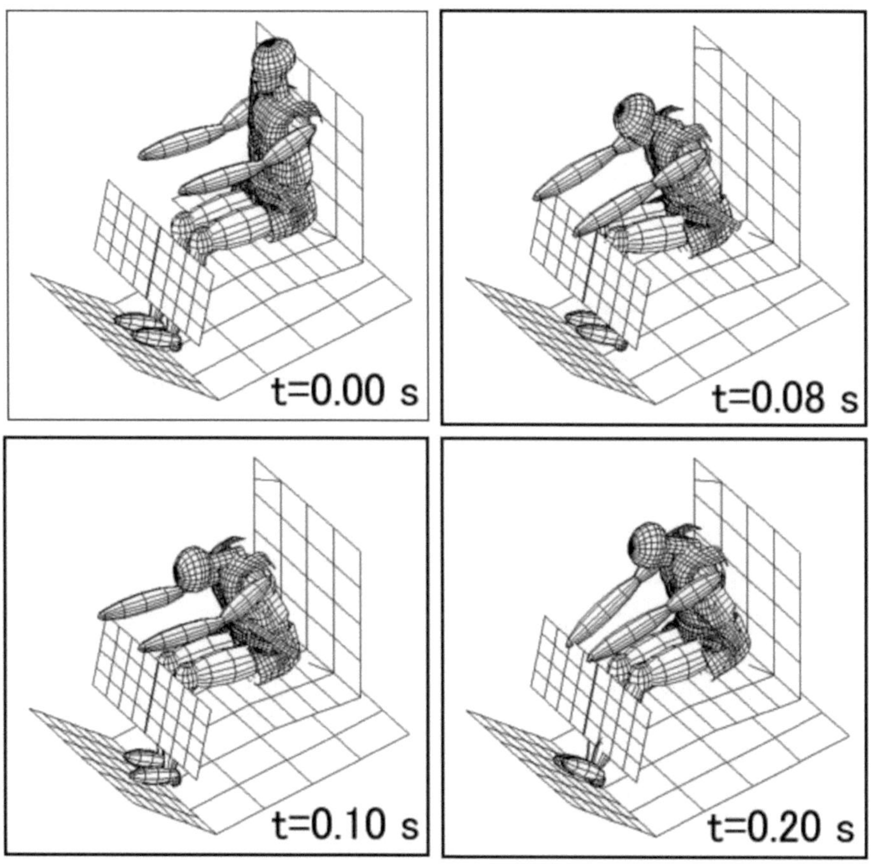

Figure 40.
Motion behavior of the dummy model at optimal design: The HIC becomes the maximum when the jaw contacts the chest around 100 ms. but the maximum HIC is rather small because the contact between the jaw and the chest becomes loose around 100 ms by the operation of the load limiter.

This means the optimal point A is never found without the MPOD's intrinsic extrapolation ability. Such result is never found only with the conventional interpolation unless the trial points are included also on the boundary. In such style, it becomes very efficiency. It is shown the motion behavior of the dummy model at optimal design in **Figure 40**. The HIC becomes the maximum when the jaw contacts the chest around 100 ms as shown in this figure. But the maximum HIC is rather small because the contact between the jaw and the chest becomes loose around 100 ms by the operation of the load limiter. Generally speaking, lots of studies for response surface optimal problem have been done for only one peak problem. On the other hand, MPOD gives us plural optimal combination simultaneously without the knowledge such as how many and where the maximum points exist. Now present vehicle also has air back in addition to seat belt as restraint device. And so it has more local maximum points in the optimal combination problem among vehicle and restraint device than this problem mentioned here. That means the best combination has not been realized between vehicle performance and restraint device one. Only the MPOD can be applied to this multi-peak optimal problem.

6. Conclusion

Here, it is presented a novel method to solve the global optimal design within a feasible design space using response surface methodology based on HNN activation. The sequential approximation procedure and extrapolation of the HNN are proposed for irregular design space and arbitrary designs. The detailed summaries of the issues raised in this paper are as follows.

1. RSM has become more and more important especially for non-linear problems. There are 2 important issues in RSM such as the selection of activation function where quadratic polynomials are most used and the selection of the design trial points where CCD is most utilized. Also artificial neural network function is applied where the multi-layered one is generally applied. In contrast we have developed MPOD where both of the selections of design trial points and the activation function are different from others.

2. In MPOD the number of design trial points can be reduced much more than others. And it is very efficient when the calculation time for one design trial point is long in real big and non-linear problems. Generally the more layers there are the higher the precision becomes in multi-layer neural network. In contrast it is shown that higher orders of the HNN can be realized by the products of fundamental terms to improve the approximation capability.

3. In higher order case with HNN, it becomes often necessary to use pseudo inverse matrix where we have developed a superior method using penalty function to Moore–Penrose generalized inverse matrix.

4. It is estimated that the HNN is more feasible to approximate the multiple peaks function than the polynomials.

5. The extrapolation can be performed by adjusting the transformation coefficient which expresses the slope of the transformation function of the HNN.

6. It is illustrated that the proposed sequential approximation procedure involves two steps. Step 1 aims to conduct a rapid scan of the entire design

space with fewer designs, Step 2 focuses on increasing and confirming the approximation accuracy near the optimal design determined in Step 1. The proposed sequential optimal design method can be applied to design problems with regular and irregular boundaries of design space by the benefit of extrapolation.

7. Rapid approximation of the entire design space can be performed by presetting the threshold d_{min} to be larger, however, it can also feasibly be used to search for relatively small clusters by adjusting the deviation of random seeds and presetting the threshold d_{min} to be smaller in Step 1. However, the deviation of random seeds is assigned smaller values in Step 1 to increase the probability of solving the design within the small cluster. In other words, a trade-off between the accuracy over the entire design space and computation expense can be comprehensively achieved.

8. Comparing with GA, the proposed method can suppress the number of samples and has the same level of accuracy as GA. That means MPOD is much superior to GA in the case of long calculation time problem for one sample such as non-linear problem.

9. Applications to the vehicle problems such as idling vibration, booming noise, crashworthiness design and crash occupant restraint problem with multiple peaks reveals that the proposed method is a feasible and practical approach.

10. Some points still need to be discussed, such as the robustness of the optimal design and the confident regions of estimated coefficients of holographic neural network, which will be addressed in the future.

Acknowledgements

This research has been done by many co-authors. Especially this author thanks to Dr. Shi, Q., Dr. Kamada, Y., Dr. Tokura, S., Mr. Takasima, F. and Mr. Fukushima, H.

Author details

Ichiro Hagiwara
Research and Intellectual Property Strategic Organization, Meiji Institute for Advanced Study of Mathematical Sciences (MIMS), Meiji University's Institute of Autonomous Driving (MIAD), Tokyo, Japan

*Address all correspondence to: ihagi@meiji.ac.jp

References

[1] Benaouali A, Stanisław KS. Multidisciplinary design optimization of aircraft wing using commercial software integration. Aerospace Science and Technology. September 2019;**92**:766-776

[2] Ma, Z.D. and Hagiwara, I., Sensitivity Analysis Methods for Coupled Acoustic-Structural Systems Part I: Modal Sensitivities, AIAA Journal, Volume 29, Number 11 (1991-11), pp. 1787-1795.

[3] Ma, Z.D. and Hagiwara, I., Sensitivity Analysis Methods for Coupled Acoustic-Structural Systems Part II: Direct Frequency Response and its Sensitivities, AIAA Journal, Volume 29, Number 11 (1991-11), pp. 1796-1801.

[4] Ma, Z.D. and Hagiwara, I., Sensitivity Calculation Methods for Conducting Modal Frequency Response Analysis of Coupled Acoustic-Structural Systems, JSME International Journal Series III, Vol. 35, No. 1(1992-3), pp. 14-21.

[5] Hagiwara I, Ma ZD. Arai, A. and Nagabuchi, K., reduction of vehicle interior noise using structural-acoustic sensitivity analysis methods. SAE 910208 Transaction Section. 1991;**6** (1992–4):267-276

[6] Hagiwara, I. and Ma, Z.D., Development of Eigen mode and Frequency Response Sensitivity Analysis Methods for Coupled Acoustic-Structural Systems, JSME International Journal Series III, Vol. 35,No. 2(1992-6), pp. 229-235.

[7] Hagiwara, I., Ma, Z.D., Arai, A. and Nagabuchi, K., Reducing Noise by Structural-Acoustic Sensitivity Analysis, Automotive Technology International '92 (1992-7), pp. 339-342.

[8] Hagiwara, I., Kozukue, W. and Kitagawa, Y., Advanced Sensitivity Analysis Techniques for Improving Automobile Noise and Vibration and Crashworthiness Characteristics, Design-Sensitivity Analysis, ATLANTA TECHNOLOGYPUBLICATIONS, Kleiber M. and Hisada T. ed.(1993-7).

[9] Hagiwara, I., Nagabuchi, K. and Arai, A., Study of Structure Identification Method using Sensitivity Analysis for Vibration, Finite Elements in Analysis and Design Elsevier, Vol. 14,No. 2 and 3 (1993-10), pp. 111-126.

[10] Pal, C. M. and Hagiwara, I., Development of Eigen mode Sensitivity Analysis Methods for Coupled Acoustic-Structural Systems and Its Application to Vehicle Interior Noise Problems, Finite Elements in Analysis and Design Elsevier, Vol. 14, No. 2 and 3(1993-10), pp. 225-234.

[11] Hagiwara, I., Kozukue, W. and Ma, Z.D., Development of Eigen mode Sensitivity Analysis Methods for Coupled Acoustic-Structural Systems and Its Application to Reduction of Vehicle Interior Noise, Finite Elements in Analysis and Design, Elsevier, Vol. 14, No. 2 and 3 (1993-10), pp. 235-248.

[12] Ma, Z.D. and Hagiwara, I., Development of a New Mode-Superposition Technique for Truncating-Lower and/or Higher-Frequency Modes (Application of Eigen mode Sensitivity Analysis Method for Systems with Repeated Eigenvalues), JSME International Journal Series C, Vol. 37,No. 1(1994-3), pp. 7-13.

[13] Hagiwara, I. and Ma, Z.D., Development of a New Mode-Superposition Technique for Truncating Lower- and/or Higher-Frequency Modes (Application for Eigen mode Sensitivity Analysis) JSME International Journal Series C, Vol. 37,No. 1 (1994-3), pp. 14-20.

[14] Pal, C. M. and Hagiwara, I., Optimization of Noise Level Reduction by Truncated Modal Coupled Structural-Acoustic Sensitivity Analysis, JSME International Journal Series C, Vol. 37,No. 2 (1994-6) pp. 246-251.

[15] Kozukue W, Hagiwara I. Development of sound pressure level integral sensitivity and its application to vehicle interior noise reduction. Engineering Computations. 1996;**13**(5): 91-107

[16] White KP Jr. Simulation optimization of the crashworthiness of a passenger vehicle in frontal collision using response surface methodology, SAETransaction. Sec. 1985;**3**:798-811

[17] Giunta, A.A., Noise Aerodynamic Response and Smooth Approximations in HSCT Design, AIAA-94-4376-CP, (1994), pp. 1117-1128.

[18] Şahin S, Samli R, Tan AS, Barba FJ, Chemat F, Cravotto G, et al. Solvent-free microwave-assisted extraction of polyphenols from olive tree leaves: Antioxidant and antimicrobial properties. Molecules. 2017;**7**:1054

[19] Guglielmetti A, Ghirardello D, Belviso S, Zeppa G. Optimisation of ultrasound and microwave-assisted extraction of caffeoylquinic acids and caffeine from coffee silverskin using response. Italian Journal of Food Science. 2017;**29**:409-423

[20] Aydar AY, Bağdatlıoğlu N. and Köseoğlu. O: Effect of ultrasound on olive oil extraction and optimization of ultrasound-assisted extraction of extra virgin olive oil by response surface methodology (RSM). Grasas Y Aceites = International journal of fats and oils; 2017. p. 68

[21] Elksibi I, Haddar W, Ticha MB, Mhenni MF. Development and optimisation of a non conventional extraction process of natural dye from olive solid waste using response surface methodology (RSM). Food Chemistry. 2016;**161**:345-352

[22] Agcam E, Akyıldız A, Balasubramaniam VM. Optimization of anthocyanins extraction from black carrot pomace with thermosonication. Food Chemistry. 2017;**237**:461-470

[23] Espínola F, Moya M, Fernández DG, Castro E. Modelling of virgin olive oil extraction using response surface methodology. International Journal of Food Science and Technology. 2011;**46**: 2576-2583

[24] Ghasemzadeh A, Jaafar HZ, Karimi E, Rahmat A. Optimization of ultrasound-assisted extraction of flavonoid compounds and their pharmaceutical activity from curry leaf (Murraya koenigii L.) using response surface methodology. BMC Complementary and Alternative Medicine. 2014;**14**:318

[25] Yilmaz T. Tavman Ş. ultrasound assisted extraction of polysaccharides from hazelnut skin. Food Science and Technology International. 2015;**22**(2): 112-121

[26] Hornik K et al. Multilayer feed forward networks are universal apploximations. Neural Networks. 1989; **2**:359-366

[27] Carpenter, W.C. and Barthelemy, J. M., A Comparison of Approximations and Artificial Nrural Nets as Response Surface, AIAA-92-2247-CP, (1992), pp. 2474-2482.

[28] Sutherland J. G. the holographic model of memory, learning and expression. International Journal of Neural System. 1990;**1**:259-267

[29] Hagiwara, I., Shi, Q., Kozukue, W., Development of Nonlinear Dynamic Optimization Method Using the Holographic Neural Network, Trans.

Jpn. Soc. Mech.Eng., (in Japanese), Vol. 63,No. 616,A (1997), pp. 2510-2517.

[30] Hagiwara I, Shi Q, Takashima F. Development of crashworthiness optimization method using the neural network, (2nd report, application to vehicle component)trans. Jpn. Soc. Mech. Eng., (in Japanese), Vol. 64. 1998;**626A**:2241-2247

[31] Ugwele, F. O., Aninwede, C. S., Chime, T. O., Christian, O. A. and Innocent, S., Application of response surface methodology in optimizing the process conditions for the regeneration of used mobil oil using different kinds of acids, Heliyon 2020 Oct 7; 6(10): e05062.doi:10.1016/j.heliyon. 2020. e05062.eCollection 2020 Oct.

[32] Hinton, G.E., Krizhevsky, A., Srivastava, N., Sutskever, I., and Salakhutdinov, R., Dropout: A Simple Way to Prevent Neural Networks from Overfitting, Journal of Machine Learning Research, 15, 1929–1958 (Cited 2084 Times, HIC: 142, CV: 536) (2014).

[33] Ciric, A., Krajnc, B., Heath, D. and Ogrinc, N., Response surface methodology and artificial neural network approach for the optimization of ultrasound-assisted extraction of polyphenols from garlic, Food and Chemical Toxicology, Volume 135, January 2020, 110976.

[34] Ilomuanya, M.O., Onwubuya, C.P. and Amenaghawon, A.N., Development and optimization of antioxidant polyherbal cream using artificial neural network aided response surface methodology, Journal Pharmaceutical technology, e-ISSN: 2717-7904 https://doi.org/10.37662/jpt.2020.6,pp.46-53.

[35] Ilomuanya MO, Amenaghawon NA. Odimegwu. J., Okubanjo, O.O. and Aghaizu, C., Oluwatobiloba a. formulation and optimization of gentamicin hydrogel infused with *tetracarpidium conophorum* extract via central composite design for topical delivery. Turk J Pharm Sci. 2018;**15**(3): 319-327

[36] Shi, Q., Hagiwara, I., Azetsu, S. and Ichikawa, T., Holographic Neural Network Approximations for Acoustics Optimization, JSAE Review, Japan Society of Automobile Engineering, Vol. 19 No. 4 (1998-10), pp. 361-363.

[37] Shi Q, Hagiwara I. Optimal design method using holographic neural Network's approximation to automobile problems, JJIAM (Japan journal of industrial and applied mathematics). Vol. 2000;**17-3**:240-247

[38] Shi Q, Hagiwara I, Takashima F. Global optimization to multiple local optima with response surface approximation methodology. JSME International Journal Series A. 2001; **44**(1):175-184

[39] Fukushima H, Kamada Y, Hagiwara I. Optimum engine mounting layout using MPOD. Transactions of the Japan Society of Mechanical Engineers, Series C. 2004;**70**(689):54-61 (in Japanese)

[40] Sakai T, Iwahara M, Shirai Y, Hagiwara I. Optimum designing of engine mounting for heavy duty vehicle: 5th report. Optimum Engine Mounting Layout by Genetic Algorithm, Transactions of the Japan Society of Mechanical Engineers, Series C. 2001; **63**(664):3815-3822 (in Japanese)

[41] Arai T, Kubozuka T, Gray SD. Development of an engine mount optimization method using modal parameters. SAE Technical Paper 932898. 1993

[42] Ishihama M, Satoh S, Seto K, Nagamatsu A. A., vehicle vibration reduction by transfer function phase control on hydraulic engine mounts. JSME International Journal Series C. 1994;**37**(3):536-541

[43] Ichikawa, T. and Hagiwara, I., Frequency Response Analysis of Large-Scale Damped Structuresusing Component Mode Synthesis, JSME International Journal Series III, Vol. 39, No. 3 (1996-9), pp. 450-455.

[44] Ma, Z.D. and Hagiwara, I., Development of New Mode-superposition technique for Modal Frequency Response Analysis of Coupled Acoustic-Structural Systems, Finite Elements in Analysis and Design Elsevier, Vol. 14, No. 2 and 3 (1993-10), pp. 209-223.

[45] Yashiro H, Suzuki K, Kajio Y, Hagiwara I, Arai A. An Applicationof structural-acoustic to Car body structure. SAE1985 Transactions Section. 1985;**4**:777-784

[46] Hagiwara, I., Advanced Structural Analysis Techniques for Improving Automobile Crashworthiness and Noise and Vibration Characteristic, The 40th Proceedings of Theoretical and Applied Mechanics (1991-6), pp 11-20.

[47] Ma, Z.D. and Hagiwara, I., Improved Mode-Superposition Technique for Modal Frequency Response Analysis of Coupled Acoustic-Strctural Systems, AIAA Jounal, Volume 29, Number 10 (1991-10), pp. 1720-1726.

[48] Li, D., Douglas, C.C., Kako, T., Suzuki, M. and Hagiwara, I., A Novel Perturbation Expasion Method for Coupled System of Acoustics and Structure, Computers and Mathematics with Applications (2006), Pp. 1689-1704.

[49] Hagiwara I, Tsuda M, Sato Y, Kitagawa Y. Simulation of automobile side member collapse for crash energy management, the international journal of supercomputer applications, (1990-summer). Vol. 4;**2**:107-114

[50] Hagiwara, I., Tsuda, M. and Sato, Y., Dynamic Analysis of Thin-Walled Box Columns Subjected to Axial Crushing Using Finite Element Method, JSME International Journal Series I, Vol. 33, No. 3(1990-10), pp. 444-452.

[51] Kitagawa Y, Hagiwara I, Tsuda M. Development of a collapse mode control Methodfor side members in vehicle collisions, SAE 910809. Transaction Section. 1991;**6**(1992–4):1101-1107

[52] Kitagawa, Y., Hagiwara, I. and Tsuda, M., Dynamic Analysis of Thin-Walled Columns with Arbitrary Section Geometry Subjected to Axial Crushing, JSME International Journal Series I, Vol. 35,No. 2,(1992-4), pp. 189-193.

[53] Hagiwara, I., Satoh, Y., and Tsuda, M., Study of an analytical and System for Conducting Vehicle Crash Simulations, International Journal of Vehicle Design, Vol. 11,No. 6(1990-6), pp. 564-577.

[54] LS-DYNA (1997), Livermore Software Technology Corporation.

[55] Shi, Q., Hagiwara, I., Takashima, F. and Tokura, S., A Study on Deformation Behavior of Vehicle Cabin and safety Belt Using a Most Probable Optimal Design Method, Transactions of the Society of Automotive Engineers of Japan, Vol. 31,No. 3(2000-7), pp. 107-113 (in Japanese).

Chapter 4

Response Surface Designs Robust against Nuisance Factors

Nam-Ky Nguyen, Mai Phuong Vuong and Tung-Dinh Pham

Abstract

This paper discusses an algorithmic approach to constructing trend-free and orthogonally-blocked response surface designs. The constructed designs have the main effects, 2-factor interactions and second-order effects being orthogonal or near-orthogonal to the nuisance factors such as the time-trend or the blocking factors. The paper also provides a catalogue of (near-) trend-free Box–Behnkens designs and orthogonally blocked Box–Behnkens designs arranged in rows and columns.

Keywords: Box–Behnken designs, D-optimality, Interchange algorithm, Orthogonal blocking, Trend-free designs

1. Introduction

Consider an experiment to study the effect of processing time, temperature and shear stress caused by pumping on the quality of skim milk powder. The milk for this experiment is blended and stored at 4°C and is used over a week for a series of experimental runs. As the milk quality deteriorates over the week, it is desirable for the scientist to have a design whose runs are in a particular order such that the main effects (MEs), 2-factor interactions (2FIs) and second-order effects (SOEs) are orthogonal or near-orthogonal to the time trend.

Box et al. [1], p. 486 discussed an experiment using the Box–Behnken design or BBD [2] to study the influence of four factors on nylon flake blender efficiency. The four factors are (1) particle size, (2) screw lead length, (3) screw rotation and (4) blending time. The design has 27 runs in three blocks of nine runs each. Let us assume there is a need to add an additional factor, i.e. supplier in addition to the existing blocking factor, say operator and find a design which can accommodate the new blocking factor.

The two mentioned experiments emphasise the need for a special class of response surface designs (RSDs) and experiment designs in general which are robust against nuisance factors such as the time trend and the heterogeneous environment. Particular attention will be given to the BBDs since BBDs are the most widely used 3-level designs.

2. Measure of goodness of a (near-) trend-free design or orthogonally blocked design

Consider the following second-order model for an n-run design with v time trend columns (or blocking factor columns) $z_1, \ldots, z_v$ and m factors $x_1, \ldots, x_m$, of which m_3 factors are at 3-level and the rest at 2-level:

$$y_u = \delta_1 z_{1u} + \ldots + \delta_v z_{vu} + \beta_0 + \sum_{i=1}^{m_3} \beta_{ii} x_{iu}^2 + \sum_{i=1}^{m} \beta_i x_{iu} + \sum_{i=1}^{m-1} \sum_{j=i+1}^{m} \beta_{ij} x_{iu} x_{ju} + \epsilon_u \quad (1)$$

where y_u $(u = 1, \ldots, n)$ is the response value of the uth run, z_w's are the nuisance columns/variables and ϵ_u is a random error associated with the uth run. If the nuisance columns z_w's are the time trend ones, there will be two columns, one for linear trend and one quadratic. The linear trend column (first column) is created by scaling a column of numbers $(1, \ldots, n)'$ by subtracting each value from the column mean and then dividing the resulting number by the largest one. The quadratic trend column (second column) is created by scaling a column obtained by squaring each element of the linear trend column.

If the nuisance columns z_w's are associated with the blocking factors, there will be $v\ \left(= \sum_{i=1}^{r}(b_i - 1)\right)$ columns where r is the number of blocking factors and b_i is the number of blocks/categories for each blocking factor. Before standardisation by subtracting the values of each column by the column mean, these columns are dummy variables, which take value 1 if the uth run is in the wth block and zero otherwise (see [3], Section 8 and [4], Chapter 8).

Eq. (1) can also be written in matrix form as:

$$\mathbf{y} = \mathbf{Z}\boldsymbol{\delta} + \mathbf{X}\boldsymbol{\beta} + \boldsymbol{\epsilon} \quad (2)$$

where $\mathbf{y}$ is an $n \times 1$ response vector, $\mathbf{Z}$ a matrix of size $n \times v$ containing v z_w columns in (1), $\boldsymbol{\delta}$ a $v \times 1$ column vector representing nuisance effects, $\mathbf{X}$ is the expanded design matrix of size $n \times p$, $\boldsymbol{\beta}$ a $p \times 1$ column vector of parameters to be estimated, and $\boldsymbol{\epsilon}$ an $n \times 1$ column vector of random errors. The least square solution for the unknown parameters $\boldsymbol{\delta}$ and β in (2) is thus the solution of the following equation:

$$\begin{bmatrix} \mathbf{Z}' \\ \mathbf{X}' \end{bmatrix} \mathbf{y} = \begin{bmatrix} \mathbf{Z}'\mathbf{Z} & \mathbf{Z}'\mathbf{X} \\ \mathbf{X}'\mathbf{Z} & \mathbf{X}'\mathbf{X} \end{bmatrix} \begin{bmatrix} \hat{\boldsymbol{\delta}} \\ \hat{\boldsymbol{\beta}} \end{bmatrix} \quad (3)$$

When the condition

$$\mathbf{Z}'\mathbf{X} = \mathbf{0} \quad (4)$$

is satisfied, it can be seen that the solution for $\boldsymbol{\beta}$ from (3) will be the same as the one from the equation $\mathbf{X}'\mathbf{y} = \mathbf{X}'\mathbf{X}\hat{\boldsymbol{\beta}}$, i.e. the equation for a model without the nuisance columns. The condition in (4) is called the time trend-free condition or orthogonal block condition.

To measure how good of a (near-) trend-free design or orthogonally blocked design is, we use the following fraction:

$$\left(|X'X| / (|\mathbf{Z}'\mathbf{Z}|\ \mathbf{X}'\mathbf{X}||)\right)^{1/p} \quad (5)$$

where $X = [\mathbf{Z}\ \mathbf{X}]$ includes the nuisance factors and p is the number of parameters of the model to be estimated (i.e. the number of columns of $\mathbf{X}$). A well known result of Fischer [5] states that given a positive definite hermitian matrix $\mathbf{G}$ of the form:

$$\mathbf{G} = \begin{bmatrix} \mathbf{A} & \mathbf{B} \\ \mathbf{B}' & \mathbf{D} \end{bmatrix} \quad (6)$$

then $|\mathbf{G}| \leq |\mathbf{A}||\mathbf{D}|$, and the equality holds if and only if $\mathbf{B} = \mathbf{0}$. If follows from this inequality that:

$$|X'X| \leq |\mathbf{Z}'\mathbf{Z}||\mathbf{X}'\mathbf{X}| \tag{7}$$

which implies the measure in (5) is less than or equal to 1, and it becomes 1 if and only if $\mathbf{Z}'\mathbf{X} = 0$.

3. An algorithm to attain the orthogonality condition $\mathbf{Z}'\mathbf{X} = \mathbf{0}$

Let $\mathbf{x}_i'$ and $\mathbf{x}_u'$ be two rows of $\mathbf{X}$ and $\mathbf{z}_i'$ and $\mathbf{z}_u'$ be the corresponding rows of $\mathbf{Z}$. Swapping the ith and uth rows of $\mathbf{X}$ is the same as adding the following matrix to $\mathbf{Z}'\mathbf{X}$:

$$-(\mathbf{z}_i - \mathbf{z}_u)(\mathbf{x}_i - \mathbf{x}_u)'. \tag{8}$$

We use this matrix result to develop an algorithm to achieve the condition $\mathbf{Z}'\mathbf{X} = \mathbf{0}$, i.e. for constructing (near-) trend-free designs and orthogonally blocked designs. This algorithm has two main steps:

1. Construct the nuisance matrix $\mathbf{Z}$ and the expanded design matrix $\mathbf{X}$. Randomly assign each of the n rows of $\mathbf{X}$ to each of the n rows of $\mathbf{Z}$. Calculate f, the sum of squares of the elements of $\mathbf{Z}'\mathbf{X}$.

2. Repeat searching for a pair of rows of $\mathbf{X}$ such that the swap of the positions of these two rows results in the biggest reduction in f. If the search is successful, swap their positions, update f and $\mathbf{Z}'\mathbf{X}$. This step is repeated until $f = 0$ or until f cannot be reduced further.

Table 1 (a) and **1 (b)** display the $X(= (\mathbf{Z}\ \ \mathbf{X}))$ matrices of a BBD for three factors in 15 runs: (a) ordered in the presence of both linear and quadratic trends; (b) arranged in three blocks of five runs each. This example shows how the $\mathbf{Z}$ matrices are constructed. The $\mathbf{X}$ matrix contains a column of 1's representing the intercept, followed by three columns representing SOEs, three columns representing the MEs and three columns representing 2FIs.

Remarks

1. The above two steps make up one computer try. Thousands of tries are required for each design parameters and the one with the smallest f will be chosen. Among designs with the same f, the one with the smallest fraction calculated in (5) will be chosen.

2. For a factorial or fractional factorial design (FFD), the orthogonality between MEs and nuisance variables is considered more important than the orthogonality between 2FIs and nuisance variables. For an RSD, the orthogonality between the MEs (and 2FIs) and nuisance variables are considered more important than the orthogonality between SOEs and nuisance variables. In these situations, partition $\mathbf{X}$ as $(\mathbf{X}_1\ \mathbf{X}_2 \ldots)$ where $\mathbf{X}_2$ is associated with the more important effects. Similarly, partition $\mathbf{Z}'\mathbf{X}$ as $(\mathbf{Z}'\mathbf{X}_1\ \mathbf{Z}'\mathbf{X}_2 \ldots)$. Let g be the sum of squares of the elements of $\mathbf{Z}'\mathbf{X}_2$ and f the sum of squares of the elements of $\mathbf{Z}'\mathbf{X}$ as defined previously. Step 2 of our algorithm now becomes

repeating searching for a pair of runs such that swapping their run positions results in the biggest reduction in g (or f if g cannot be reduced further). If the search is successful, swap their positions, update f and $\mathbf{Z}'\mathbf{X}$. This step is repeated until $f = 0$ or until both g and f cannot be reduced further.

z_1	z_2	x_0	x_1^2	x_2^2	x_3^2	x_1	x_2	x_3	x_1x_2	x_1x_3	x_2x_3
−1	1	1	0	0	0	0	0	0	0	0	0
−0.86	0.57	1	0	1	1	0	−1	1	0	0	−1
−0.71	0.21	1	1	0	1	1	0	−1	0	−1	0
−0.57	−0.09	1	1	0	1	−1	0	−1	0	1	0
−0.43	−0.32	1	0	1	1	0	1	1	0	0	1
−0.29	−0.48	1	1	1	0	−1	1	0	−1	0	0
−0.14	−0.58	1	1	1	0	1	1	0	1	0	0
0	−0.61	1	0	0	0	0	0	0	0	0	0
0.14	−0.58	1	1	1	0	−1	−1	0	1	0	0
0.29	−0.48	1	1	1	0	1	−1	0	−1	0	0
0.43	−0.32	1	0	1	1	0	−1	−1	0	0	1
0.57	−0.09	1	1	0	1	1	0	1	0	1	0
0.71	0.21	1	1	0	1	−1	0	1	0	−1	0
0.86	0.57	1	0	1	1	0	1	−1	0	0	−1
1	1	1	0	0	0	0	0	0	0	0	0

(a)

z_1	z_2	x_0	x_1^2	x_2^2	x_3^2	x_1	x_2	x_3	x_1x_2	x_1x_3	x_2x_3
0.67	−0.33	1	1	0	1	−1	0	1	0	−1	0
0.67	−0.33	1	1	0	1	1	0	1	0	1	0
0.67	−0.33	1	0	0	0	0	0	0	0	0	0
0.67	−0.33	1	0	1	1	0	1	−1	0	0	−1
0.67	−0.33	1	0	1	1	0	−1	−1	0	0	1
−0.33	0.67	1	0	0	0	0	0	0	0	0	0
−0.33	0.67	1	1	0	1	−1	0	−1	0	1	0
−0.33	0.67	1	1	0	1	1	0	−1	0	−1	0
−0.33	0.67	1	0	1	1	0	−1	1	0	0	−1
−0.33	0.67	1	0	1	1	0	1	1	0	0	1
−0.33	−0.33	1	1	1	0	1	−1	0	−1	0	0
−0.33	−0.33	1	1	1	0	−1	1	0	−1	0	0
−0.33	−0.33	1	1	1	0	−1	−1	0	1	0	0
−0.33	−0.33	1	1	1	0	1	1	0	1	0	0
−0.33	−0.33	1	0	0	0	0	0	0	0	0	0

(b)

Table 1.
The $X (= (\mathbf{Z}\ \mathbf{X}))$ matrices of a BBD for three factors in 15 runs (a) ordered in the presence of both linear and quadratic trends; (b) arranged in three blocks of five runs each.

The designs whose X matrices are in **Table 1 (a)** and **1 (b)** were constructed by this approach. In **Table 1 (a)**, the $\mathbf{X}_2$ matrix is associated with the MEs and in **Table 1 (b)**, the $\mathbf{X}_2$ matrix is associated with the MEs and 2FIs. Readers can verify that $\mathbf{Z}'\mathbf{X}_2 = \mathbf{0}$.

The algorithm we describe in this Section is closely aliased to the ones of Nguyen [6] and Nguyen [7]. As such, it does not require matrix inversions and therefore is much faster than the ones by other authors (See e.g. [8–11]).

4. BBDs robust against time trend and blocking factors

The algorithm in the previous Section has been used to construct BBDs for 3–7 factors which are (near-) trend-free and orthogonally blocked with two blocking factors (i.e. rows and columns).

We use the correlation cell plots (CCPs) proposed by Jones & Nachtsheim [12] to display the magnitude of the correlation between the SOEs, MEs and 2FIs with the columns/variables representing nuisance factors. The colour of each cell in these plots ranges from white (no correlation) to dark (correlation of 1 or close to 1). **Figures 1 (a)** and **1 (b)** show the CCPs of the BBDs whose X matrices are **Table 1 (a)** and **1 (b)**. In **Figure 1 (a)**, it can be seen that the MEs are orthogonal to both the linear and quadratic trends. In **Figure 1 (b)**, it can be seen that both the MEs and 2FIs are orthogonal to the block effects. For both Figures, the SOEs are slightly correlated with the nuisance effects.

Appendix 1 tabulates the BBDs which are near- trend-free. These BBDs are for three factors in 15 runs, four factors in 27 runs, five factors in 46 runs and six factors in 54 runs. The BBD for seven factors in 62 runs can be found in the link given in the next Section. These BBDs have the MEs being orthogonal to the time trends and have CCPs very similar to the one in **Figure 1 (a)**. Let TF (time factor) be the fraction computed in (5) for these designs. The TF values of these five designs are 0.91, 0.959, 0.986, 0.974 and 0.976 respectively.

Appendix 2 tabulates the BBDs which are near-orthogonally blocked. These BBDs are for three factors in 16 runs arranged in two rows and two columns, four factors in 28 runs arranged in two rows and two columns, five factors in 48 runs arranged in two rows and three columns, and six factors in 54 runs arranged in two rows in three columns. The BBD for seven factors in 60 runs arranged in two rows

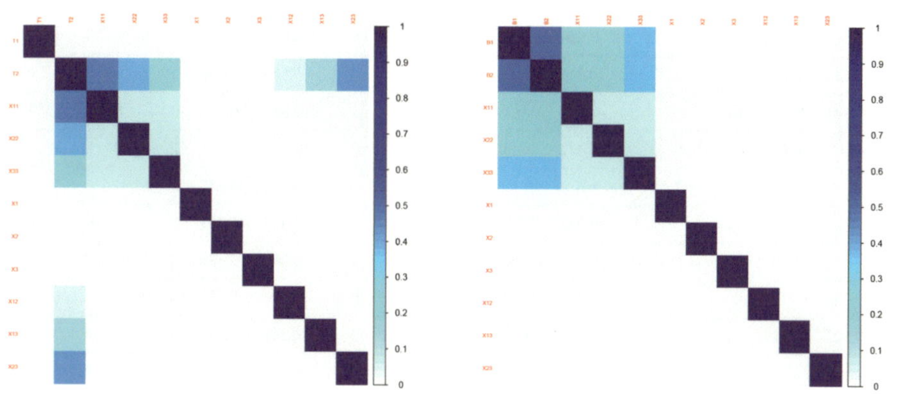

(a) (b)

Figure 1.
CCPs of the BBDs whose X matrices are ***Table 1*** *(a) and 1 (b).*

and three columns can be found in the link given in the next Section. With the exception of the first BBD, these BBDs have both the MEs and 2FIs being orthogonal to the block effects and have CCPs very similar to the one in **Figure 1 (b)**. Let BF (blocking factor) be the fraction computed in (5) for these designs. The BF values of these five designs are 0.944, 1, 0.992, 0.927 and 0.962 respectively. Note that for the 4-factor BBD, all effects are orthogonally blocked.

5. Conclusion

This paper describes an algorithmic approach to arrange the runs of an experimental design in general and an RSD in particular so that it is robust against nuisance factors such as time trend and blocking factors. Designs constructed by this approach can supplement the existing catalogue of designs in the literature. Although 3-level designs are used in this paper to illustrate our blocking approach, ours can also be used with 2-level designs (the factorial and fractional factorial designs) or mixed-level designs [13–15] or mixture designs [16].

The link of the supplemental material which contains the Java implementation of the algorithm in Section 3 and additional examples is: https://drive.google.com/drive/folders/14g7E3I4F8KIL2rcZMovlvJmsaM7iC7pJ?usp=sharing.

Appendix 1: BBDs for 3–6 factors with the trend-free main effects

-1	**1**	0	0	0		**-1**	**1**	-1	0	-1	0	0	**-1**	**1**	0	0	0	0	0	0
-0.857	**0.571**	-1	0	-1		**-0.956**	**0.867**	1	0	-1	0	0	**-0.962**	**0.887**	-1	1	0	0	0	-1
-0.714	**0.209**	0	1	1		**-0.911**	**0.739**	0	0	0	1	1	**-0.925**	**0.778**	1	1	0	0	0	1
-0.571	**-0.088**	0	-1	1		**-0.867**	**0.618**	0	0	0	0	0	**-0.887**	**0.673**	0	0	0	1	-1	1
-0.429	**-0.319**	1	0	-1		**-0.822**	**0.503**	0	1	0	0	-1	**-0.849**	**0.573**	0	-1	0	0	1	-1
-0.286	**-0.484**	1	-1	0		**-0.778**	**0.394**	0	-1	0	-1	0	**-0.811**	**0.478**	1	0	0	-1	1	0
-0.143	**-0.582**	1	1	0		**-0.733**	**0.291**	-1	0	1	0	0	**-0.774**	**0.386**	0	-1	-1	-1	0	0
0	**-0.615**	0	0	0		**-0.689**	**0.194**	1	0	0	0	-1	**-0.736**	**0.299**	-1	0	0	-1	1	0
0.143	**-0.582**	-1	-1	0		**-0.644**	**0.103**	0	0	0	0	0	**-0.698**	**0.216**	0	0	0	1	-1	-1
0.286	**-0.484**	-1	1	0		**-0.6**	**0.018**	0	0	1	1	0	**-0.66**	**0.138**	1	0	-1	0	-1	0
0.429	**-0.319**	-1	0	1		**-0.556**	**-0.061**	0	0	1	0	1	**-0.623**	**0.064**	-1	0	1	0	-1	0
0.571	**-0.088**	0	1	-1		**-0.511**	**-0.133**	1	-1	0	0	0	**-0.585**	**-0.006**	0	0	0	0	0	0
0.714	**0.209**	0	-1	-1		**-0.467**	**-0.2**	0	0	0	-1	1	**-0.547**	**-0.071**	0	-1	0	0	1	1
0.857	**0.571**	1	0	1		**-0.422**	**-0.261**	0	0	0	0	0	**-0.509**	**-0.132**	1	-1	0	1	0	0
1	**1**	0	0	0		**-0.378**	**-0.315**	0	-1	0	0	-1	**-0.472**	**-0.189**	0	1	-1	0	-1	0
						-0.333	**-0.364**	0	1	1	0	0	**-0.434**	**-0.241**	1	-1	1	0	0	0
						-0.289	**-0.406**	-1	0	0	-1	0	**-0.396**	**-0.289**	-1	0	0	-1	0	1
-1	**1**	0	-1	-1	0	**-0.244**	**-0.442**	1	1	0	0	0	**-0.358**	**-0.332**	-1	0	0	1	1	0
-0.923	**0.769**	1	1	0	0	**-0.2**	**-0.473**	-1	1	0	0	0	**-0.321**	**-0.372**	0	1	1	0	1	0
-0.846	**0.557**	1	0	0	-1	**-0.156**	**-0.497**	0	0	1	-1	0	**-0.283**	**-0.406**	0	-1	0	0	-1	-1
-0.769	**0.363**	0	0	1	1	**-0.111**	**-0.515**	0	0	0	0	0	**-0.245**	**-0.437**	-1	1	0	0	0	1
-0.692	**0.188**	-1	1	0	0	**-0.067**	**-0.527**	0	-1	0	1	0	**-0.208**	**-0.463**	0	1	1	-1	0	0
-0.615	**0.031**	-1	0	1	0	**-0.022**	**-0.533**	0	0	-1	1	0	**-0.17**	**-0.485**	0	0	1	1	0	-1
-0.538	**-0.108**	0	0	-1	1	**0.022**	**-0.533**	0	0	-1	0	-1	**-0.132**	**-0.502**	1	1	0	0	0	-1

-0.462	-0.228	0	0	1	-1	0.067	-0.527	0	1	-1	0	0	-0.094	-0.515	-1	-1	0	1	0	0
-0.385	-0.329	0	0	0	0	0.111	-0.515	0	0	0	0	0	-0.057	-0.524	0	0	0	0	0	0
-0.308	-0.412	-1	-1	0	0	0.156	-0.497	1	0	0	0	1	-0.019	-0.528	0	0	-1	0	1	1
-0.231	-0.477	-1	0	-1	0	0.2	-0.473	-1	0	0	0	1	0.019	-0.528	-1	0	-1	0	-1	0
-0.154	-0.523	0	0	0	0	0.244	-0.442	0	0	1	0	-1	0.057	-0.524	0	0	1	1	0	1
-0.077	-0.551	0	-1	0	-1	0.289	-0.406	-1	-1	0	0	0	0.094	-0.515	1	1	0	1	0	0
0	-0.56	0	-1	0	1	0.333	-0.364	0	0	-1	-1	0	0.132	-0.502	1	0	0	-1	0	1
0.077	-0.551	1	0	0	1	0.378	-0.315	0	0	-1	0	1	0.17	-0.485	0	0	0	0	0	0
0.154	-0.523	0	1	0	-1	0.422	-0.261	0	0	0	-1	-1	0.208	-0.463	0	0	0	0	0	0
0.231	-0.477	0	1	1	0	0.467	-0.2	0	1	0	1	0	0.245	-0.437	1	0	0	-1	0	-1
0.308	-0.412	1	-1	0	0	0.511	-0.133	-1	0	0	1	0	0.283	-0.406	-1	0	0	-1	0	-1
0.385	-0.329	1	0	-1	0	0.556	-0.061	0	-1	-1	0	0	0.321	-0.372	0	1	1	0	-1	0
0.462	-0.228	0	1	-1	0	0.6	0.018	1	0	0	1	0	0.358	-0.332	0	-1	1	-1	0	0
0.538	-0.108	0	1	0	1	0.644	0.103	1	0	1	0	0	0.396	-0.289	0	0	-1	0	1	-1
0.615	0.031	-1	0	0	-1	0.689	0.194	0	0	0	1	-1	0.434	-0.241	0	-1	0	0	-1	1
0.692	0.188	0	0	0	0	0.733	0.291	1	0	0	-1	0	0.472	-0.189	0	0	1	0	1	-1
0.769	0.363	1	0	1	0	0.778	0.394	0	-1	0	0	1	0.509	-0.132	0	0	-1	1	0	-1
0.846	0.557	0	-1	1	0	0.822	0.503	0	-1	1	0	0	0.547	-0.071	0	0	-1	1	0	1
0.923	0.769	0	0	-1	-1	0.867	0.618	0	1	0	-1	0	0.585	-0.006	1	-1	-1	0	0	0
1	1	-1	0	0	1	0.911	0.739	0	1	0	0	1	0.623	0.064	-1	-1	-1	0	0	0
						0.956	0.867	0	0	0	0	0	0.66	0.138	0	0	0	-1	-1	1
						1	1	-1	0	0	0	-1	0.698	0.216	0	1	-1	0	1	0
													0.736	0.299	0	1	-1	-1	0	0
													0.774	0.386	0	0	1	0	1	1
													0.811	0.478	0	0	0	-1	-1	-1
													0.849	0.573	0	0	0	0	0	0
													0.887	0.673	1	0	0	1	1	0
													0.925	0.778	-1	1	0	1	0	0
													0.962	0.887	1	0	1	0	-1	0
													1	1	-1	-1	1	0	0	0

Appendix 2: BBDs for 3–6 factors arranged in rows and columns

1	1	-1	-1	0	1	1	-1	0	0	1	0	1	1	0	0	1	0	1	1
1	1	1	0	-1	1	1	0	0	1	0	1	1	1	1	-1	-1	0	0	0
1	1	0	0	0	1	1	0	-1	0	0	-1	1	1	1	1	0	0	0	1
1	1	0	1	1	1	1	0	0	0	0	0	1	1	1	1	0	0	0	-1
1	2	-1	0	1	1	1	0	0	1	0	-1	1	1	0	0	1	0	1	-1
1	2	0	1	-1	1	1	0	0	0	0	0	1	1	1	0	1	0	-1	0
1	2	1	-1	0	1	1	0	1	0	-1	0	1	1	0	0	0	0	0	0
1	2	0	0	0	1	1	1	0	-1	0	0	1	1	1	-1	1	0	0	0

2	**1**	0	-1	1		**1**	**2**	0	1	0	0	-1	**1**	**1**	-1	0	1	0	-1	0
2	**1**	-1	0	-1		**1**	**2**	0	0	0	0	0	**1**	**2**	0	1	1	0	-1	0
2	**1**	0	0	0		**1**	**2**	0	0	-1	-1	0	**1**	**2**	0	-1	0	0	-1	-1
2	**1**	1	1	0		**1**	**2**	1	0	0	0	-1	**1**	**2**	0	1	-1	-1	0	0
2	**2**	1	0	1		**1**	**2**	0	-1	0	-1	0	**1**	**2**	0	-1	-1	-1	0	0
2	**2**	0	0	0		**1**	**2**	0	0	-1	1	0	**1**	**2**	0	-1	0	0	-1	1
2	**2**	0	-1	-1		**1**	**2**	0	0	0	0	0	**1**	**2**	-1	-1	0	1	0	0
2	**2**	-1	1	0		**1**	**2**	1	0	0	0	1	**1**	**2**	0	0	0	0	0	0
						1	**3**	-1	0	-1	0	0	**1**	**2**	0	1	-1	0	-1	0
						1	**3**	0	-1	1	0	0	**1**	**2**	-1	1	0	1	0	0
1	**1**	0	0	-1	-1	**1**	**3**	0	0	0	1	1	**1**	**3**	-1	0	0	-1	0	1
1	**1**	-1	0	0	-1	**1**	**3**	-1	-1	0	0	0	**1**	**3**	0	1	-1	0	1	0
1	**1**	1	1	0	0	**1**	**3**	0	0	0	-1	1	**1**	**3**	0	0	-1	1	0	1
1	**1**	-1	0	0	1	**1**	**3**	1	0	0	1	0	**1**	**3**	-1	0	0	-1	0	-1
1	**1**	0	-1	1	0	**1**	**3**	0	1	1	0	0	**1**	**3**	0	-1	0	0	1	-1
1	**1**	1	-1	0	0	**1**	**3**	-1	1	0	0	0	**1**	**3**	0	0	-1	1	0	-1
1	**1**	0	1	0	1	**2**	**1**	1	0	1	0	0	**1**	**3**	0	1	1	0	1	0
1	**2**	0	1	1	0	**2**	**1**	0	1	-1	0	0	**1**	**3**	0	0	0	0	0	0
1	**2**	1	0	-1	0	**2**	**1**	0	-1	0	0	1	**1**	**3**	0	-1	0	0	1	1
1	**2**	0	0	0	0	**2**	**1**	0	0	0	0	0	**2**	**1**	-1	-1	-1	0	0	0
1	**2**	0	0	1	-1	**2**	**1**	0	0	0	0	0	**2**	**1**	0	0	0	0	0	0
1	**2**	0	-1	0	1	**2**	**1**	0	1	0	1	0	**2**	**1**	0	0	-1	0	1	1
1	**2**	0	0	0	0	**2**	**1**	0	-1	-1	0	0	**2**	**1**	0	0	-1	0	1	-1
1	**2**	-1	0	-1	0	**2**	**1**	-1	0	0	-1	0	**2**	**1**	-1	0	-1	0	-1	0
2	**1**	1	0	1	0	**2**	**2**	0	0	0	0	0	**2**	**1**	-1	1	0	0	0	1
2	**1**	-1	0	1	0	**2**	**2**	-1	0	0	0	-1	**2**	**1**	-1	-1	1	0	0	0
2	**1**	0	0	-1	1	**2**	**2**	-1	0	0	0	1	**2**	**1**	-1	1	0	0	0	-1
2	**1**	0	-1	-1	0	**2**	**2**	0	0	0	0	0	**2**	**1**	1	0	-1	0	-1	0
2	**1**	0	1	0	-1	**2**	**2**	0	-1	0	1	0	**2**	**2**	1	0	0	1	1	0
2	**1**	0	0	0	0	**2**	**2**	0	1	0	0	1	**2**	**2**	0	-1	1	-1	0	0
2	**1**	0	0	0	0	**2**	**2**	0	0	1	-1	0	**2**	**2**	-1	0	0	1	1	0
2	**2**	0	-1	0	-1	**2**	**2**	0	0	1	1	0	**2**	**2**	1	1	0	1	0	0
2	**2**	0	1	-1	0	**2**	**3**	0	0	-1	0	-1	**2**	**2**	0	0	0	0	0	0
2	**2**	1	0	0	-1	**2**	**3**	0	0	-1	0	1	**2**	**2**	1	-1	0	1	0	0
2	**2**	-1	1	0	0	**2**	**3**	1	-1	0	0	0	**2**	**2**	0	1	1	-1	0	0
2	**2**	1	0	0	1	**2**	**3**	-1	0	1	0	0	**2**	**2**	-1	0	0	-1	1	0
2	**2**	-1	-1	0	0	**2**	**3**	1	0	0	-1	0	**2**	**2**	1	0	0	-1	1	0
2	**2**	0	0	1	1	**2**	**3**	0	0	0	-1	-1	**2**	**3**	1	0	0	-1	0	1
						2	**3**	1	1	0	0	0	**2**	**3**	0	0	0	1	-1	1
						2	**3**	0	0	0	1	-1	**2**	**3**	0	0	1	1	0	1
													2	**3**	0	0	1	1	0	-1

	2	**3**	0	0	0	-1	-1	1
	2	**3**	0	0	0	0	0	0
	2	**3**	0	0	0	1	-1	-1
	2	**3**	0	0	0	-1	-1	-1
	2	**3**	1	0	0	-1	0	-1

Author details

Nam-Ky Nguyen[1*], Mai Phuong Vuong[2] and Tung-Dinh Pham[3]

1 Vietnam Institute for Advanced Study in Mathematics, Hanoi, Vietnam

2 Hanoi University of Science and Technology, Vietnam

3 VNU University of Science, Vietnam National University, Hanoi

*Address all correspondence to: nknam@viasm.edu.vn

References

[1] Box, G.E.P., Hunter, J.S. & Hunter, W.G. (2005) Statistics for Experiments 2nd ed., New York: Wiley.

[2] Box, G.E.P., & Behnken, D. (1960) Some new three level designs for the study of quantitative variables. Technometrics, 2, 455–475.

[3] Box, G.E.P. & Hunter, J.S. (1957) Multi-factor experimental designs for exploring response surfaces, The Annals of Mathematical Statistics, 28, 159-241.

[4] Khuri, A.I. & Cornell, J.A. (1996) Response Surfaces, Designs and Analyses. 2nd ed. (New York, Marcel Dekker).

[5] Fischer, E. (1908) Über den Hadamardschen Determinantensatz. Archiv Der Mathematik und Physik, 13, 32-40.

[6] Nguyen, N-K (2001) Cutting Experimental Designs into Blocks. Austral. & New Zealand J. of Statistics 43, 367-374.

[7] Nguyen, N-K. (2014) Making Experimental Designs Robust Against Time Trend. Statistics & Applications, 11, 79-86.

[8] Atkinson, A.C. & Donev, A.N. (1996) Experimental designs optimally balanced for trend. Technometrics 38, 333-341.

[9] Cook, R.D. & Nachtsheim, C.J. (1989) Computer-aided blocking of factorial and response surface designs. Technometrics 31, 339-346.

[10] Gilmour, S.G. & Trinca L.A. (2003) Row-Column Response Surface Designs. Journal of Quality Technology 35, 184-193.

[11] Jones, B. & Nachtsheim, C. J. (2016) Blocking Schemes for Definitive Screening Designs, Technometrics 58, 74-83.

[12] Jones, B. and Nachtsheim, C. J. (2011). A Class of Three Levels Designs for Definitive Screening in the Presence of Second-Order Effects. Journal of Quality Technology 43, 1-15.

[13] Jones, B. & Nachtsheim, C. J. (2013) Definitive Screening Designs with Added Two-Level Categorical Factors. Journal of Quality Technology 45, 121-129.

[14] Nguyen, N-K., Pham, D-T. and Vuong, M.P. (2020) Constructing D-Efficient Mixed-Level Foldover Designs Using Hadamard Matrices. Technometrics, 62, 48-56,

[15] Nguyen, N-K., Kenett, R.S., Pham, D-T. and Vuong, M.P. (2021) D-efficient Mixed-Level Foldover Designs for Screening Experiments. To appear in Springer Handbook of Engineering Statistics, 2nd ed, Hoang Pham (Editor).

[16] Cornell, J.A. (2002) Experiments with Mixtures: Designs, Models and the Analysis of Mixture data. 3rd ed. New York: John Wiley & Sons,Inc.

Chapter 5

Central Composite Design for Response Surface Methodology and Its Application in Pharmacy

Sankha Bhattacharya

Abstract

The central composite design is the most commonly used fractional factorial design used in the response surface model. In this design, the center points are augmented with a group of axial points called star points. With this design, quickly first-order and second-order terms can be estimated. In this book chapter, different types of central composite design and their significance in various experimental design were clearly explained. Nevertheless, a calculation based on alpha (α) determination and axial points were clearly described. This book chapter also amalgamates recently incepted central composite design models in various experimental conditions. Finally, one case study was also discussed to understand the actual inside of the central composite design.

Keywords: central composite design, response surface method, circumscribed design, the uncoded value value of alpha (α), two-factor central composite design

1. Introduction

Any optimization process is achieving by going through certain phases, i.e., Screening; where identification of significant and important factor is important [1]; Improvement; where factors need to be identified which is near to optimum, Response surface design [2]; where optimum or best product has been designing by response surface method (RSM) by quantifying the relationship between one or more measured responses and vital input factor [3]. It is always been a tedious tasks to choice a suitable experimental design, which can easily explain many response variables. Such variables often end as quadratic surface model. For such kind of interpretation central composite design can be an excellent choice. In the process of Optimization and finding the best possible product from the ongoing batches, an experimental design called the central composite design (CCD) concept has emerged [4]. The CCDmodel is an integral part of response surface mythology. The biggest advantage of this type of optimization model is, it is more accurate, and no need for a three-level factorial experiment for building a second-order quadratic model [5]. After excising the CCDmodel within the experiment, a linear regression model has been used to construct the model, and coadded values have been used [6]. The CCDmodel is otherwise called A Box-Wilson Central Composite Design. In this design, the center points are eventually augmented with the group of "star points" that allows estimation of curvature [7]. If the distance from the center of the design space to a factorial point is ± 1 unit for each factor, the distance from the center of the design space to a star point is $\pm\alpha$ with |α8] 1 < |]. The precise value of α depends on certain specific properties required for the design. Since there are many factors available in the CCD model, therefore, the possibility of more than two or many star points within the model is more palpable. The star points represent lower and higher extreme values. The CCD model allows to extends 2 level factors, which have been widely used in response surface modeling and Optimization. As far as pharmaceutical research is a concerned, much scientific research has been carried out in recent times in this direction. As per Krishna Veni *et al* (2020), environment-sensitive Eudragit coated solid lipid nanoparticles can be prepared using a central composite design (CCD) model [9]. In another study, Ye, Qingzhuo, et al.(2020) prepared puerarin nanostructured lipid carriers by central composite design, where 5 levels 3 factors central composite design was used to utilized to anticipate response variables and to constrats 3D plots [10].

However, in this book chapter, an attempt was made to highlight the basics of the CCD model and to corelate the concepts of CCD with suitable case studies, which could increase the readers' inquisitiveness.

2. Essential steps in responses surface methodology

a. After necessary Screening, the various factors and subsequent interactions of the experiment were identified [11].

b. The priority was given to the established various level of characteristics

c. Upon Optimization, the best suitable model has been selected [12].

d. The appropriate model, which is ideal for experimental design, can also be chosen [13].

e. To performed experimental studies, it is necessary to incept tangible factors and values which are needed to analyze systematically [14]

f. The selected model can be validated

g. There is a provision where if the data are not satisfactory, then another model of the experimental equation and experimental design is preferred. While pursuing the study, the aforementioned point c,d, and f need to be repeated until a suitable model is obtained, which is an acceptable representation of the data [15].

h. If required, a graphical representation of the surface is generated.

2.1 Models of the model used in the optimization process

The first-order model for the Optimization can be depicted as:

$$Y = \beta_o + \beta_1 X_1 + \beta_2 X_2 + \epsilon \tag{1}$$

For quadratic or second-order model, if nonlinearity was reported, then the following equation was incorporated:

$$Y = \beta_o + \beta_1 X_1 + \beta_2 X_2 + \beta_{12} X_1 X_2 + \beta_{11} {X_1}^2 + \beta_{22} {X_2}^2 + \epsilon \tag{2}$$

The factor must be very at level three while activating to fit the second-order model [16]. It was observed that, during the dictation of center point and two-level design, the quadratic terms can be identified, but it cannot be adequately estimated [17]. In **Figure 1,** the condition at which the Optimization can occur was explained, i.e., Optimization can be confirmed when second order model can be optioned from statistical outcomes and which coincide with the optimum value. During the factorial design experiment, it is preferable to avoid three-level designs as chances of an increase in the number of runs would be more [16].

For CCDDesign and Box–Behnken Design, second-order models are widely used. The analyzing aspect of these two designs can be explained by the following equation:

$$Y = b_o + b_1 X_1 + \ldots. + b_k X_k + b12 X_1 X_2 + b_{13} X_1 X_3 + \ldots + b_{k-1}, k\, X_{k-1} X_k + b_{11} {X_1}^2 + \ldots + b_{kk} {X_k}^2 + \epsilon \tag{3}$$

The above equation represents the quadratic model, which is near to the Optimization.

In this equation, Y = Dependent variables or Outcome variables or estimated responses, X_1 = independent variables, b_0 = overall mean response or intercept constant, b_1 = regression model coefficients, K = number of independent variables, ϵ = error.

Put into words, a mathematical model to the observed values of the dependent variables y, that indicates:

1. Main effects for factor $X_1 \ldots \ldots. X_k$

2. Their interactions (X_1X_2, $X_1X_{3 \ldots.,}$ $X_{k\,-\,1}$, $X_{k)}$

3. Their quadratic components (${X_1}^2, \ldots \ldots {X_k}^2$). No assumptions are made concerning the levels of the factors, and you can analyze any set of continuous values for the factors.

Based on the outcomes and empirical models from various experimental design, the central composite design gives us a direction to logically think and exercised multivariable analysis [18]. Three design points are prerequisite to establishing a

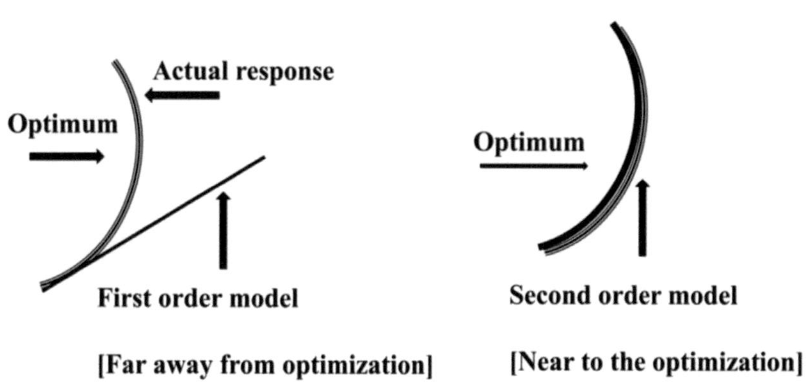

Figure 1.
Optimization condition.

second-order polynomial equation in CCDmodel [19]. When two levels of fractional factorial design need to be established, then 2^k should have possible +1 and − 1 levels of factors. In similar patterns, 2 k needs to be calculated, which can be otherwise called star points, and α forms the center to generate quadratic terms. The center point of the CCD., the model, provides an excellent independent estimation of experimental error.

$$N = k^2 + 2k + n, \tag{4}$$

Where N is the actual number of experiments, n is a number of repetition and k is the number of different factors which were incorporated within the study. Eventually, the CCD model can be best explained by the design of an expert (Version 11.0) software. The various steps involved in central composite design (CCD) was discussed in **Figure 2.**

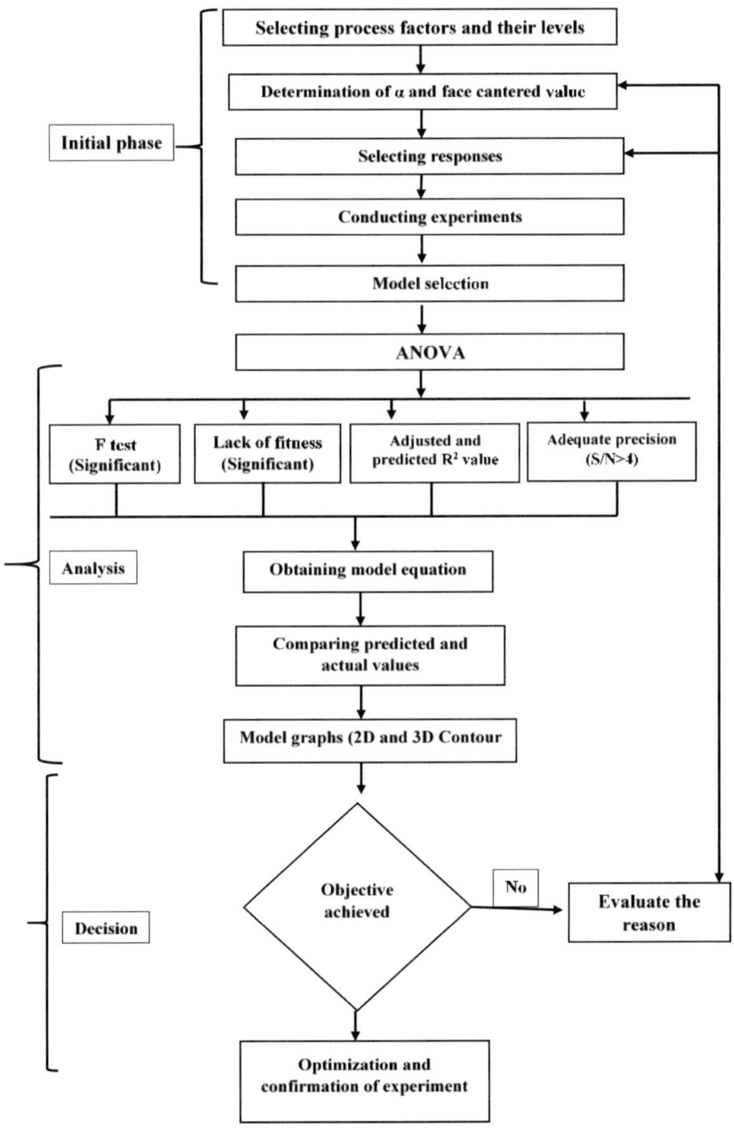

Figure 2.
Central composite design flow diagram.

To determine the local axial point, it is necessary to identify the alpha value in the CCD model. Depending on the alpha vale design can face cantered, rotatable, orthogonal. The alpha value can be calculated using the following equation:

$$\alpha = \left(2^k\right)^{0.25}.$$

If α value comes equals 1, the position of axial points stands within the factorial region. This is otherwise called a face-centered design, with three levels of factors that need to be kept in the design matrix. To calculate and analyze experimental results from response surface methodology, a polynomial equation needs to be implemented to study the correlation between dependent and independent variables.

$$Y = \beta_+ \beta_1 X_1 + \beta_2 X_2 + \beta_{12} X_1 X_2 + \beta_{11} {X_1}^2 + \beta_{22} {X_2}^2 + E \quad (5)$$

3. Types of central composite design

The Box and Wilson design or CCD model comprising of factorial[1], factorial[2], and factorial[3] design [20]. The star point outside the domine and the center point, representing the experimental domine, helps determine the response surface plot [21]. By estimating the precision of surface responses, the value of α can be determined; where star design is ± α. There are three types of CCD; the α can be determined according to the calculation possibilities and the required precision, which can be obtained from surface responses. The α value's positioning determines the quality of the design or estimation. The rate by design is identified by determining the position of the points [22]. The precision of the estimation influence by the number of trials at the center of the domine. The quality by design approach is necessary to estimate the coefficients' variability and responses [23]. One key aspect is rotatability or iso-variance per-rotation, which means that the prediction error is identical from all the points to the center points from the same distance [24]. Eventually, the center composite design was classified into three types:

3.1 Circumscribed design (CCC)

In central composite design, the levels of the factors eventually stand on the edge.

The CCD model **(Figure 3)** is always magnate with corner points, which was represented in red dots. From the center point (blue), the extract points are constrained from the sides (green dots). In this CCD model, each factor would have 5 levels. The star points are establishing new extremes for the low and high settings for all factors. These designs having circular, spherical or hyperspherical symmetry and required 5 levels for each factor. Supplementing an already existing factor or factorial design with a start point can produce the design. The Circumscribed (CCC) was found to be a rotatable design [25].

3.2 Inscribed design (CCI)

When the limit is specified for factor settings, the CCI design utilized the factor setting as star points and created a factorial design within those limits [26, 27]. In other words, CCI design is a modified version of CCC design, where CCC design has been divided by α to generate the CCI model. Eventually, CCC and CCI were found to be a rotational model **(Figure 4)**.

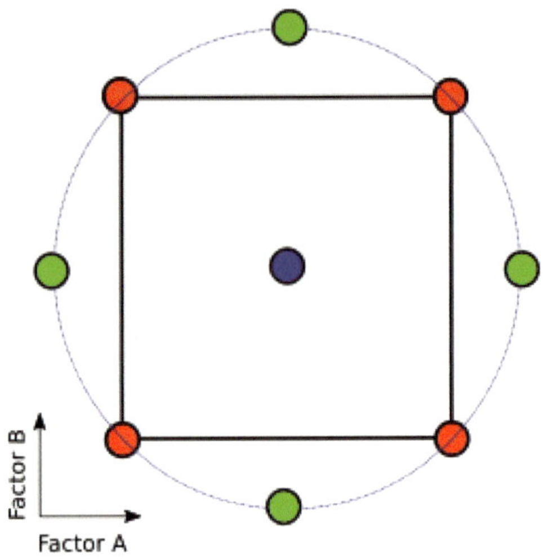

Figure 3.
CCDmodel.

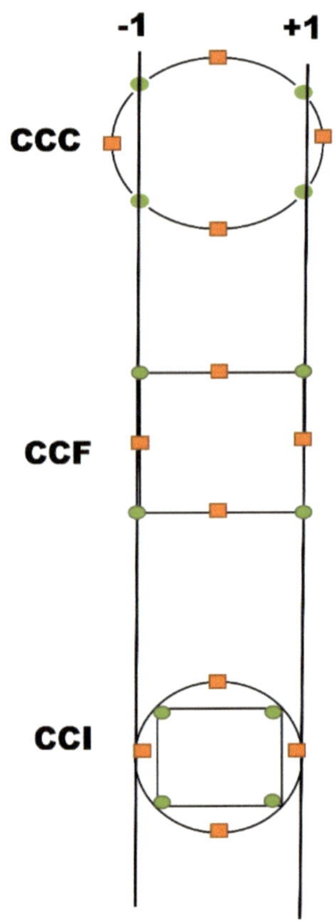

Figure 4.
Comparison of the three types of central composite designs.

3.3 Face cantered (CCF)

In this design, for each face of the factorial space, star points are the center point. Therefore. $\alpha = \pm 1$. This variable requires 3 levels of each factor [28]. The face cantered designs (CCF) are a non-rotatable design **(Figure 4)**.

In **Figure 4**, with two factors, three types of center composite design are used. From this design, one thing is clearly evidenced that; CCI explores the smallest process space, and CCC enjoys the largest process space. The CCC models looking like a sphere rotates around the factorial cube.

4. The parts of box and Wilson composite design

4.1 Determination of α value

Alpha (α) value can be defined as the calculated distance of each individual axial point (star point) from the center in the center composite design [29]. If Alpha (α) is less than 1, which indicates the axial point must be a cube, and if it is greater than 1, it indicates it is outside the cube. In central composite design, each factor has five levels, i.e., Extreme high or otherwise called a star point, higher point, center point, low point, and finally, extreme low star point. **Figures 5** and **6** describe how to select the total number of experimental runs for the CCD model as well as how to design two factors factorial design **(Table 1).** Coming to the Alpha (α) determination; which can be determined by the following equation:

α = (Number of factorial runs)$^{1/4}$.

= (2^k or 2^{k-r}) ^{¼}.

If K (number of factors) =2.

Alpha (α) = $(2^2)^{1/4} = 2^{2/4} = 2^{1/2} = 1.414$.

4.2 Uncoded value value of alpha (α)

To determine the uncoded value α value, the following equation can be used:

uncoded value value = (Coated value x L) + C; Where, L = Length expressed in real units between centre points and + 1 value of factor and C = Centre point value expressed in real units.

For temperature:

Uncoded value of – α = (Coded value × L) + C.

= (−1.414 × 30) + 60 = 17.58.

Uncoded vale of + α = (Coded value× L) + C = (1.414 × 30) + 60 = 102.42 **(Table 2).**

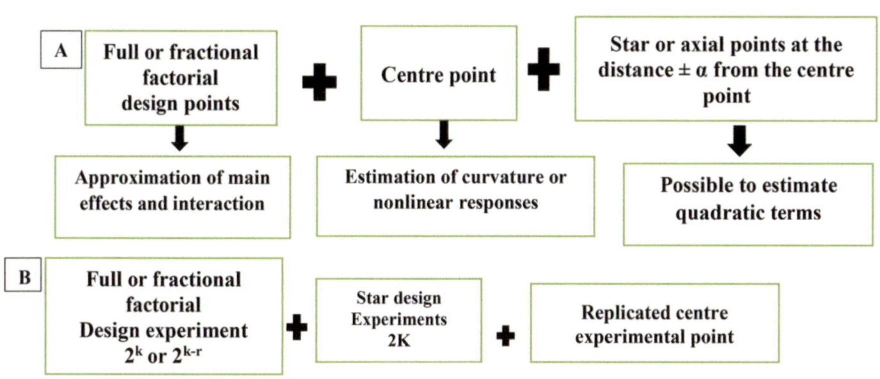

Figure 5.
(A) Parts of CCD (B) Total number of experimental runs required in the CCDmodel; where K = number of variables and r = fraction of full factorial. Thus, two-factor central composite design, the number of experimental runs is; $2^2 + 2(2) + 1 = 9$.

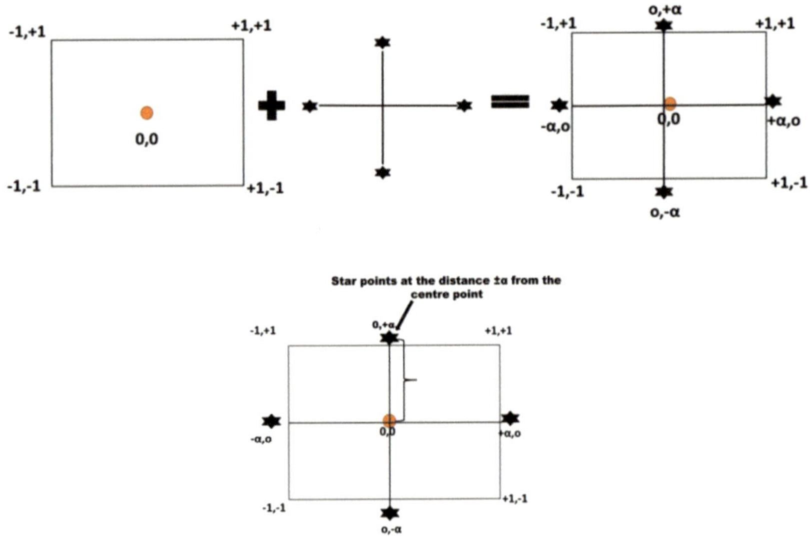

Figure 6.
Schematic presentation of two-factor central composite design.

Number of factors	α value related to ±1
2	±1.414
3	±1.682
4	±2
5	±2.378
6	±2.828

Table 1.
Number of factors and α value.

Level of factor	The temperature in ° C	Pressure in bar
-α (Lowest)	17.58	2.93
−1 (Lower)	30	5
0 (Centre point)	60	10
+1 (High)	90	15
+α (Highest)	102.42	17.07

Table 2.
α value of one experiment.

5. Advantages of center composite design

- It turns out to be the extension of 2 level factorial or fractional factorial design [21]
- To estimate nonlinearity of responses in the given data set
- Helps to estimate curvature in obtained continuous responses
- Maximum information in a minimum experimental trial

No.		Factor A	Factor B
1.	**Factorial runs 2^2 = 4**	−1	-1
2.		1	-1
3.		-1	1
4		1	1
5.	**Axial or star point runs 2(2) = 4**	−1.44	0
6.		1.414	0
7.		0	−1.414
8.		0	1.414
9.	**Centre point**	0	0

Table 3.
Design matrix for 2 factors central composite design [30].

No.		Factor A	Factor B	Factor C
1.	**Factorial runs 2^3 = 8**	−1	−1	−1
2.		1	−1	−1
3.		−1	1	−1
4.		1	1	−1
5.		−1	−1	1
6.		1	−1	1
7.		−1	1	1
8.		1	1	1
9.	**Axial or star point runs 2(3) = 6**	−1.682	0	0
10.		1.682	0	0
11.		0	−1.682	0
12.		0	1.682	0
13.		0	0	−1.682
14.		0	0	1.682
15.	**Centre point**	0	0	0

Table 4.
Design matrix for three factors central composite design [31].

- Reduction in the number of trials required to estimate the squared terms in the second-order model

- They have been widely used in response to surface modeling and Optimization (**Tables 3** and **4**)

6. Limitation of central composite design

- It was observed that the star points are outside the hypercube, so the number of levels that have to be adjusted for every factor is five instead of three, and sometimes it is not easy to achieve the adjusted values of factors [32].

- Depending upon the Design, the squared terms in the model will not be orthogonal to each other.

- Inability to estimate individual interaction terms, i.e., linear by quadratic or quadratic by quadratic.

Same examples of CCDoptimization in recent experimental conditions are mentioned in **Table 5.**

Title of the research	Author and Year	Model utilized	Variables taken	Findings	References
Development and optimization of baicalin-loaded solid lipid nanoparticles prepared by coacervation method using central composite design	Jifu Hao .et. a.l (2012)	central composite design	A two-factor five-level central composite design (CCD) was introduced.	The composition of optimal formulation was determined as 0.69% (w/v) lipid and 26.64% (w/w) drug/lipid ratio. The results showed that the optimal formulation of baicalin-loaded SLN had entrapment efficiency (EE) of 88.29%, particle size of 347.3 nm and polydispersity index (PDI) of 0.169.	[33]
Formulation, Development and Optimization of Propranolol Mucoadhesive Bilayer Tablets by Using Central Composite Design and its In Vitro Studies	Asif MASSUD		Angle of repose, compressibility index, bulk and tapped densities, Hausner's ratio for powders and granules were performed.	Mucoadhesive tablets with adequate mucoadhesion by adopting a new oral drug delivery concept Power was successfully developed to prevent liver degradation and propranololol enhancement Bioavailability Availability	[34]
Statistical Analysis of the Tensile Strength of Coal Fly Ash Concrete with Fibers Using Central Composite Design	Barbuta Marinela, et al. (2015)	CCD, Response Surface Method (RSM.)	Length, percentage, The total number of tests were statistically established taking into account the number of independent variables, the type of analyze that was done and	DOE is a structured, organized method that is used to determine the relationship between the different factors affecting a process and the output of that process. Analysis of	[35]

Title of the research	Author and Year	Model utilized	Variables taken	Findings	References
			the type of experimental plan that was chosen	variance (ANOVA) is a common method used to compare the relative strength of two related models	
Response surface modeling of lead (II) removal by graphene oxide-Fe3O4 nanocomposite using a central composite design	Khazaei Mohammad et al. (2016)	CCD, RSM (Response Surface Methodology)	4 independent variables: initial pH of Solution, nanocomposite dosage, contact time, initial lead ion concentration	Quadratic and reduced models were examined to correlate the variables with the removal efficiency of Magnetic Graphene Oxide. According to ANOVA, influential factors were pH and contact time.	[36]
Optimization of ferulic acid production from banana stem waste using central composite design	Sharif Nurul Shareena Aqmar Mohd (2017)	CCD, RSM (Response Surface Methodology)	Ratio of water to Banana stem waste (BSW), incubation time (hrs)	The RSM-CCD method optimize the hydrolysis conditions for maximum ferulic acid production. Hence, B.S.W. is proven useful and highly feasible for producing good quality natural products.	[37]
Central Composite Design Optimization of Zinc Removal from Contaminated Soil, Using Citric Acid as Biodegradable Chelant	Asadzadeh Farrokh et al. (2018)	CCD, RSM.	Citric acid concentration, pH, washing time	RSM based CCD is a promising tool for modeling and optimizing Zn removal from the contaminated soil using citric acid. It was found that pH and citric acid conc. Are the significant parameters in Zn removal process.	[38]
Removal of reactive red-198 dye using chitosan as an adsorbent: Optimization by Central composite design coupled with response surface methodology	Haffad Hassan et al. (2019)	C.C.D., Langmuir isotherm model, pseudo-second-order equation	pH, Concentration, temperature	Chitosan material based shrimp cells have countless opportunities in use of waste water treatment. The major effect is played by pH	[39]

Title of the research	Author and Year	Model utilized	Variables taken	Findings	References
Mixture optimization of high-strength blended concrete using central composite design	Hassan Wan Nur Firdaus Wan et al. (2020)	CCD, RSM.	Micro and Nano Palm Oil Fuel Ash (POFA.)	Mixture optimization of high-strength blended concrete using central composite Design, Run 1, containing 10% micro POFA and 2% nano POFA, showed the highest flexural strength	[40]

Table 5.
Examples of CC demployed for the optimization.

7. Case study-1

As per S. Bhattacharya., (2020) studies [41] varius independent variables viz., entrapment efficacy percentage, zeta potential, particle size, percentage of calmative drug release of a polymeric nanoparticle formulation was evaluated and optimized using central composite design (CCD); which was interpreted by Design Expert (Stat-Ease; version 11.0) software. Upon considering the alpha point at 1.68179, in this 21 baches experimental design, 4 factors, and 2 levels were considered **(Table 6)**. Based on the optimization surface plot batch with desired particle size, zeta potential, cumulative drug release (%) entrapment efficacy (%) were selected for further characterization studies. **Table 7** indicating critical quality attributes and necessary process attributes that affect the outcomes of the nanoparticles. By using polynomial equations and a 3-dimensional surface plot, the

S. No	Factors	Low Value	High Value
1	Homogenization speed (rpm)	10000	15000
2	**Homogenization Time (min)**	10	15
3	Surfactant Concentration (%)	1	1.25
4	Polymer concentration (mg/mL)	3	6

Table 6.
Critical process parameters that influence various critical quality attributes.

S. No	Critical Quality Attributes	Desired constrained
1	Particle Size (nm)	Finest
2	Zeta Potential (mV)	Finest
3	Cumulative drug release (%)	Moderately high
4	Entrapment efficacy (%)	Supreme

Table 7.
Desired construction of critical quality attributes.

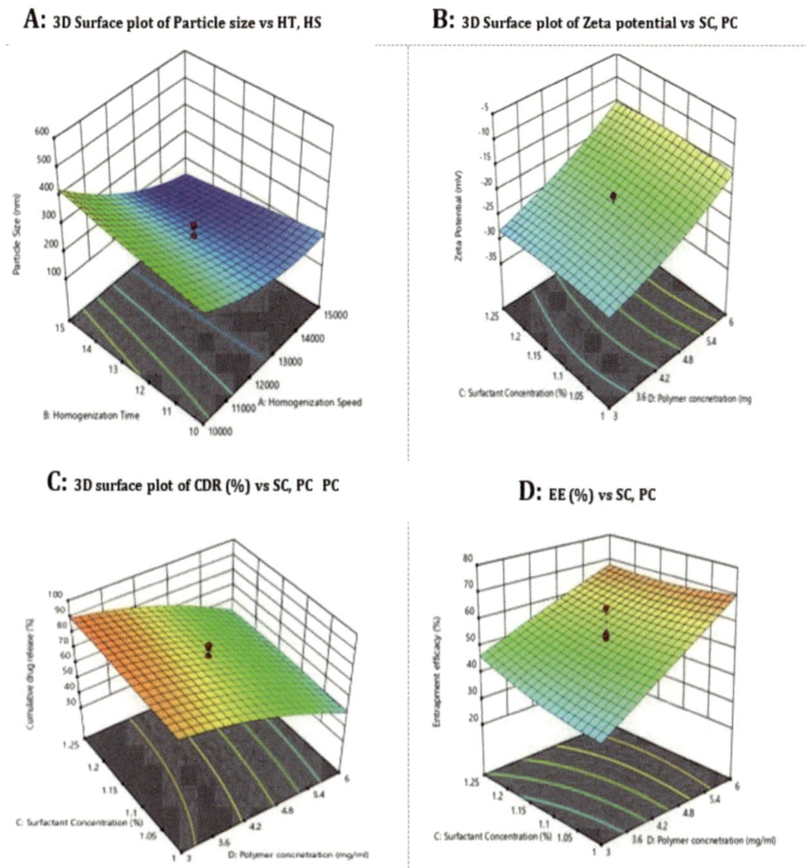

Figure 7.
(A-D) represents the surface plot identifying the effects of critical process parameters on essential attributes of quality.

effects of critical process parameters on essential attributes of quality were examined **Figure 7.**

From this CCDmodel, the following polynomial equations can be derived:

$$\begin{aligned}\textbf{Particle Size}\ (\textbf{nm}) = {} & 236.054 - 101.38\text{A} - 3.6663\text{B} - 8.85986\text{C} + 0.594604\text{D} \\ & - 40.7804\text{AB} - 24.375\text{AC} - 16.375\text{AD} + 2.625\text{BC} \\ & - 30.2549\text{BD} - 18.375\text{CD} + 45.2505\ \text{A}^2 + 1.23307\text{B}^2 \\ & + 11.1326\text{C}^2 + 0.879517\text{D}^2\end{aligned} \tag{6}$$

$$\begin{aligned}\textbf{Zeta Potential}\ (\textbf{mV}) = {} & -21.8583 + 1.11191\text{A} - 0.772985\text{B} - 0.19847\text{C} + 6.18685\text{D} \\ & - 1.42565\text{AB} + 0.77\text{AC} - 0.242985\text{AD} - 1.0125\text{BC} \\ & + 0.829409\text{BD} + 0.385\text{CD} - 0.644786\text{A}^2 - 1.44735\ \text{B}^2 \\ & - 0.805653\text{C}^2 + 1.23081\text{D}^2\end{aligned} \tag{7}$$

$$\begin{aligned}& \textbf{Cumulative drug release}\ (\%)\ \textbf{at}\ \textbf{80}^{\textbf{th}}\ \textbf{hours} \\ & = 75.2947 + 2.52409\text{A} - 1.66192\text{B} + 2.1758\text{C} - 17.5081\text{D} + 2.73315\text{AB} \\ & \quad - 0.65625\text{AC} - 0.605667\text{AD} + 0.71625\text{BC} + 2.56284\text{BD} - 0.4687\ \text{CD} \\ & \quad + 0.534547\text{A}^2 - 1.0918\ \text{B}^2 - 1.11301\text{C}^2 - 5.01271\ \text{D}^2\end{aligned} \tag{8}$$

$$\textbf{Entrapment efficacy}\ (\%) = 53.323 + 0.184327A + 3.54978B + 2.29125C + 14.0832D + 5.83819AB + 1.2475AC + 1.6522AD + 3.1875BC - 4.10817BD - 3.25CD + 2.30194A^2 + 0.088697\,B^2 + 1.78045C^2 - 0.192378\,D^2 \tag{9}$$

By considering A as homogenization speed, B as homogenization time, C as surfacetant concentration (%) and D as polymeric concentration; respectively, the polynomial Eqs. 6,7,8, & 9 can be interpreted. From these polynomial equations, critical process parameters of qualifiable effects on essential attributes can be determined. It can easily predict from the polynomial Eq. 6, that particle size of the polymeric nanoparticles can be increased, when homogenization time & speed and surfactant concentration decrease. The elevated negative co-efficient in homogenization speed of polynomial Eq. 6, indicates it has a significant influence on particle size. Higher shearing stress during elevated homogenization time & speed could lead to mass transfer between the particles, ultimately resulting in nucleation and smaller particle size. From Eq. 7, it can be predict that homogenization time & surfactant concentration has antagonistic effects on zeta potential and homogenization speed has an agonistic effect on zeta potential. In a similar fashion equation, 8 shows homogenization speed & surfactant concentration has an agonistic effect on cumulative drug release (%) at 80th hours. From Eq. 9 it was clear that entrapment efficacy (%) increases with increases of homogenization speed, homogenization time & surfactant concentration.

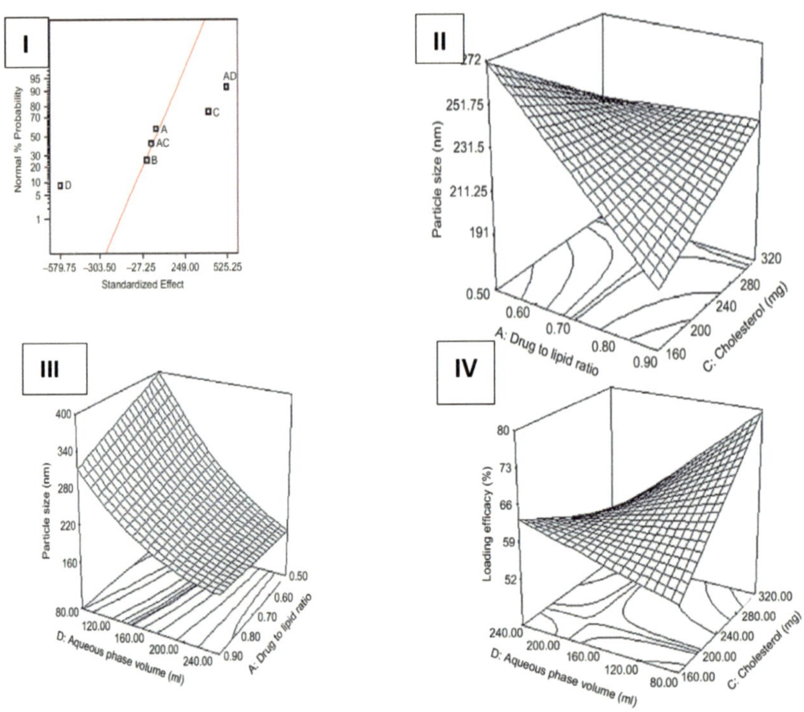

Figure 8.
(I) *Normal probability plot for broadcast, the most critical variables are influencing the particle size. The ratio of drug to lipid (a), surfactant type (B), amount of lipid phase (C), and volume of aqueous phase (D).* ***(II)*** *the effect of drug-to-lipid ratio (a) and amount of cholesterol (C) on particle size.* ***(III)*** *the effect of drug-to-lipid ratio (a) aqueous-phase volume (D) on particle size.* ***(IV)*** *the impact of aqueous-phase volume (D) and the amount of cholesterol (C) on loading efficacy(%).*

8. Case study 2

As per Jaleh Varshosaz *et al.* (2010) [42] research, amikacin solid lipid nanoparticles can be prepared by using a central composite design. In this research, central composite design (CCD) was utilized to identify a suitable formula with minimum particle size; where, three independent variables were considered, i.e., the ratio of drug to lipid (A), amount of lipid phase (B), the volume of aqueous phase (D). The alpha value of the experiment was found to be −1.682; the alpha value helps in determining rotatability and orthogonality within this design. In this experiment, a total of 20 experimental designs have been incepted, along with 8 factorial points and 6 axial points were considered. The best-fitted model can be assumed after quadratic model analysis by ANOVA and F-value determination. From the **Figure 8(II&III)** it was clarly evident that, decrease concentration of drug to lipid ratio, aquous phase valume whould certlay decrease particle size; which indicates agonistic effects on particle size, where else, from the **Figure 8(IV),** it was evident that, increase concentration of cholesterol would increases the drug loading capacity. Therfore, by resolving all the polynomial equation obtained from **Figure 8 graph**, it was identified that, at 0.5 drugs to lipid ratio, 314 mg cholesterol, 229 mL of aqueous phase an optimized formulation would possibly be constructed with lower particle size and higher drug loading efficacy (%). Therefore, At these levels of independent variables, predicted amikacin particle size and loading efficiency were calculated to be 153 nm and 86%.

9. Conclusion

This book chapter's main agenda was to enlighten the present approaches and recent optimization research activities based on the CCD model, as specially for pharmaceutical product development. The CCDmodel is useful for modeling and analyzing programs in which the response of interest influences several variables. The CCDmodel can be considered as a robust statistical tool for process optimization. The best part of CCDis, as compared to Plackett–Burman design, a limited number of experiments are required with less computational experience. The biggest challenge of the CCDmodel is finding the critical factor. Central composite designs are beneficial in sequential experiments because you can often build on previous factorial experiments by adding axial and center points.

Acknowledgements

The author is like to acknowledge the help and motivation of Dr. R.S. Gaud, Director, SVKM's NMIMS. Deemed-to-be University, Shirpur Campus, for providing excellent research facilities and profound inspiration while drafting this book chapter. The author is also like to acknowledge his own original research publication entitled "Fabrication and characterization of chitosan-based polymeric nanoparticles of Imatinib for colorectal cancer targeting application"; for taking inputs from that article while drafting this book chapter.

Conflict of interest

The author declares that the author has no competing interests.

Author details

Sankha Bhattacharya
School of Pharmacy and Technology Management, Department of Pharmaceutics, SVKM'S NMIMS Deemed-to-be University, Shirpur, Maharashtra, India

*Address all correspondence to: sankhabhatt@gmail.com

References

[1] Cooper, R.G., S.J. Edgett, and E.J.J.R.-T.M. Kleinschmidt, *Optimizing the stage-gate process: what best-practice companies do—I*. 2002. **45**(5): p. 21–27.

[2] Del Castillo, E., *Process optimization: a statistical approach*. Vol. 105. 2007: Springer Science & Business Media.

[3] Kilickap, E., M. Huseyinoglu, and A. J.T.I.J.o.A.M.T. Yardimeden, *Optimization of drilling parameters on surface roughness in drilling of AISI 1045 using response surface methodology and genetic algorithm*. 2011. **52**(1–4): p. 79–88.

[4] Beg, S., et al., *Application of design of experiments (DoE) in pharmaceutical product and process optimization*, in *Pharmaceutical quality by design*. 2019, Elsevier. p. 43–64.

[5] Granato, D., et al., *The use and importance of design of experiments (DOE) in process modelling in food science and technology*. 2014. **1**: p. 1–18.

[6] Ellis, D.I., et al., *Rapid identification of closely related muscle foods by vibrational spectroscopy and machine learning*. 2005. **130**(12): p. 1648–1654.

[7] Bevilacqua, A., et al., *Design of experiments: a powerful tool in food microbiology*. 2010: p. 1419–1429.

[8] Coman, G. and G.J.A.o.m. Bahrim, *Optimization of xylanase production by Streptomyces sp. P12–137 using response surface methodology and central composite design*. 2011. **61**(4): p. 773–779.

[9] Veni, D.K., N.V.J.I.J.o.P.M. Gupta, and P. Biomaterials, *Development and evaluation of Eudragit coated environmental sensitive solid lipid nanoparticles using central composite design module for enhancement of oral bioavailability of linagliptin*. 2020. **69**(7): p. 407–418.

[10] Ye, Q., et al., *Development and evaluation of puerarin-loaded controlled release nanostructured lipid carries by central composite design*. 2020: p. 1–37.

[11] Imandi, S.B., et al., *Application of statistical experimental designs for the optimization of medium constituents for the production of citric acid from pineapple waste*. 2008. **99**(10): p. 4445–4450.

[12] Candioti, L.V., et al., *Experimental design and multiple response optimization. Using the desirability function in analytical methods development*. 2014. **124**: p. 123–138.

[13] Zolgharnein, J., M. Bagtash, and N.J. J.o.E.C.E. Asanjarani, *Hybrid central composite design approach for simultaneous optimization of removal of alizarin red S and indigo carmine dyes using cetyltrimethylammonium bromide-modified TiO2 nanoparticles*. 2014. **2**(2): p. 988–1000.

[14] Schmid, R.F., et al., *The effects of technology use in postsecondary education: A meta-analysis of classroom applications*. 2014. **72**: p. 271–291.

[15] Gefen, D., D. Straub, and M.-C.J.C. o.t.a.f.i.s. Boudreau, *Structural equation modeling and regression: Guidelines for research practice*. 2000. **4**(1): p. 7.

[16] Jones, B. and C.J.J.J.o.Q.T. Nachtsheim, *A class of three-level designs for definitive screening in the presence of second-order effects*. 2011. **43**(1): p. 1–15.

[17] Bezerra, M.A., et al., *Response surface methodology (RSM) as a tool for optimization in analytical chemistry*. 2008. **76**(5): p. 965–977.

[18] Esbensen, K.H., et al., *Multivariate data analysis: in practice: an introduction to multivariate data analysis and experimental design*. 2002: Multivariate Data Analysis.

[19] Chu, P.C. and C.J.J.o.C.P. Fan, *A three-point combined compact difference scheme*. 1998. **140**(2): p. 370–399.

[20] Yolmeh, M., S.M.J.F. Jafari, and B. Technology, *Applications of response surface methodology in the food industry processes*. 2017. **10**(3): p. 413–433.

[21] Tarley, C.R.T., et al., *Chemometric tools in electroanalytical chemistry: methods for optimization based on factorial design and response surface methodology*. 2009. **92**(1): p. 58–67.

[22] Wang, G.G.J.J.M.D., *Adaptive response surface method using inherited latin hypercube design points*. 2003. **125** (2): p. 210–220.

[23] Xu, X., M.A. Khan, and D.J.J.I.j.o.p. Burgess, *A quality by design (QbD) case study on liposomes containing hydrophilic API: II. Screening of critical variables, and establishment of design space at laboratory scale*. 2012. **423**(2): p. 543–553.

[24] Pradhan, M., et al., *Quality by design and formulation optimization using statistical tools for safe and efficient bioactive loading*, in *Advances and Avenues in the Development of Novel Carriers for Bioactives and Biological Agents*. 2020, Elsevier. p. 555–594.

[25] Singh, B., R. Kumar, and N.J.C.R.i.T. D.C.S. Ahuja, *Optimizing drug delivery systems using systematic" design of experiments." Part I: fundamental aspects*. 2005. **22**(1).

[26] Leiviskä, K.J.U.o.O.C.E.L., *Introduction to experiment design*. 2013.

[27] Cavazzuti, M., *Design of experiments*, in *Optimization methods*. 2013, Springer. p. 13–42.

[28] Hasanien, H.M., A.S. Abd-Rabou, and S.M.J.I.T.o.E.C. Sakr, *Design optimization of transverse flux linear motor for weight reduction and performance improvement using response surface methodology and genetic algorithms*. 2010. **25**(3): p. 598–605.

[29] Asghar, A., A.A. Abdul Raman, and W.M.A.W.J.T.S.W.J. Daud, *A comparison of central composite design and Taguchi method for optimizing Fenton process*. 2014. **2014**.

[30] Ait-Amir, B., P. Pougnet, and A. El Hami, 6 - Meta-Model Development, in Embedded Mechatronic Systems 2, A. El Hami and P. Pougnet, Editors. 2015, Elsevier. p. 151-179.

[31] Rajmohan T, Palanikumar K. Application of the central composite design in optimization of machining parameters in drilling hybrid metal matrix composites. Measurement. 2013 May 1;46(4):1470-81.

[32] Ferreira, S.L.C., et al., *Statistical designs and response surface techniques for the optimization of chromatographic systems*. 2007. **1158**(1–2): p. 2–14.

[33] Hao, J., et al., *Development and optimization of baicalin-loaded solid lipid nanoparticles prepared by coacervation method using central composite design*. 2012. **47**(2): p. 497–505.

[34] Massud, A., et al., *Formulation, development and optimization of propranolol mucoadhesive bilayer tablets by using central composite design and its in-vitro studies*. 2015. **34**(8): p. 1637–44.

[35] Barbuta, M., et al., *Statistical analysis of the tensile strength of coal fly ash concrete with fibers using central composite design*. 2015. **2015**.

[36] Khazaei, M., et al., *Response surface modeling of lead (II) removal by graphene oxide-Fe 3 O 4 nanocomposite using central composite design*. 2016. **14**(1): p. 2.

[37] Mohd Sharif, N.S.A., et al., *Optimization of ferulic acid production from banana stem waste using central composite design*. 2017. **36**(4): p. 1217–1223.

[38] Asadzadeh, F., et al., *Central composite design optimization of zinc removal from contaminated soil, using citric acid as biodegradable chelant.* 2018. **8**(1): p. 1–8.

[39] Haffad, H., et al., *Removal of reactive red-198 dye using chitosan as an adsorbent: optimization by Central composite design coupled with response surface methodology.* 2019: p. 1–13.

[40] Hassan, W.N.F.W., et al., *Mixture optimization of high-strength blended concrete using central composite design.* 2020. **243**: p. 118251.

[41] Bhattacharya, Sankha. "Fabrication of poly (sarcosine), poly (ethylene glycol), and poly (lactic-co-glycolic acid) polymeric nanoparticles for cancer drug delivery." Journal of Drug Delivery Science and Technology (2020): 102194.

[42] Varshosaz, J., et al., *Development and optimization of solid lipid nanoparticles of amikacin by central composite design.* 2010. **20**(2): p. 97–104.

Section 3

Application of Response Surface Methodology in Engineering

Chapter 6

Response Surface Methodology Optimization in Asphalt Mixtures: A Review

Aliyu Usman, Muslich Hartadi Sutanto, Madzlan Bin Napiah and Nura Shehu Aliyu Yaro

Abstract

The application of statistical modeling and optimization approaches such as response surface methodology (RSM) is important for the excellent potential to tackle different constraints and goals and the analysis of the relationships between independent factors influencing a particular response. This chapter provides a simple yet detailed literature review on the utilization of RSM for the design of experiments, modeling, and optimization of virgin and alternative materials into asphalt binder and mixtures for sustainability. Meanwhile, an in-depth analysis based on the literature reviewed in terms of asphalt binder modification employing RSM with various independent parameters were summarized. Also, a critical review of the application of RSM to optimize the engineering and mechanical performance characteristics of asphalt concrete mixtures is presented in this chapter. The current chapter concluded that the use of RSM statistical analysis in a highway materials perspective provides a broader understanding of the factors that control pavement performance throughout the pavement service life.

Keywords: asphalt binder, asphalt concrete mixture, response surface methodology, prediction, optimization

1. Introduction

In order to get the most benefits from a process and maximize its efficiency, optimization of the process is needed. Optimization refers to picking the ideal factor from a collection of potential solutions (independent variables). In certain applications, the value of the variable is enough to achieve the optimal output or best possible outcome. The common practice used to assess optimum design parameters is to evaluate the effect of each variable independently on the response [1, 2]. In this process, interactive effects among parameters are not regarded. Consequently, this approach does not display the full impact of these variables on the outcome. Another primary disadvantage of this method is the increase in the overall number of tests needed for the investigation, leading to an increase in costs and materials, and time [3, 4]. Owing to these drawbacks, researchers have been exploring alternative approaches. For the past decades, response surface methodology (RSM) is among the most effective approaches used for modeling and optimization. Myers, et al. [5] defined response surface methodology (RSM) as a statistical method that is commonly used to optimize engineering design and operations. The majority of RSM applications require multiple response variables. In a typical RSM analysis, the experimenter will create an empirical model like the second-order model to every response using these models to evaluate the configuration of the design variables that generate optimum or at least appropriate response values [2, 6]. RSM has great relevance in the design, development, and introduction of new material and the enhancement of material designs. RSM describes the influence of independent variables, either alone or in synergetic effect, during the process. It also helps in evaluating the influence of independent variables, this experimental approach produces a statistical model that explains the mechanism and processes [2, 5, 7]. While RSM has several benefits, it can be concluded that it applies to all modeling and optimization analyses of various aspect of engineering, RSM has now been commonly and successfully used to optimize asphalt binder and asphaltic concrete mixtures to maximize their performances and promote sustainability. The current research examined the application of RSM for modeling and optimization in asphalt pavement studies. In this chapter, several recently published RSM studies have been examined to have an overview and access the application of RSM in the pavement industry.

2. Design of experiment (DoE)

Design-Expert is a software for experimental design, statistical analysis, modeling, and optimization. It provides a variety of programs such as fractional factorial design, surface response, full factorial, mixing, and D-optimal designs [1, 2, 8]. RSM experimental designs are developed using the Design-Expert program. The program is also utilized to analyze the data obtained. Regression is applied to data obtained where the measured variable (response) is estimated based on a functional relationship between the predicted input variables. Experimental values that can be altered independently of one another are referred to as factors or independent variables [2, 9]. Variable levels are the various stages of the factors at which the experiments are to be performed. After conducting the experiment, the data obtained are called responses or dependent variables. While residual is the discrepancy between the experimental and predicted values for the determined range of experimental criteria also model with low residual values indicates a strong fitting of the experimental data [8, 9].

3. Response surface methodology (RSM)

Box and Wilson [10] introduced RSM in 1951 and they proposed using a second-degree polynomial model. RSM has recently been used for process parameter optimization. RSM can be considered as a systematic calculation technique for the optimization problem. This method provides an appropriate experimental method that incorporates all the independent variables and utilizes the experiment's input data to subsequently create a set of equations that can offer an output's theoretical value [11]. The findings are achieved from a well-designed regression analysis that examines the relationship between independent variables'-controlled values. Based on the new values of independent variables, the dependent variable can then be forecast [11]. RSM is an effective statistical tool for both the modeling and optimization of multiple variables with a minimum number of experiments to forecast the optimum performance parameters [7, 12]. By employing the RSM technique in the optimization process, the testing of all the variables relating to the product assessment requires just a short time, making the laboratory test stage more efficient [11].

Furthermore, the estimation of parameters that profoundly influence the model can determine which allows researchers to concentrate on those specific variables to improve the performance of the process [11]. In a set of experimental designs, one factor or process variable can depend strongly on or be dependent on another variable. In an attempt to discover the output–input relationship, understanding the interaction between the variables is critical, that is why taking a single factor at a time approach is seldom used to evaluate interrelationships between parameters. RSM can determine the relationship as well as interactions between the multiple parameters using quantitative data by creating a model equation. In RSM implementation, there are three steps; (i) experiment design, i.e., Box Behnken (BBD) and Central Composite Design (CCD); (ii) statistical and regression analysis to build model equations that describe the modeling of the response surface; and (iii) optimization of parameters/variables carried out via model Equation [13]. A statistical experimental design is presented in RSM. Different experimental designs could be carried out based on the special criteria and the choice of experimental points and numbers. In addition to randomizing the experimental error to every experimental point, operating with a statistical experimental design often implies the distributions of experimental points in the examined set of independent variables. These improve the reliability of the model Equation [5].

3.1 Central composite design (CCD)

The most widely and frequently employed and effective design technique is central composite design (CCD). A minimum of two numerical inputs is required and varied in the CCD approach at a range of alpha (α) over three (−1, 0, +1) or five (−α, −1, 0, +1, +α) stages. Three features are contained in the CCD model: (i) a complete factorial or fractional factor design; (ii) an additional design, mainly a star design with experimental points from the centre at alpha; and (iii) a central point. There are various ways to CCD depending on the alpha (α)-value, such as face-centred central composite (FCCD), rotational, spherical, orthogonal quadratic, and practical. Five levels are utilized for the rotational and spherical approaches depending on the number of variables. The value of alpha (a) for the FCCD is always unity. This means that the axial points are not positioned on the spheres, but rather on the centre of the faces, so only three variants of each parameter are involved [2]. **Figure 1**(**a** and **b**) illustrate the full factorial central composite model for two and three parameters optimization.

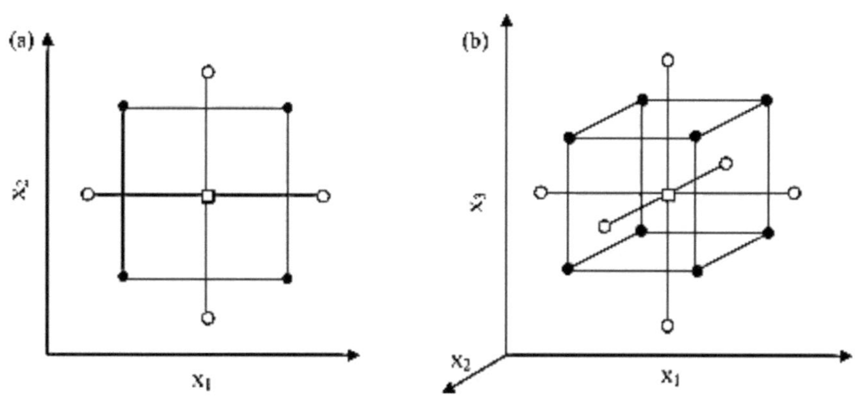

Figure 1.
Optimization of two and three parameters using CCD (a) two parameters (X_1 and X_2) where alpha (α) = 1.41 (b) three parameters (X_1, X_2, and X_3) where alpha (α) = 1.68 [12].

3.2 Box Behnken design (BBD)

Box–Behnken Design (BBD) suggested how to select points from a three-level parameter model to permit the first- and second-level coefficients of the statistical model to be effectively assessed. In this way, these designs, especially for many variables, are more efficient and cost-effective than their respective 3^k design models. A minimum of three numerical factors are required in BBD and vary across three levels. In Box–Behnken models, the experimental points are situated on a hypersphere spaced uniformly from the central point. BBD's key characteristics are: (1) it requires an experimental number depending on the Eq. $N = 2k(k - 1) + cp$, where k is the number of variables and (cp) the number of central points; (2) it is appropriate to change all factor levels at three levels only (−1, 0, + 1) with intervals spaced equally. BBD has been used successfully as a physical and chemical technique for different optimization processes. Other RSM design techniques, such as one factor, optimal design, miscellaneous design, user-defined, and historical data designs, are available [2].

3.3 Motivation for the chapter

Due to the challenges facing the conventional pavement materials such as asphalt binder and mixtures and the paradigm shifts towards green and sustainable construction. The application of RSM is one of the promising ways used by pavements engineers as it has shown to be of great importance in the design, development, and incorporation of alternative green materials for the improvement of the pavement material design. Modeling and optimization of the synergetic impact of different parameters that affect the engineering properties and performance output of the asphalt binder and mixtures utilization help to provide more insight into the influences of the various variables. This experimental method generates a numerical method that helps elaborate on the mechanism and procedures that involve a lesser

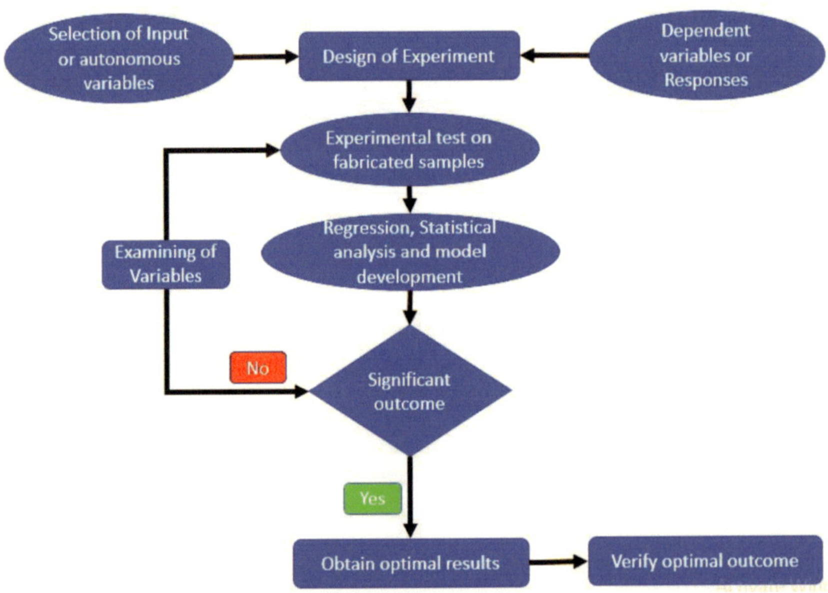

Figure 2.
General RSM design flow chart.

number of experimental runs. Thus, the utilization of RSM is of pivotal use for pavement engineers.

RSM is an effective instrument for engineering new bituminous blends. By forecasting the response of the materials on the basis of experimental plans drawn up with a scientific and statistical method [14]. Also, another RSM's key goal is to evaluate the optimal settings for control variables that lead to the maximum or minimum output over a certain area of study. The generated equations can be used for interpolation to obtain the maximum (or minimum) predictable results within the levels of the analyzed variables [14–16].

The general design flow chart of the RSM statistical technique framework is presented in **Figure 2**.

4. Asphalt mixing process for incorporating new materials

Three methods are employed to incorporate new sustainable materials into asphalt binders and mixtures for optimization. These methods are selected based on the type and properties of the new material to be incorporated. These methods are wet process, dry process, and modified dry process.

4.1 Wet process

The wet method comprises blending the newly added material with the heated asphalt binder to form the modified asphalt binder, The optimization is commonly done on the mixing parameters or the conventional properties of the mixes and the mixture is formed by blending hot aggregates with the modified asphalt binder. This method is suitable for materials having low melting points such as low-density polyethylene (LDPE), and polypropylene (PP). This method requires a high-shear mixer and sufficient time for blending the newly added material and asphalt binder [15].

4.2 Dry process

In the dry method, before adding asphalt binder to the mixture, the new sustainable green material is initially incorporated into the heated aggregates and mixed [16]. In this technique, some of the filler is lost as dust when blending the new material with the aggregates, which is not suitable [15]. This method is suitable for materials with a higher melting point above the asphalt mixture mixing temperatures such as polyethylene terephthalate (PET) and high-density polyethylene (HDPE). The optimization process using involves the mixing process and mechanical properties of the asphalt mixtures.

4.3 Modified dry process

In the modified dry method, new added sustainable green materials are incorporated while the heated asphalt binder and aggregates are thoroughly mixed [17]. The added new materials particles are ensured to be well coated by the asphalt binder. It is hypothesized that minor changes in the shape and properties of added waste materials during mixing will result from the modified dry process [15, 18]. Some experts modified the blending techniques to obtain a good distribution of the waste materials particles in the modified blends. The optimization process using involves the mixing process and mechanical properties of the asphalt mixtures.

5. Utilization of RSM optimization techniques in asphalt binder modification

RSM has been utilized in the modification of asphalt binder and optimization of the modification mixing parameters to enhance sustainability and improve performance of various bitumen modifiers using different dependent and independent variables associated with the bitumen, modification process, and performance of the bitumen. Assessment and investigation of the interactive effects on the response of process variables can thus be studied and analyzed using RSM. The model equation also quickly clarifies the effects with the various combination of the independent parameters. In this section the application of RSM for the modification of asphalt binder utilizing several types of alternative materials. Most of the prior study includes studies of the mixing conditions of asphalt binder and its conventional properties are discussed below.

A study conducted by Liu, et al. [19] evaluated the effects of mixing variables parameters such as mixing speed, time, temperature, and the modifier (diatomite and crumb rubber) on modified asphalt binder properties (penetration, softening point, penetration index, viscosity, elastic recovery, and ductility) were examined and optimization using RSM. The findings showed that with the increase in crumb rubber concentration, softening points, viscosity, elastic recovery, and penetration index increased, while penetration and ductility decreased. With the rise in diatomite concentration, the softening points, viscosity, and penetration index increase, while penetration and ductility decrease, which has little effect on elastic binder recovery. The shear temperature has had major impacts on penetration, ductility, softening point, and viscosity. Because of its similar mechanism of action, shear velocity, shear duration, and storage time have similar influences on binder properties. The optimum speed, time, and temperature of mixing were achieved at 55 min and 5300 rpm, 55 minutes, and 183 °C, respectively, based on the optimization. The effects of the various mixing parameters on the convention properties of the diatomite and crumb rubber modified bitumen is illustrated in the 3D surface plots in **Figures 3-5**.

Recently, the effects of various crumb rubber (CR) contents and their interactions with high-temperature ranges on the rheological activity of asphalt binder were also studied by Badri, et al. [20] At temperatures ranging from 46 °C to 82 °C, a temperature sweep test was performed using a dynamic shear rheometer (DSR) on modified binders with 5, 7, 10, 12 and 15% crumb rubber content. Based on central composite design (CCD) and considering the rheological parameters complex modulus (G^*) and phase angle (δ) as response variables, RSM statistical analysis was performed. The outcome of the ANOVA test showed that rheological parameters were significantly affected by temperature and rubber content with p less than 0.05. Independent effects of temperature and rubber content were analyzed, and the results showed that both responses were affected by the interaction of both independent variables but were more impacted by rubber content than temperature as depicted in the synergetic effect of the variables on the CR modified bitumen are shown in the 2D and 3D contour diagrams in **Figures 6** and 7. It can therefore be inferred that RSM is an efficient tool for investigating rheological characteristics. For all responses, a percentage error of <5% was achieved, suggesting that the optimization process by RSM is a very efficient and effective technique.

Also, the RSM was used to evaluate both the volumetric and mechanical properties of asphalt mixtures after modifying the mix variables. A study conducted by Nassar, et al. [21] optimize the design mix variables, namely bitumen emulsion content (BEC), pre-wetting water content (PWC), and curing temperature (CT), the Response Surface Methodology (RSM) was used. The purpose of this work was

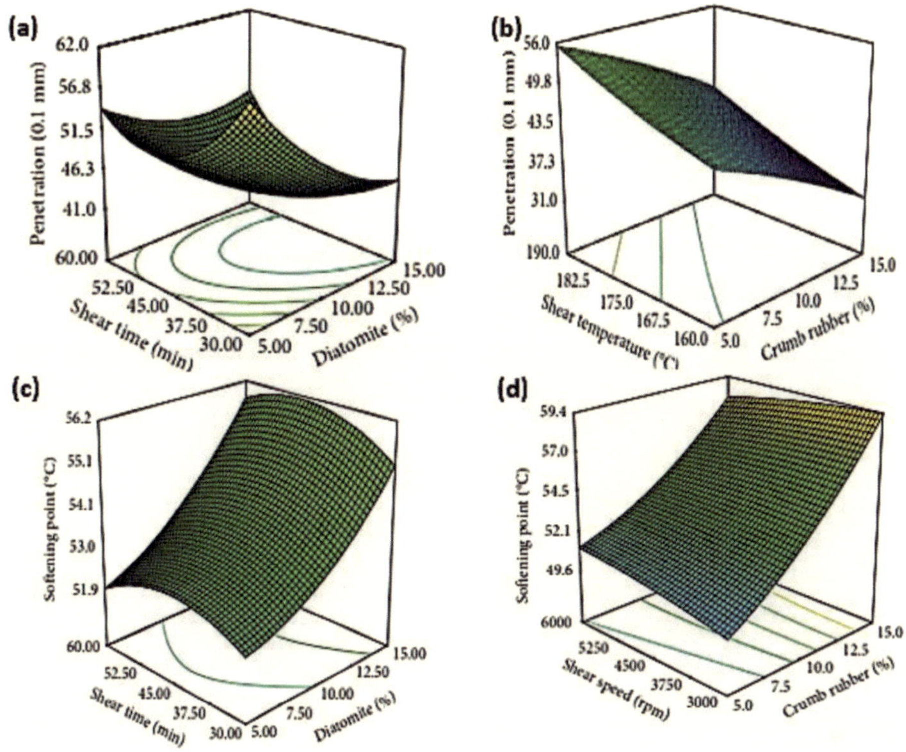

Figure 3.
Response surface plots for the preparation parameters on penetration values (a) shear time and diatomite (b) shear temperature and crumb rubber and for softening point. (c) shear time and diatomite content, (d) shear speed and crumb rubber content [19].

to evaluate the relationship influence on the mechanical and volumetric properties of cold bitumen emulsion (CBEMs) between these parameters. To determine the mechanical response, indirect tensile strength (ITS) and indirect tensile stiffness modulus (ITSM) tests were performed while air voids and dry density were calculated to obtain volumetric responses. Besides, the total fluid content was used in the typical mix design technique, the individual effects of BEC and PWC are significant. The findings show a lower CBEM strength/stiffness at the same overall fluid content and with varying ratios of BEC/PWC. The stiffness modulus assessment of the CBEM after 10 days is anticipated to provide the designer with sufficient information to optimize the CBEM mix model in a reasonable period. The overall 3D surface plots showing the interaction between the three autonomous factors on the output parameters.

Another research by Varanda, et al. [22] utilized RSM in the formulation of bitumen mixtures from the refining process of base oils utilizing asphaltic residue, vacuum residue, and three aromatic extracts (by-products). The asphaltic residue (A), vacuum residue (B), and the three different aromatic extracts (hereafter denoted by Extract-1 (C), Extract-2 (D) and Extract-3 (E)) from base oil refining, all derived from the same crude oil source (Arabian Light), To control bitumen formulation, a constrained mixture methodology was employed. The projected and calculated responses show that both models are reliable with average deviations of just 4.67% and 1.53% respectively for the penetration and the softening stage. Mixture design has been shown to be an effective and important instrument for bitumen formulation in general especially if a significant number of components are included in the blends. In general, in contrast to aromatic extracts, both asphaltic and vacuum residues impart greater stability to bitumen, which ensures that the

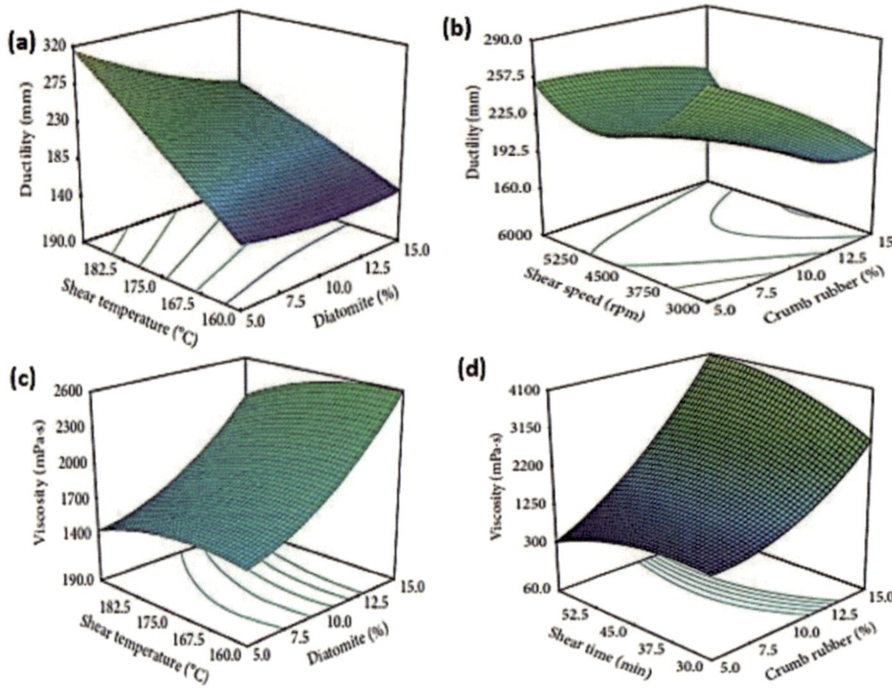

Figure 4.
Response surface plots for the preparation parameters on Ductility. (a) Shear time and diatomite (b) shear speed and Crumb rubber content and Viscosity values on (c) shear temperature and diatomite (d) shear time and crumb rubber [19].

softening point increases while penetration reduces. The asphaltic residue, however, transmits more difficult properties to bitumen n vacuum residue.

Wang and Fan [23] utilize Box–Behnken model RSM was to obtain the optimum values of process parameters for calcium sulphate whisker (CSW) modified bitumen. To access the mechanism of modification, three input variables, stirring time, stirring temperature, and production temperature was chosen. Three performances were evaluated as reactions by testing in the, including high-temperature efficiency, low-temperature efficiency, and deformation resistance related to the bitumen properties. The study found that mixing time of 32 min, mixing temperature of 175 °C, and production temperature of 175 °C were the optimum process parameters within the range of this analysis, with accurate precision compared to actual experimental results. In contrast, the influence of three process parameters on bitumen properties was also investigated, and stirring time was found to have a more important impact on the softening point, penetration, and ductility. CSW has a reasonable dispersion in the bitumen matrix under the optimization process parameters and has been shown to increase the physical performance of the bitumen.

del Barco Carrión, et al. [14] use RSM to develop a modified bitumen with a mixture of Liquid Rubber (LR), a uniform mix of 50–70% pre-processed Recycled Tyre Rubber (RTR) in conjunction with wax and heavy oils blend. To forecast the response of different combinations of LR and Ethylene Bis Stearamide (EBS) in terms of temperature (high and low) properties and expenses, the RSM was used to optimize both conventional and rheological properties of Polymer Modified Bitumen (PMBs) commonly utilized in bitumen blends and roofing membranes. Both two modifiers show improve bitumen elasticity and stiffness, thus complementing each other. In general, the LR oils decrease the stiffness of the neat bitumen whereas the EBS has been shown to increase the stiffness and elasticity of

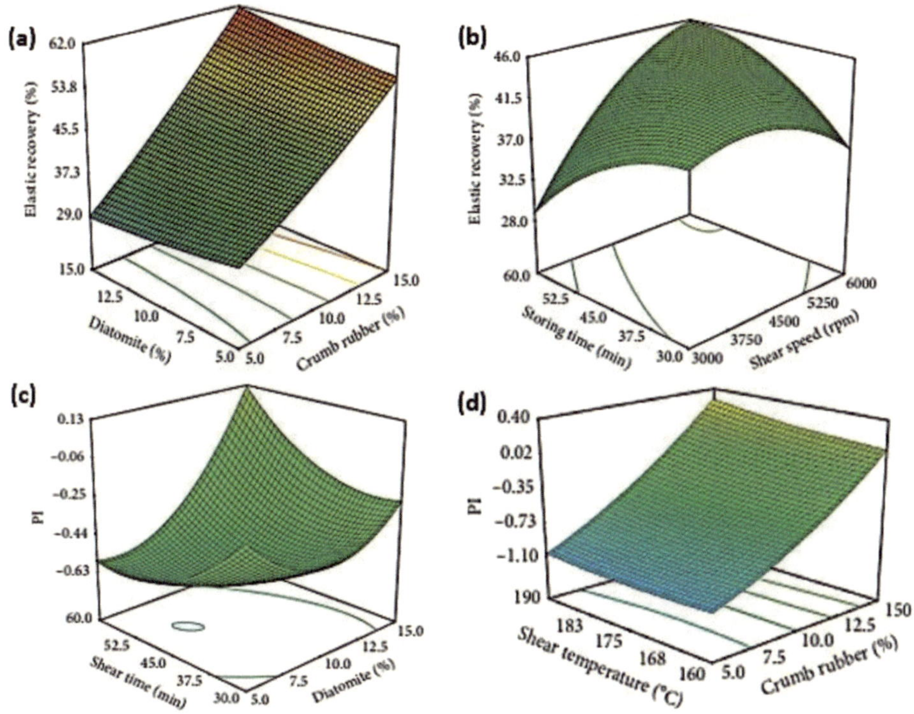

Figure 5.
Response surface plots for the preparation parameters on Elastic recovery. (a) Diatomite and crumb rubber (b) storing time and shearing speed and PI values of (c) Shearing time and diatomite content (d) shear temperature and crumb rubber content [19].

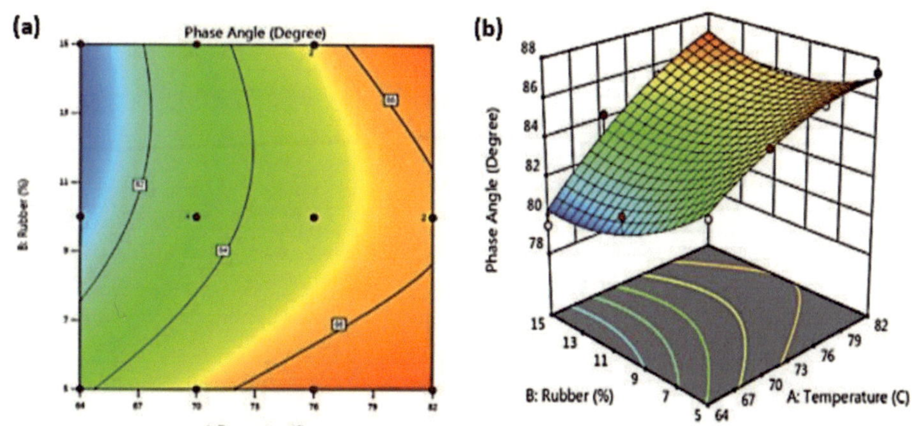

Figure 6.
Effects of input factors on phase angle (a) 2D contour for the synergistic influence of rubber content and temperature (b) 3D surface diagram for the synergistic influence of rubber content and temperature.

the blends significantly, thus enabling the use of higher LR quality to comply with disadvantages. LR contents in the range of 30–40% by weight of neat bitumen are the optimum composition of blends, thus reducing the need for virgin bitumen in terms of overall modification for both pavement and roofing.

Recently, Memon, et al. [24] utilizes RSM for modeling and optimizing the bitumen physical characterization, the mixing conditions of petroleum sludge modified bitumen (PSMB) using the penetration, softening point, penetration index, and

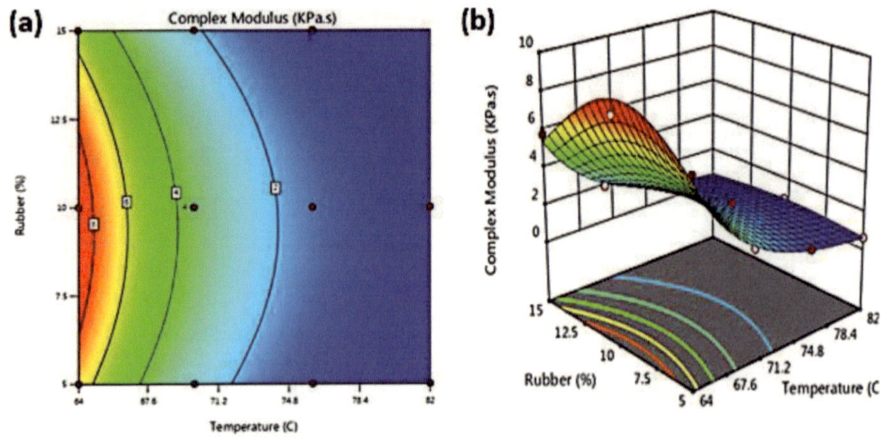

Figure 7.
Simultaneous effect of input parameters on complex modulus (a) Plot of 2D contour on the synergistic influence of rubber content and temperature (b) 3D plot for the synergistic influence of rubber content and temperature.

Authors	Autonomous factors	Responses
Bala, et al. [25]	Nanosilica content, and temperature	Complex modulus, Phase angle, Complex viscosity
Phan, et al. [26]	Hydrated lime content and asphalt binder concentration	Bitumen Linear viscoelastic properties and Fatigue resistance
Bala, et al. [25]	LDPE and binder content	Complex modulus, Phase angle, and Viscosity
Chen, et al. [27]	Test temperature, polystyrene dosage, and polystyrene molecular weight	$G^*/\sin(\delta)$ for unaged blends, $G^*/\sin(\delta)$ for short-term aged blends, and stiffness, m-value
Varanda, et al. [22]	Aromatic extracts vacuum residue and asphalt residue	Softening point and penetration,
Al-Sabaeei, et al. [28]	Crude palm oil (CPO) and temperature	Complex modulus (kPa), and phase angle (δ) for both aged and unaged bitumen
Jamshidi, et al. [29]	Solution temperature, sasobit, and test temperature	Unaged viscosity, STA viscosity, LTA viscosity
Mohammed, et al. [30]	Bitumen and rice husk warm mix asphalt	Penetration and softening point
Yıldırım, et al. [31]	Number of blows, temperature, additive rate, and bitumen content	Optimum bitumen content (OBC)
Khairuddin, et al. [32]	Polyurethane (PU) and bitumen	Penetration, softening point, and viscosity
Solatifar, et al. [33]	Mixing temperature, mixing time, mixing speed, and crumb rubber content percentage	Rutting parameter.

Table 1.
Summary of previous works of literature on the utilization of RSM for asphalt binder optimization.

storage stability. The findings show that the stiffness of PSMB was strengthened by the synergistic effects of mixing temperature and speed, whereas the mixing time initially reduces and then increases the stiffness of PSMB. The PSMB met the storage stability and the penetration index requirements for bitumen modification under optimum mixing conditions. This enhancement in stiffness at higher mixing conditions is due to the increased ratio of maltenes to asphaltenes in PSMB. While at high mixing conditions, the softening point influences were reduced, whereas it was noted that the lightweight oil component of the PS was also responsible for the softening point decline of the PSMB. The mixing time, speed, and temperature for PSMB were evaluated to be 53 min, 1292 rpm, and 149 °C, respectively, based on the RSM optimization. The overview of prior works on the optimization of various autonomous parameters on bitumen conventional properties is displayed in **Table 1**.

6. Application of RSM in modeling and optimization of materials for asphalt mixtures modification

Several RSM studies have been published, this study emphasizes the utilization of RSM for optimizing virgin or waste materials as alternative asphaltic mixtures materials for sustainability. In this section, the application of RSM for the optimization of the virgin as well as waste secondary materials will be discussed depending on the work of existing literature.

The application of Response Surface Methodology (RSM) for the prediction of Marshal volumetric properties is being explored by Bala, et al. [34] Polyethylene and nanosilica were used as the independent factors in the analysis, while the air

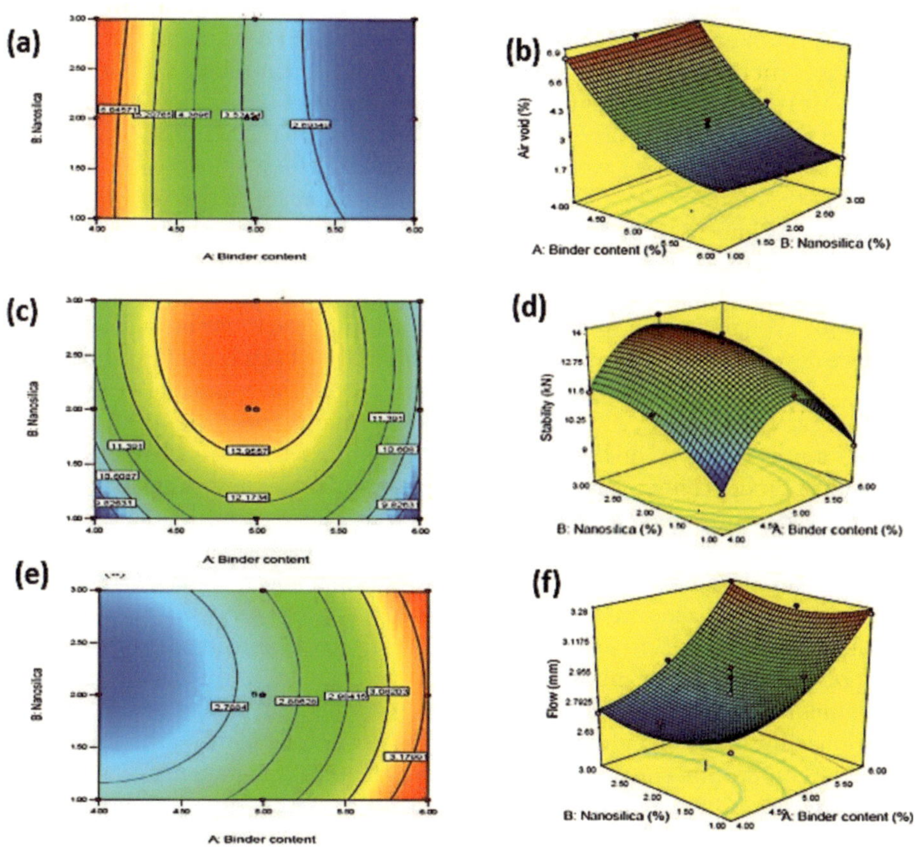

Figure 8.
2D and 3D RSM contour plots for the optimization of nanosilica and binder content (a-b) Air void (c-d) Stability (e-f) Flow [34].

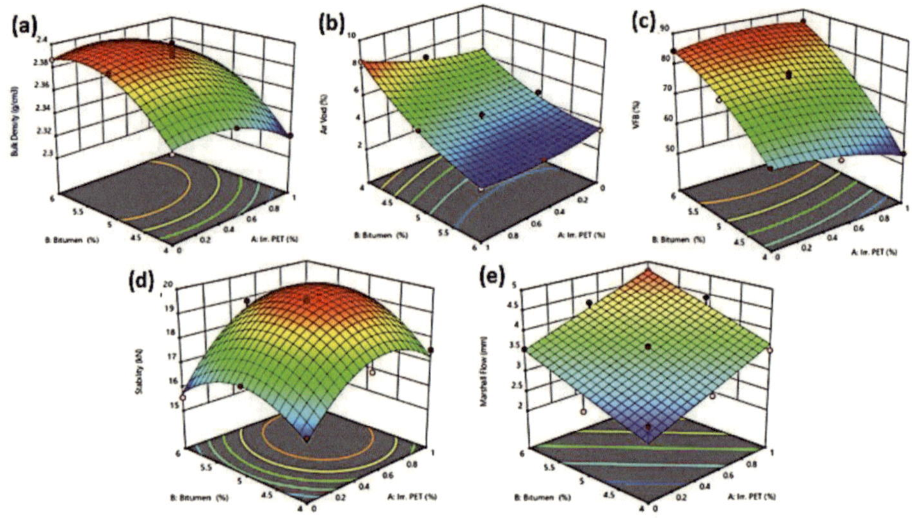

Figure 9.
RSM 3D contour for the optimization of irradiated PET fiber and binder content (a) Density (b) Air void (c) VFB (d) Stability (e) Marshall flow [35].

void, flow, and Marshall stability were the responses. Findings show that RSM-based statistical analysis confirms that it is possible to use a quadratic model built with a high degree of correlation and predictive capacity to predict the Marshal volumetric properties of the mixture. **Figure 8(a)** and **(b)** demonstrates that the binder content has more impact on the air void than the nanosilica content, whereas **Figure 8(c)** and **(d)** illustrates that both independent factors have an influential effect on the Marshall stability of the asphalt mixture. On the Marshall flow values for the nanosilica modified asphalt blend, a combined effect of both variables was noted but the binder content has a more pronounced impact on the flow values as presented in **Figure 8(e)** and **(f)**.

Recently, Usman, et al. [35] optimize irradiated waste PET fiber and binder contents utilizing RSM on the volumetric and strength properties of fiber-reinforced asphalt mixes. **Figure 9(a-e)** illustrates the interactive impacts of the mixture design parameters on the dependent variables. It was hinted that both independent factors have a significant positive effect on the bulk specific density (BSD), Marshall stability, and the Marshall flow values for the fiber-reinforced asphalt mixes. However, on the air void (AV) and voids filled with bitumen (VFB), the asphalt binder content has a more pronounced influence compared to the irradiated waste PET fiber content. The investigation concluded that the use of fibers in asphalt mixture production improves the strength and performance characteristics of asphalt mixes and based on multi-objective optimization analysis, 5.25% and 0.53% as the optimized contents for binder and irradiated waste PET, respectively.

Likewise, in 2020, Omranian, et al. [36] investigate the effects of short-term aging on asphalt concrete mixture compactibility and volumetric characteristics. Three independent parameters, including aging temperature, aging period, and duration in humidity and ultraviolent chambers, were considered in this analysis, while the compaction energy index (CEI) and volumetric characteristics, were considered as responses. Significant impacts of aging temperature and duration on compactibility, air voids (AV), mineral aggregate voids (WMA), and asphalt filled voids (WFA) are revealed in the research findings. However, duration in the environment chamber did not exhibit any significant effect on responses, as shown in **Figure 10(a-d)**. Finally, the analysis shows the ability of the RSM to predict

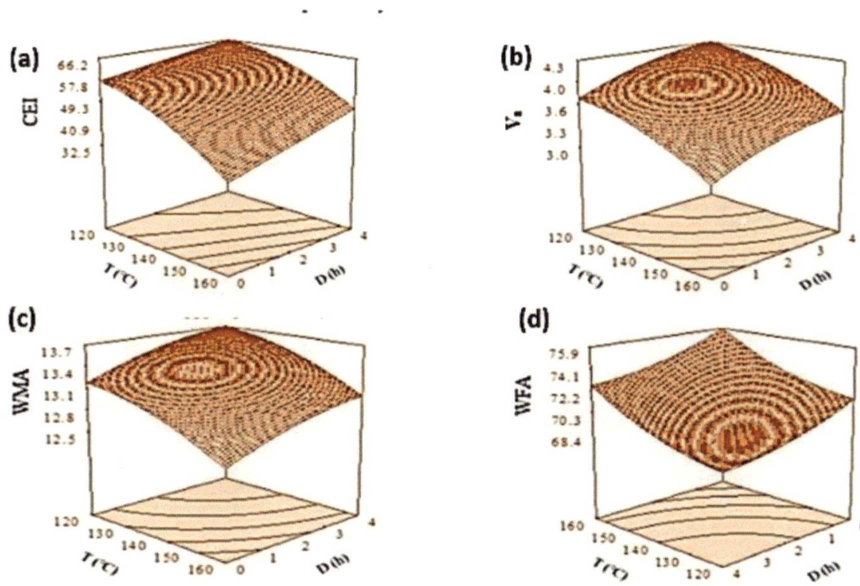

Figure 10.
Graphical representation of RSM 3D contour (a) CEI (b) Va (c) VMA (d) WFA [36].

changes from mathematical equation responses that correlate with the empirical findings with good precision. This finding concluded that to achieve the desired requirements, pavement designers should use RSM statistical technique to predict the pavement density and adjust pressure as well as the number of rolling passes.

Khan, et al. [37] explore the applicability of utilizing RSM to investigate the relationship between autonomous and dependent parameters for the formulation of cementitious grout. In the study, regular and irradiated waste PET and fly ash contents were optimized. A high coefficient of determination (R^2) with an adequate precision (AP) of greater than 4 from the analysis of variance was reported. It was further revealed that gamma irradiation exposure of waste PET resulted in the usage of a higher percentage of the waste PET in comparison to the regular waste PET without compromising the properties considered in the study. Moreover, the investigation concluded that waste PET treatment with gamma rays could be an innovative and effective way to recycle waste PET in the formulation of cementitious grout for semi-flexible pavement application as cement replacement material and can be an important advancement to attaining the zero-waste plastic goal of the united nation sustainable development.

The permanent deformation property of asphalt concrete incorporating various concentrations of recycled asphalt pavement (RAP) and quantities of the waste engine and cooking oil was evaluated by Taherkhani and Noorian [38] using response surface methodology. The study substituted 25, 50, and 75% of total aggregates with RAP and rejuvenated each with 5, 10, and 15 percent (by total binder weight) of waste engine oil (WEO) and waste cooking oil (waste cooking oil) (WCO). A polynomial quadratic model was reported to be accurate to predict the permanent strain employing RSM in the Design-Expert software. RAP and oil content, squared oil content, and oil types were significant terms for the prediction of permanent strain. Results show that with increasing RAP content, permanent strain decreases, but for each RAP content, the lowest permanent strain is reached at a specific oil content amount. The finding concluded that through RSM optimization, the concentration of oils was obtained to achieve asphalt mixture with a comparable control mix deformation property.

Authors	Autonomous factors	Responses
Hamzah, et al. [49]	Mixing temperature and test temperature	Direct tensile strength (DTS), adhesion, broken aggregate, and fracture energy
Bala, et al. [50]	Polyethylene, polypropylene, and nanosilica	Fatigue life
Usman, et al. [51]	PET fiber and temperature	Resilient modulus
Moghaddam, et al. [52]	PET and binder contents	VIM, VMA, BSG, stability, and flow
Yıldırım and Karacasu [53]	Temperature, waste rubber content, glass fiber content, and bitumen content	PSG, voids, VFA, Marshall stability
Soltani, et al. [54]	PET, stress level, and temperature	Fatigue cycles
Moghaddam, et al. [55]	PET, stress level, and temperature	Stiffness
Haghshenas, et al. [56]	Grading, bitumen content, and lime content	Indirect tensile strength (dry and saturated), TSR
Khodaii, et al. [57]	Grading and lime content	Indirect tensile strength (dry and saturated), TSR
Hamzah, et al. [58]	Compaction temperature, binder content, and recycled asphalt content	VFA, air void, G_{mb}, resilient modulus, stability, and flow
Nassar, et al. [21]	Curing time, Pre-wetting water and Bitumen emulsion content	Air void, indirect tensile, strength (wet and dry), and indirect tensile stiffness modulus
Golchin and Mansourian [59]	RAP content, asphalt binder type, and Loading strain	50% of initial stiffness, fatigue life (Number of cycles), and final stiffness
Santos, et al. [60]	Temperature, plastic percentage, and size of particles	Bulk Specific Gravity

Table 2.
Summary of previous works of literature on the utilization of RSM for asphalt mixtures optimization.

Bala, et al. [39] employed RSM to optimize two independent parameters, namely nanosilica and binder content effects on the response factors VMA, Marshall stability, flow, fatigue life, and indirect tensile strength (ITS). It was observed that VMA decreases substantially for binder content from 4–5%, beyond that it experiences a decreasing trend for mixtures fabricated with binder content between 5–6%. The findings also reported that nanosilica content has less influence on the VMA compared to the binder content. Also, the contour plot revealed that on the Marshall stability, the nanosilica particles have a marginal effect than the binder content. For the flow, fatigue life, and ITS, both nanosilica and binder contents have significant impacts on the nanocomposite modified asphalt mixes. The finding concluded that 2.67% and 4.65% as the optimized nanosilica and binder contents for an improved mixture performance property.

Several studies have been performed on the effect of different alternative materials on asphalt concrete mixtures [19, 40–48]. **Table 2** presents the summary of some selected work done utilizing RSM for different independent factors on several mechanical performance properties of asphalt concrete mixes.

7. Conclusions

RSM has effectively been utilized for various applications in the pavement industry. The current book chapter provides an overview of RSM applications for bitumen, asphaltic mixture modification, and performance properties. From the review, it was observed that RSM has many benefits in improving the knowledge about the synergetic effect of various modification variables on the bitumen performance. RSM technique provides an in-depth understanding of the influence of other responses with the variability each mix design factor would have on the asphalt mixture performance. RSM also has shown the benefit of analyzing various autonomous variables concurrently. For several optimization research, the implementation of RSM to achieve optimum dependent variable outcome leads to considerably reduced costs of running experiments while at the same time optimizing the performance properties of both the bitumen and asphaltic mixtures. Thus, with the need to optimized mix design, RSM plays a crucial role in the pavement industry to establish a performance-based mix design about the above criteria. Generally, RSM can also be regarded as an effective alternative to optimize and understand the pavement mix design parameters effectively to produce optimal mixtures with less cost and sample runs. It was shown in this chapter that the RSM exhibits the great ability to determine the compactness and mechanical efficiency properties of asphalt binder and mixtures under different conditions quickly and accurately. With appropriate precision, the model developed by the RSM always fits into the experimental observations. This means that in predicting the effects of the autonomous parameters on the response variables, these models are accurate and realistic. RSM's proposed models can be used by the highway industry as a valuable and effective way to controlling and preparing the construction method to achieve the best pavement performance characteristics, which can significantly increase pavement performance and longevity.

Acknowledgements

The authors would like to appreciate the support provided by Universiti Teknologi PETRONAS for carrying out this review.

Conflict of interest

The authors declare no conflict of interest.

Author details

Aliyu Usman*, Muslich Hartadi Sutanto, Madzlan Bin Napiah
and Nura Shehu Aliyu Yaro
Department of Civil and Environmental Engineering, Universiti Teknologi
PETRONAS, Bandar Seri-Iskandar, Malaysia

*Address all correspondence to: aliyu_17005209@utp.edu.my

References

[1] Bashir MJ, Aziz HA, Aziz SQ, Amr SA. An overview of wastewater treatment processes optimization using response surface methodology (RSM). In The 4th International Engineering Conference–Towards engineering of 21st century, Malaysia 2012 Jan (pp. 1-11).

[2] Montgomery DC. Design and analysis of experiments. John wiley & sons; 2017.

[3] Arslan-Alaton I, Tureli G, Olmez-Hanci T. Treatment of azo dye production wastewaters using Photo-Fenton-like advanced oxidation processes: Optimization by response surface methodology. Journal of photochemistry and Photobiology A: Chemistry. 2009 Feb 25;202(2-3):142-153.

[4] Baş D, Boyacı IH. Modeling and optimization, I: Usability of response surface methodology. Journal of food engineering. 2007 Feb 1;78(3):836-845.

[5] Myers RH, Montgomery DC, Anderson-Cook CM. Response surface methodology: process and product optimization using designed experiments. John Wiley & Sons; 2016 Jan 4.

[6] Shah HK, Montgomery DC, Carlyle WM. Response surface modeling and optimization in multiresponse experiments using seemingly unrelated regressions. Quality Engineering. 2004 Jan 4;16(3):387-397.

[7] Baş D, Boyacı İH. Modeling and optimization II: comparison of estimation capabilities of response surface methodology with artificial neural networks in a biochemical reaction. Journal of Food Engineering. 2007 Feb 1;78(3):846-854.

[8] Lundstedt T, Seifert E, Abramo L, Thelin B, Nyström Å, Pettersen J, Bergman R. Experimental design and optimization. Chemometrics and intelligent laboratory systems. 1998 Aug 24;42(1-2):3-40.

[9] Anderson MJ, Whitcomb PJ. Design of experiments. Kirk-Othmer Encyclopedia of Chemical Technology. 2000 Dec 4:1-22.

[10] Box GE, Wilson KB. On the experimental attainment of optimum conditions. Journal of the royal statistical society: Series b (Methodological). 1951 Jan;13(1):1-38.

[11] Said KA, Amin MA. Overview on the response surface methodology (RSM) in extraction processes. Journal of Applied Science & Process Engineering. 2015;2(1).

[12] Bezerra MA, Santelli RE, Oliveira EP, Villar LS, Escaleira LA. Response surface methodology (RSM) as a tool for optimization in analytical chemistry. Talanta. 2008 Sep 15;76(5):965-977.

[13] Steppan DD. Essential regression and experimental design for chemists and engineers. http://www. geocities. com/SiliconValley/Network/1900/ index. html. 1998.

[14] del Barco Carrión AJ, Subhy A, Rodriguez MA, Presti DL. Optimisation of liquid rubber modified bitumen for road pavements and roofing applications. Construction and Building Materials. 2020 Jul 20;249:118630.

[15] Taherkhani H, Arshadi MR. Investigating the mechanical properties of asphalt concrete containing waste polyethylene terephthalate. Road Materials and Pavement Design. 2019 Feb 17;20(2):381-398.

[16] Wu S, Montalvo L. Repurposing waste plastics into cleaner asphalt

pavement materials: A critical literature review. Journal of Cleaner Production. 2020 Sep 28:124355.

[17] Ahmadinia E, Zargar M, Karim MR, Abdelaziz M, Shafigh P. Using waste plastic bottles as additive for stone mastic asphalt. Materials & Design. 2011 Dec 1;32(10):4844-4849.

[18] Musa NF, Aman MY, Shahadan Z, Taher MN, Noranai Z. Utilization of synthetic reinforced fiber in asphalt concrete–a review. International Journal of Civil Engineering and Technology. 2019;10(5):678-694.

[19] Liu H, Zhang M, Jiao Y, Fu L. Preparation parameter analysis and optimization of sustainable asphalt binder modified by waste rubber and diatomite. Advances in Materials Science and Engineering. 2018 Jan 1;2018.

[20] Badri RM, Sutanto M, k Alobaidi M. Investigating the rheological properties of asphalt binder incorporating different crumb rubber contents based on a response surface methodology. Journal of King Saud University-Engineering Sciences. 2020 Oct 24.

[21] Nassar AI, Thom N, Parry T. Optimizing the mix design of cold bitumen emulsion mixtures using response surface methodology. Construction and Building Materials. 2016 Feb 1;104:216-229.

[22] Varanda C, Portugal I, Ribeiro J, Silva AM, Silva CM. Optimization of bitumen formulations using mixture design of experiments (MDOE). Construction and Building Materials. 2017 Dec 15;156:611-620.

[23] Wang XS, Fan TT. Optimization in Process Parameters of Calcium Sulfate Whisker Modified Asphalt Using Response Surface Methodology. Iranian Journal of Science and Technology, Transactions of Civil Engineering. 2020 Nov 23:1-11.

[24] Memon AM, Sutanto MH, Napiah M, Khan MI, Rafiq W. Modeling and optimization of mixing conditions for petroleum sludge modified bitumen using response surface methodology. Construction and Building Materials. 2020 Dec 20;264:120701.

[25] Bala N, Kamaruddin I, Napiah M, Danlami N. Polyethylene polymer modified bitumen: Process optimization and modeling of linear viscoelastic rheological properties using response surface methodology. Journal of Engineering and Applied Sciences. 2018;13(9):2818-2827.

[26] Phan CV, Di Benedetto H, Sauzéat C, Lesueur D, Pouget S, Olard F, Dupriet S. Complex modulus and fatigue resistance of bituminous mixtures containing hydrated lime. Construction and Building Materials. 2017 May 15;139:24-33.

[27] Chen C, Podolsky JH, Williams RC, Cochran EW. Determination of the optimum polystyrene parameters using asphalt binder modified with poly (styrene-acrylated epoxidised soybean oil) through response surface modelling. Road Materials and Pavement Design. 2019 Apr 3;20(3):572-591.

[28] Al-Sabaeei AM, Napiah M, Sutanto M, Alaloul W, Ghaleb AA. Prediction of Rheological Properties of Bio-asphalt Binders Through Response Surface Methodology. In IOP Conference Series: Earth and Environmental Science 2020 May 1 (Vol. 498, No. 1, p. 012012). IOP Publishing.

[29] Jamshidi A, Hamzah MO, Zahed MA. Rheological evaluation and modeling of Sasobit®-modified asphalt binder at high temperatures. Petroleum science and technology. 2013 Aug 3;31(15):1574-1584.

[30] Mohammed KM, Abdalla IK, Mohammed AH, Khairuddin FH,

Ibrahim AN, Rosyidi SA, Abdullah ME, Yusoff NI. Determining The Effects of Rh-Wma on The Engineering Properties of Bitumen. Jurnal Teknologi. 2019 Jan 22;81(2).

[31] Yıldırım ZB, Karacasu M, Okur V. Optimisation of Marshall Design criteria with central composite design in asphalt concrete. International Journal of Pavement Engineering. 2020 Apr 15;21(5):666-676.

[32] Khairuddin FH, Yusoff NM, Badri K, Koting S, Choy PN, Misnon NA, Osmi SC. Design and optimization of polyurethane modified bitumen (PUMB) using response surface method. In IOP Conference Series: Earth and Environmental Science 2020 Apr 1 (Vol. 476, No. 1, p. 012061). IOP Publishing.

[33] Solatifar N, Azadedel R, Khalili M, Rahbarnia M. Optimization of Rutting Parameter for Crumb Rubber Modified Binder Using Response Surface Methodology (RSM). Journal of Transportation Research. 2019 May 22;16(1):245-258.

[34] Bala N, Napiah M, Kamaruddin I. Performance evaluation of composite asphalt mixture modified with polyethylene and nanosilica. International Journal of Civil Engineering and Technology. 2017;8(9):616-628.

[35] Usman A, Sutanto MH, Napiah M, Zoorob SE, Abdulrahman S, Saeed SM. Irradiated polyethylene terephthalate fiber and binder contents optimization for fiber-reinforced asphalt mix using response surface methodology. Ain Shams Engineering Journal. 2020 Aug 3.

[36] Omranian SR, Hamzah MO, Pipintakos G, Vuye C, Hasan MR. Effects of Short-Term Aging on the Compactibility and Volumetric Properties of Asphalt Mixtures Using the Response Surface Method. Sustainability. 2020 Jan;12(15):6181.

[37] Khan MI, Sutanto MH, Napiah MB, Zoorob SE, Yusoff NM, Usman A, Memon AM. Irradiated polyethylene terephthalate and fly ash based grouts for semi-flexible pavement: design and optimisation using response surface methodology. International Journal of Pavement Engineering. 2020 Dec 2:1-16. doi: https://doi.org/10.1080/10298436.2020.1861446.

[38] Taherkhani H, Noorian F. Investigating Permanent Deformation of Recycled Asphalt Concrete Containing Waste Oils as Rejuvenator Using Response Surface Methodology (RSM). Iranian Journal of Science and Technology, Transactions of Civil Engineering. 2020 Oct 18:1-13.

[39] Bala N, Napiah M, Kamaruddin I. Nanosilica composite asphalt mixtures performance-based design and optimisation using response surface methodology. International Journal of Pavement Engineering. 2020 Jan 2;21(1):29-40.

[40] Khan MI, Sutanto MH, Napiah MB, Zahid M, Usman A. Optimization of Cementitious Grouts for Semi-Flexible Pavement Surfaces Using Response Surface Methodology. In IOP Conference Series: Earth and Environmental Science 2020 May 1 (Vol. 498, No. 1, p. 012004). IOP Publishing.

[41] Adnan AM, Luo X, Lü C, Wang J, Huang Z. Physical properties of graphene-oxide modified asphalt and performance analysis of its mixtures using response surface methodology. International Journal of Pavement Engineering. 2020 Aug 18:1-15.

[42] Kofteci S, Masbah AK, Firat MZ, Gurcan T. Prediction of the Asphalt Mixture Performance Prepared with Recycled Fine Aggregate by using Response Surface Analysis. Acta Physica Polonica, A. 2019 May 1;135(5).

[43] Saha G, Biligiri KP. Cracking performance analysis of asphalt mixtures using response surface methodology: experimental investigations and statistical optimization. Materials and Structures. 2017 Feb 1;50(1):33.

[44] Lapian FE, Ramli MI, Pasra M, Arsyad A. Opportunity applying response surface methodology (RSM) for optimization of performing butonic asphalt mixture using plastic waste modifier: a preliminary study. In IOP Conference Series: Earth and Environmental Science 2020 Jan (Vol. 419, No. 1, p. 012032). IOP Publishing.

[45] Lv S, Yuan J, Peng X, Cabrera MB, Guo S, Luo X, Gao J. Performance and optimization of bio-oil/Buton rock asphalt composite modified asphalt. Construction and Building Materials. 2020 Dec 20;264:120235.

[46] Zhang P, Cheng YC, Tao JL, Jiao YB. Molding process design for asphalt mixture based on response surface methodology. Journal of Materials in Civil Engineering. 2016 Nov 1;28(11):04016120.

[47] Omranian SR, Hamzah MO, Valentin J, Hasan MR. Determination of optimal mix from the standpoint of short term aging based on asphalt mixture fracture properties using response surface method. Construction and Building Materials. 2018 Aug 10;179:35-48.

[48] Wang W, Cheng Y, Tan G. Design optimization of SBS-modified asphalt mixture reinforced with eco-friendly basalt fiber based on response surface methodology. Materials. 2018 Aug;11(8):1311.

[49] Hamzah MO, Teh SY, Golchin B, Voskuilen J. Use of imaging technique and direct tensile test to evaluate moisture damage properties of warm mix asphalt using response surface method. Construction and Building Materials. 2017 Feb 1;132:323-334.

[50] Bala N, Napiah M, Kamaruddin I, Danlami N. Optimization of Nanocomposite Modified Asphalt Mixtures Fatigue Life using Response Surface Methodology. In IOP Conference Series: Earth and Environmental Science 2018 Apr (Vol. 140, No. 1). Institute of Physics Publishing.

[51] Usman N, Masirin MI, Ahmad KA, Wurochekke AA. Reinforcement of asphalt concrete mixture using recycle polyethylene terephthalate fibre. Indian Journal of Science and Technology. 2016 Dec;9(46):107143.

[52] Moghaddam TB, Soltani M, Karim MR, Baaj H. Optimization of asphalt and modifier contents for polyethylene terephthalate modified asphalt mixtures using response surface methodology. Measurement. 2015 Oct 1;74:159-169.

[53] Yıldırım ZB, Karacasu M. Modelling of waste rubber and glass fibber with response surface method in hot mix asphalt. Construction and Building Materials. 2019 Dec 10;227:117070.

[54] Soltani M, Moghaddam TB, Karim MR, Baaj H. Analysis of fatigue properties of unmodified and polyethylene terephthalate modified asphalt mixtures using response surface methodology. Engineering Failure Analysis. 2015 Dec 1;58:238-248.

[55] Moghaddam TB, Soltani M, Karim MR. Stiffness modulus of Polyethylene Terephthalate modified asphalt mixture: A statistical analysis of the laboratory testing results. Materials & Design. 2015 Mar 5;68:88-96.

[56] Haghshenas HF, Khodaii A, Khedmati M, Tapkin S. A mathematical model for predicting stripping potential of Hot Mix Asphalt. Construction

and Building Materials. 2015 Jan 30;75:488-495.

[57] Khodaii A, Haghshenas HF, Tehrani HK, Khedmati M. Application of response surface methodology to evaluate stone matrix asphalt stripping potential. KSCE Journal of Civil Engineering. 2013 Jan 1;17(1):117-121.

[58] Hamzah MO, Gungat L, Golchin B. Estimation of optimum binder content of recycled asphalt incorporating a wax warm additive using response surface method. International Journal of Pavement Engineering. 2017 Aug 3;18(8):682-692.

[59] Golchin B, Mansourian A. Evaluation of fatigue properties of asphalt mixtures containing reclaimed asphalt using response surface method. International Journal of Transportation Engineering. 2017 Apr 1;4(4):335-350.

[60] Santos F, Alzona A, Taguba MA, Santos MJ. Optimization of the Bulk Specific Gravity of Plastic Waste Bituminuous Binder using Box-Behnken Design. In2019 IEEE 11th International Conference on Humanoid, Nanotechnology, Information Technology, Communication and Control, Environment, and Management (HNICEM) (pp. 1-5). IEEE.

Chapter 7

Application of Response Surface Method for Analyzing Pavement Performance

Seyed Reza Omranian

Abstract

Hot mix asphalt (HMA) is a common material that has been largely used in the road construction industries. The main constituents of HMA are asphalt binder, mineral aggregate, and filler. The asphalt binder bounds aggregate and filler particles together and also waterproofs the mixture. The aggregate acts as a stone skeleton to impart strength and toughness to the structure, while the filler fills pores in the mixture which can improve adhesion and cohesion as well as moisture resistance. The HMA behavior depends on individual component properties and their combined reaction in the mixture. Asphalt binder properties change due to different factors. Over the years, asphalt pavement materials age, causing binder embrittlement which adversely affects pavement service life. Response Surface Method (RSM) is a set of techniques that are used to develop a series of experiment designs, determining relationships between experimental factors and responses, and using these relationships to determine the optimum conditions. Incorporating RSM in pavement technologies can beneficially help researchers to develop a better experimental matrix and give them the opportunity to analyze the changes in pavement performance in a faster, more effective, and reliable way.

Keywords: hot mix asphalt, road construction, short-term aging, pavement performance, optimization

1. Introduction

Infrastructure plays a pivotal role in all countries' social and economic development. The road construction industry consumes a huge amount of energy and non-renewable materials. According to the literature, the United States followed by China, Canada and Australia annually produced approximately 500, 150, 45, and 8 million tons of asphalt mixtures, respectively, which lead to the dedication of a significant amount of funds not only to pavement construction but also on pavement maintenance and rehabilitation (M & R) [1–3]. Incorporation of high-quality construction materials, proper construction strategy and equipment, and consequently evaluation of pavement performance along with proper M&R activities, can prevent premature failures in road networks and supply safety and convenience for road users.

Hot Mix Asphalt (HMA) has been widely used for road construction. HMA mainly consists of asphalt binder, mineral aggregate, and filler. The HMA behavior depends on individual component properties and their combined reaction in the mixture. However, the binder is the dominant constituent that controls asphalt mixtures' overall performance. Asphalt binder is a complex organic material. The primary factor affecting the durability of the asphalt mixtures is binder age hardening. Aging in itself is a complex physico-chemical phenomenon. Aging by changing asphalt binder chemical and rheological properties cause binder embrittlement which adversely affects pavement service life [4, 5]. Aging can be divided into short-term and long-term stages. Short-term aging is a result of binder volatilization and oxidation which increase binder viscosity and results in stiffer mixture. Although aging enhances the load-bearing capacity and permanent deformation resistance of pavements by producing stiffer mixtures, it can also cause or accelerate several distresses such as fatigue, low-temperature cracking, and moisture damage by reducing pavement flexibility.

A study to clearly understand the effects of short-term aging on asphalt binder and mixture properties may help to predict asphalt mixtures performance and design a better mixture which can lead to longer-lasting pavements that require less maintenance cost and time while being in service. In this regard, different conditioning scenarios and failure aspects were studied to understand the binder and mixtures' performance before and after aging. Response Surface Method (RSM) as one of the promising methods was therefore employed to assist the experimental matrix design and evaluation procedures.

RSM can establish the relationships between experimental factors and responses by combining and analyzing a series of experiment designs. RSM has been widely used in several disciplines such as environment, material and chemical sciences as well as pavement engineering. Khodaii et al. (2012) used RSM to evaluate the effects of aggregate gradation and lime content on the tensile strength ratio of dry and saturated hot mix asphalt [6]. Jamshidi et al. (2013) determined the changes in the rheological properties of asphalt binder which was modified by different amount of Sasobit at high temperatures using RSM [7]. Kavussi et al. (2014) used the same technique to evaluate the effects of aggregate gradation, hydrated lime and Sasobit content on the indirect tensile strength of warm mix asphalt (WMA) [8]. Hamzah et al. (2015) utilized the RSM to study the effects of elongated short-term aging on the rheological properties of binder at intermediate temperatures [9]. Hamzah and Omranian (2016) studied the effects of aging on the asphalt binder behavior at high temperatures by applying RSM [10]. Saha and Biligiri (2017) used RSM to optimize the asphalt mixtures fracture toughness characteristics that affect cracking performance [11]. RSM was also employed by Omranian et al. (2018, 2020) to study the impact of short-term aging on mixtures fracture and volumetric properties, respectively [12, 13]. Li et al. (2018) evaluated the incorporation of recycled pavement concrete on mixtures mechanical properties. It was found that the three RSM relation models were fit well and effectively represent the mixtures characteristics after a series of fatigue and freeze–thaw cycles [14]. Long et al. (2019) studied the impacts of corrosion, fatigue, and fiber content on the pavement concrete mechanical properties and revealed that RSM model fits well with pavement performance [15]. In other material study, Hou et al. (2020) clearly demonstrated the great potential of RSM in developing magnesium phosphate cement in patch repair and maintenance works [16]. Bala et al. (2020) successfully used RSM to optimize nano-silica and binder content for nanocomposite-modified asphalt mixtures for replacing and reducing the application of polymer-modified binders [17]. Lapian et al. (2021) also determined the optimum conditions of incorporating plastic waste in asphalt mixing, using RSM, to improve the performance in terms of mixtures failure resistance under repetitive loading [18]. The successful application of RSM technique in the previous studies indicates its capability to characterize the complex behavior of asphalt binders and mixtures.

In this article, the effects of aging on the asphalt binder and mixtures performance will be discussed. The aging conditions and experimental plans were designed using the RSM. RSM was also used to develop regression models and predict the binder and mixture's behavior subjected to short-term aging.

2. Materials and methods

In this study, four different binders were used for both binder and three of them were selected for mixture testing. Following the previous publications and for ease of reference, binders and mixtures are designated according to their source, type and constituents [9, 12]. Binders A1 and A2 refer to the conventional penetration grade 80/100 and 60/70 binders from Source A, respectively. Binders B1 and B2 refer to the conventional penetration grade 80/100 and 60/70 binders from source B, respectively. The basic properties of the binders are summarized in **Table 1**. Mixtures were also designated in accordance with source and binder grade. To simplify the nomenclature, mixtures produced with binders 60/70 and 80/100 from source A are referred to as A60 and A80, respectively, while mixtures produced with binder 80/100 from source B is designated as B80.

Granite aggregates, and filler are other mixture constituents that were used in this study. The median aggregate gradation in accordance with Malaysian Public Works Department (PWD) specifications (as shown in **Table 2**) for mixture type AC14 was used [19].

The effects of aging on binders were evaluated from the differences between their un-aged and aged rheological properties. The Rolling Thin Film Oven (RTFO) was used to produce a homogenous artificial short-term age asphalt binder following the procedures outlined by Hamzah and Omranian (2016) [10]. To prepare mixtures, binders and batched aggregates with fillers were mixed at a temperature between 160 °C and 170 °C (based on the viscosities obtained from the Rotational Viscometer test). Loose mixtures were placed in a conventional oven which was set to the compaction temperature to simulate short term aging. The short-term aged loose mixtures were then compacted using the Servopac gyratory compactor to 4% air voids. Exposure of the pavement to the environmental condition while in service

Binder type	Specific gravity (g/cm3)	Aging state	Penetration at 25 °C (dmm)	Softening point (°C)	Ductility at 25 °C (cm)
A1	1.020	Un-Aged	80	46	>100
		85 min Aged	—	—	—
A2	1.030	Un-Aged	63	49	>100
		85 min Aged	—	—	—
B1	1.020	Un-Aged	81	47	>100
		85 min Aged	—	—	—
B2	1.030	Un-Aged	62	50	>100
		85 min Aged	—	—	—

Table 1.
Conventional binder properties.

Sieve size (mm)	Min-Max passing limitation (%)	Selected median gradation (%)
20	100	100
14	90–100	95
10	76–86	81
5	50–62	56
3.35	40–54	47
1.18	18–34	26
0.425	12–24	18
0.150	6–14	10
0.075	4–8	6

Table 2.
Aggregate gradation [12].

was finally simulated by placing the compacted samples in a humidity (H) and ultraviolet (UV) chamber. This procedures are in line with the study conducted by Omranian et al. (2020) [13].

Two sets of experiment was designed using the central composite method to separately characterize behavior of asphalt binders and mixtures as shown in **Table 3**. In the case of binder evaluation, test temperature, binder type and aging duration were selected as the independent variables (IVs), while the complex modulus and viscosity are defined as the responses. In the case of mixture evaluation, the IVs used include

No.	Aging parameters or IVs for mixtures			No.	Aging parameters or IVs for binders	
	Aging temperature (°C)	Aging duration (h)	H & UV chamber (h)		DSR testing temperature (°C)	Aging duration (h)
1	160	4	4	1	82	92.5
2	160	0	0	2	46	0.0
3	140	2	2	3	82	185.0
4	140	2	2	4	46	185.0
5	140	4	2	5	64	92.5
6	140	2	0	6	64	92.5
7	140	0	2	7	46	92.5
8	120	0	4	8	64	0.0
9	140	2	2	9	64	185.0
10	160	2	2	10	82	0.0
11	140	2	2	11	64	92.5
12	120	4	4	12	64	92.5
13	120	4	0	13	64	185.0
14	120	2	2	14	64	0.0
15	160	4	0	15	64	92.5
16	120	0	0	16	82	0.0
17	140	2	2	17	82	92.5

No.	Aging parameters or IVs for mixtures			No.	Aging parameters or IVs for binders	
	Aging temperature (°C)	Aging duration (h)	H & UV chamber (h)		DSR testing temperature (°C)	Aging duration (h)
18	140	2	4	18	46	92.5
19	160	0	4	19	46	0.0
				20	46	185.0
				21	82	185.0
				22	64	92.5

Table 3.
Matrix of experimental plan.

aging temperature, aging duration in a conventional oven and duration that samples were conditioned in the humidity and ultraviolet chamber, while the compaction energy index (CEI) and fracture toughness or stress intensity factor (K) were selected as responses or dependent variables (DVs).

3. Results and discussion

ANOVA analysis was performed to develop models. The significant values that influence the responses were first selected for the model development. Different models were studied to fit the experimental results and the most accurate was nominated based on the higher R-square. Eventually, the mathematical model to predict the responses was developed by RSM. Since these procedures were repeated for both binders and mixtures samples, **Table 4** only presents the ANOVA, selected model type and mathematical regression models developed for complex modulus of binders.

Steps	Factor	Sum of squares	DF[a]	F value	Prob > F		
ANOVA procedures	A[b]	30573.37	1	229.74	< 0.0001		
	B[c]	5021.94	1	37.74	< 0.0001		
	C[d]	2845.49	3	7.13	0.0010		
	A^2	9289.86	1	69.81	< 0.0001		
	B^2	73.62	1	0.55	0.4630		
	AB	6302.33	1	47.36	< 0.0001		
	AC	3883.88	3	9.73	0.0001		
	BC	541.44	3	1.36	0.2757		
Proposed models	Factor	Sum of Squares	DF[a]	Mean Square	F Value	Prob > F	Model Type
	Model	59108.52	10	5910.85	43.59	< 0.0001	Quadratic (Sig)
	Residual error	4474.36	33	135.59			
	Lack of fit	4473.51	25	178.94	1677.49	< 0.0001	(Sig)
	R-squared	0.93					

Steps	Factor	Sum of squares	DF[a]	F value	Prob > F
Developed regression models	Binder type	Equations			
	A1	$y^{e} = +393.59482 - 12.52088 * A + 0.91926 * B + 0.095669 * A^2 - 0.011920 * A * B$			
	A2	$y = +481.57055 - 13.68645 * A + 0.91926 * B + 0.095669 * A^2 - 0.011.920 * A * B$			
	B1	$y = +379.40260 - 12.33686 * A + 0.91926 * B + 0.095669 * A^2 - 0.011.920 * A * B$			
	B2	$y = +501.65104 - 13.95958 * A + 0.91926 * B + 0.095669 * A^2 - 0.011.920 * A * B$			

[a]*Degree of Freedom.*
[b]*Test Temperature.*
[c]*Aging Duration.*
[d]*Binder Type.*
[e]*Complex modulus.*
In order to propose accurate regressions, factors B^2 and BC were eliminated due to "Prob > F" greater than 5%.

Table 4.
Procedures to develop models for complex modulus [9].

After testing and preliminary analysis of model development, **Figure 1** illustrates the contour plot of relationship between aging duration and test temperature effects on the complex modulus of binders A1, A2, B1 and B2. It can be seen that growth in test temperature declines the complex modulus. In addition, prolonged aging duration escalates the complex modulus. As an example, the complex modulus of binder A2 declines by approximately 400%, when temperature increases from 52 °C

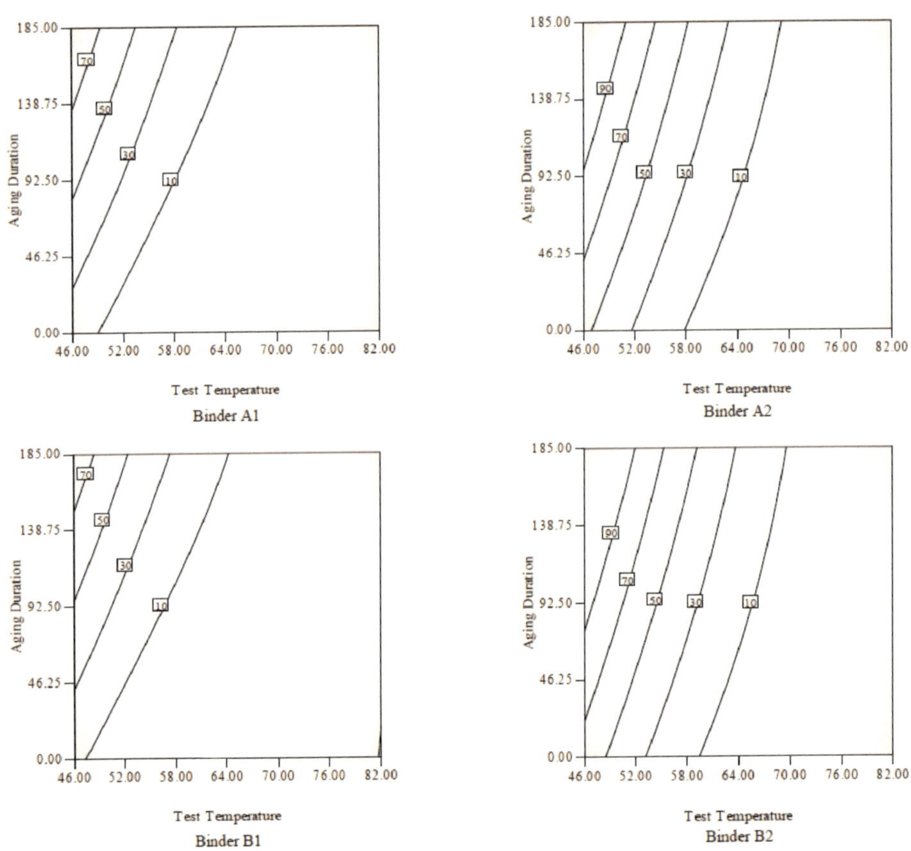

Figure 1.
Binders' complex modulus pattern (kPa).

to 64 °C. However, the corresponding vale of binder A2 at 52 °C approximately 28% increases when aging duration is extended approximately 70 minutes. The results can be correlated to the impacts of extended aging on volatilization and oxidation of the binders which increases complex modulus. On the other hand, temperature increment relates to the binder tendency to behave as a viscous material, hence, the complex modulus decreases. Changes in complex modulus at different conditions show higher test temperature effects on complex modulus compared to the aging duration. The complex modulus is also reliant on the different chemical composition of binder type. The impacts of origin at 52 °C and 70 °C are also determined. For instance, the results show that complex modulus of binder B2 is approximately 10% higher compared to the complex modulus for binder A2 at 52 °C when they are aged 92.5 minutes. While the corresponding values of both A2 and B2 binders exhibit no significant differences at 70 °C. Binders exhibit different rheological behavior when the temperature varies. From the results, it can therefore be concluded that the RSM exhibited a great potential to estimate the changes in binder behavior by fitting the developed models into the experimental outcomes [9]. According to Wang et al. (2019) changes in aging temperature can significantly influence rheological response at both short and long term aging levels [20]. However in this study, the impact of aging temperature is only explored on behavior of asphalt mixtures. It is, therefore, recommended that RSM, due to its capability, can be employed to perform such effects on complex modulus of binders in the future.

The relationship between aging duration and test temperature effects on the viscosity of binders are shown in the form of contour plots in **Figure 2**. The results

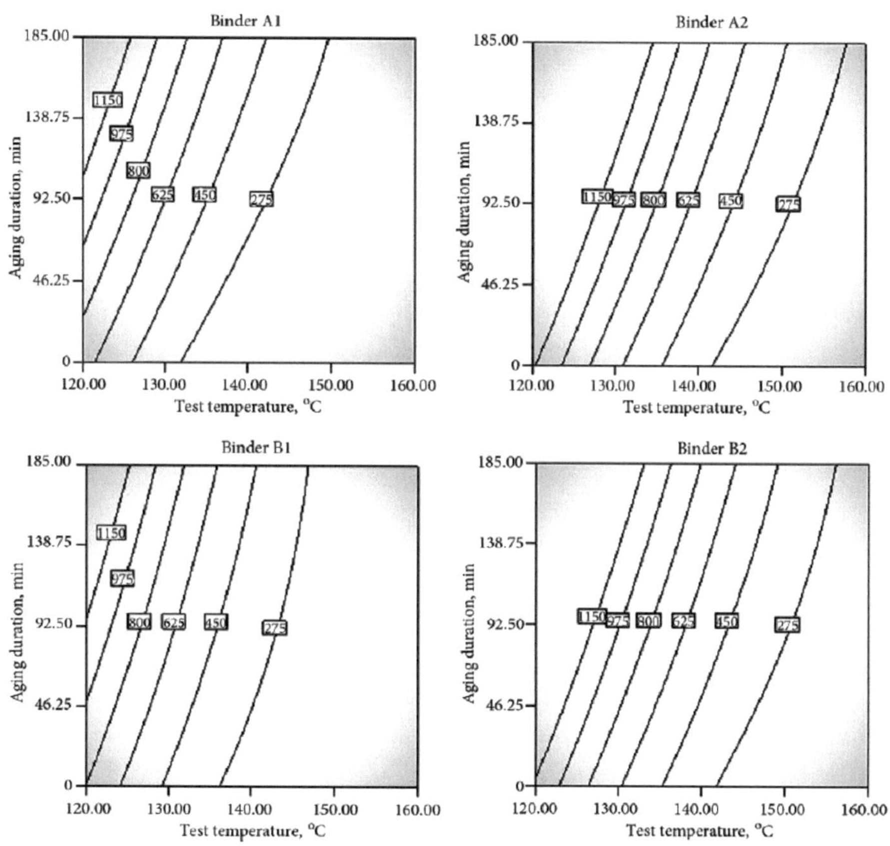

Figure 2.
Binders' viscosity pattern.

show that the viscosity decreases by temperature increment. On the other hand, the viscosity increases when aging duration is extended. For example, the viscosity of binder A1 decreases more than 70% when temperature increases from 130 °C to 150 °C. On the contrary, the corresponding value at 140 °C increases by approximately 53% when aging is extended from 70 to 140 minutes. From these result it can be understood that the aging duration impacts are lower compared to the test temperature. The figures obtained from RSM show that the maximum changes by temperature fluctuation are on 620%, 715%, 565% and 710% for binders A1, A2, B1 and B2, respectively, while the maximum changes by aging extension are 120%, 96%, 65% and 84% for binders A1, A2, B1 and B2, respectively. This testifies that RSM has a great potential to identify the different IVs impacts on the DVs. It can also be found that test temperature exhibits more significant viscosity of binders compare to the aging duration. The RSM outcomes also show that the viscosity reduces radically at lower temperatures, whereas asymptotes at higher temperatures. Conversely, viscosity increases radically at lower aging duration and then asymptotes by extending aging duration. The viscosity of binders are directly dependent on their types and sources. The comparison between the viscosity of binders from different sources was performed and the results indicated that viscosity of binder A2 at 130 °C is almost 5% higher compared to the corresponding values of binder B2. This difference lowered to 3% at higher temperature (150 °C). According to Yan et al. (2017) increasing temperature reduces the viscosity and makes binder fluid. It also causes higher binder molecular activities that results in more chemical reaction with oxygen [21]. Both procedures have significant effects on aging which was confirmed by RSM outcomes. Base on the RSM results it can be clearly concluded that the binders with same penetration grade exhibit different behavior at lower temperatures but these discrepancies are reduced by increasing the test temperature [10].

The impacts of aging temperature, aging duration in a conventional oven and duration that samples were conditioned in the humidity and ultraviolet chamber as IVs on Compaction Energy Index (CEI) are presented in the form of 3D counter plots in **Figure 3**. This figure also presents the relation between DVs and CEI based on normal plots of residuals and the actual versus predicted plots. The 3D counter plots indicate that CEI increased by aging duration increment. Conversely, the corresponding value decreased by increasing aging temperature. The CEI fluctuations differ by aging condition variation. For example in the case of mixtures produced using binder A60, the maximum discrepancies of aging temperature effects on CEI is 35.5%, while the maximum discrepancies of aging duration effects on the corresponding value is 27%. These results indicate the RSM great capability to differentiate between IVs impacts and great potential to find the relations between the outcomes. The 3D counter plots also show that extension of aging duration at higher temperature causes higher impacts on CEI compared to lower temperatures such as 120 °C which can found according to the steeper slope of aging duration effects at 160 °C compared to the corresponding value at 120 °C. The effectiveness of higher aging temperature can be correlated to the existence of lighter oily fraction volume in the binders, which accelerates the volatilization as clearly was detected by RSM. These results are in line with the results outlined by Omranian et al. (2018) [22]. The RSM great capability to estimate the CEI with respect to the IVs can be also understood from the even or normal distribution of residuals along the fitting lines as shown in **Figure 3**. Furthermore **Figure 3** shows all predicted DVs from mathematical equations fit into the experimental observation with excellent accuracy based on the actual versus predicted results plots. These findings clarify the RSM robustness and reliability to predict effects of IVs on the CEI [13].

The 3D counter plots showing the effects of aging temperature, aging duration in a conventional oven and duration that samples were conditioned in the humidity

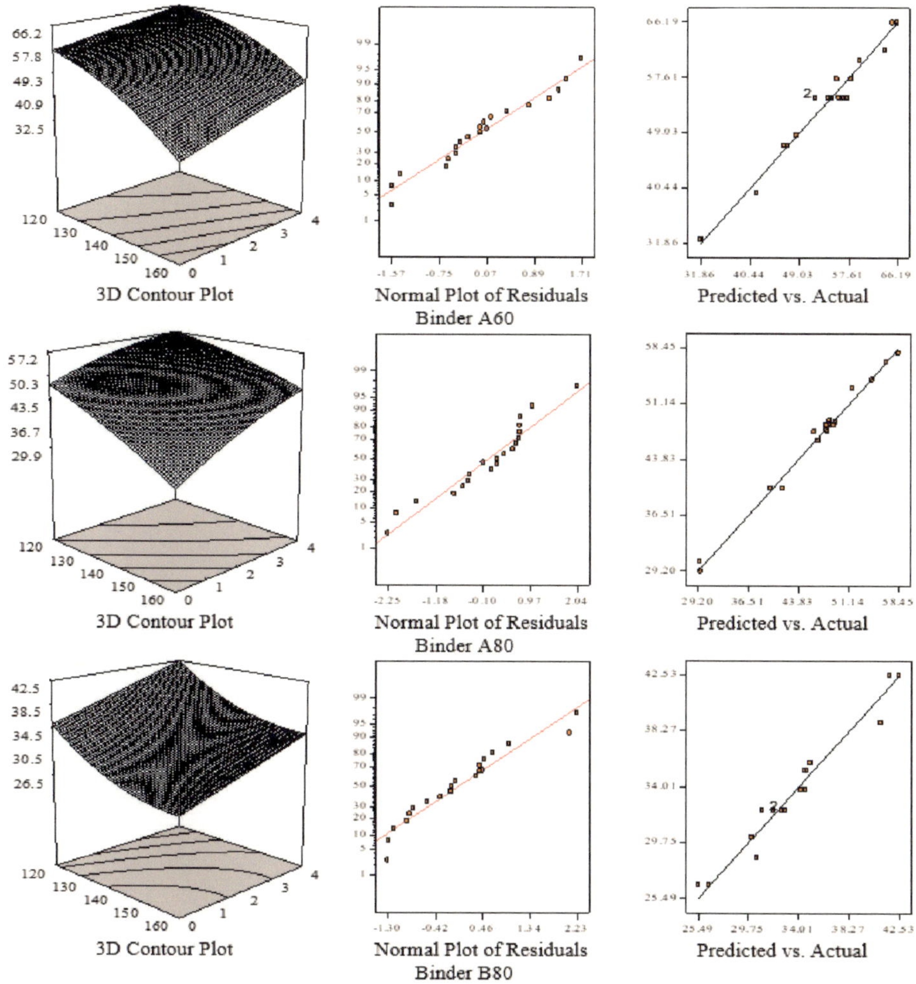

Figure 3.
Mixtures' compaction energy index pattern.

and ultraviolet chamber as IVs on fracture toughness (K) are displayed in **Figure 4**. The results show that test temperature increment reduces the K. It can also be seen that the corresponding value escalates when aging temperature and aging duration increases. However, the changes in the fracture toughness can be related to the binder type or binder content. According to Chen and Solaimanian (2019) although binder content did not significantly influence the aging index, aging significantly changed flexibility index and stiffness of samples [23]. The aging temperature exhibits fairly low impacts on the slight elevation of K for lower aged mixtures. On the other hand, the corresponding value significantly increases by aging temperature increment particularly when the mixtures were aged for 4 h in the oven. The effects of extended aging duration at 120 °C is inferior compared to the corresponding value at 160 °C, which can be clearly observed from the steeper slope of aging duration raise obtained from RSM. The results can be correlated to the higher volatilization and oxidation rates at higher temperature which was clearly detected by RSM. The fracture toughness fluctuates more significantly at 10 °C which result in more obvious changes in the K pattern in the case of mixtures produced using binder A60. According to the literature, softer binders are more vulnerable to aging, while the stiffer binder consists less light oily fraction, hence, their aging

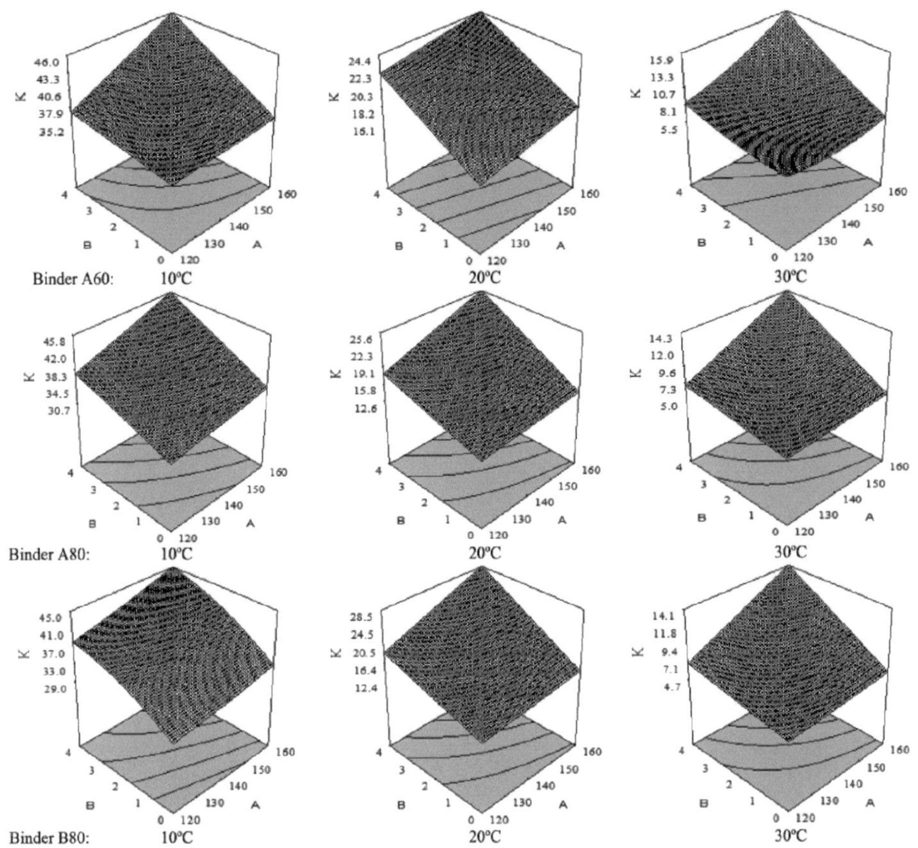

Figure 4.
Mixtures' fracture toughness pattern.

susceptibility is lower. Softer binders may lose higher proportion of lighter oily fraction both at low and high temperatures, hence, faster volatilization. Furthermore, It can be seen that mixtures contain binder A60 and extremely aged exhibit approximately similar fracture toughness at 10 °C compared to the corresponding values of the mixtures with the same conditions and produced by binder A80. These results are aligned with the mixtures' brittleness performance. The harder binders are more brittle particularly at lower temperatures, while the are stiffer at higher temperatures. Hence, the mixtures produced using binder A60 exhibit higher fracture toughness at higher testing temperatures (20°C and 30°C) compared to the corresponding values of the mixture produced using softer binders. These variations and different IVs impact on the fracture toughness of mixtures are detected by RSM, which indicates the software robustness and applicability [12].

4. Conclusion

Extended aging duration and test temperature effects on binders were quantified in terms of their complex modulus and viscosity using response surface method. The RSM exhibited the ability to develop precise regression models with R-square higher than 90%. The experimental results indicate that extending the aging duration increases the binder complex modulus and viscosity, while increasing test temperature leads to the corresponding values reduction. The effects of test temperature on the viscosity are higher compare to the effects of aging duration. It was found based on the extreme changes in the results. For instance, the maximum viscosity changes by temperature fluctuation was approximately 6 to 7 times higher than the corresponding value maximum changes by aging extension.

This work also evaluated the effects of different aging scenarios on the mixtures' compaction energy index and fracture properties using RSM. Similar to binders, RSM exhibited great capability to predict the mixtures performance in terms of CEI and fracture toughness by precise regression models development. The overall results indicate that IVs (individually and together) significantly affect CEI and fracture properties. Extended aging resulted in a higher K and CEI. Mixtures produced using stiffer binders exhibited higher energy requirements for compaction, which resulted in a higher CEI. Test temperature increment declined K, which can be contributed to the reduction in the binder viscosity due to the test temperature elevation. Although the magnitudes of changes in the responses, for both binder and mixture samples, varied depending on the variation in binder sources and types, RSM accurately detected the changes in the responses. RSM also determined the changes in the changes in the aging rate at higher and lower aging temperatures.

Employing the experimental design obtained from RSM reduced the sample size. The sample size reduction resulted in time, energy, and money saving for the entire project. Although the required number of samples significantly reduced, yet RSM detected the IVs and DVs relation and the IVs influence on the responses with an excellent accuracy. Hence, The RSM technique exhibits the ability to quickly and precisely determine the behavior of binders at various conditions. These advantageous impacts of RSM were also concluded in other studies [9, 12, 17, 24].

Acknowledgements

The author would like to thank Universiti Sains Malaysia and University of Antwerp for supporting this project.

Conflict of interest

There is no conflict of interest.

Author details

Seyed Reza Omranian
Faculty of Applied Engineering, EMIB Research Group, University of Antwerp, Antwerp, Belgium

*Address all correspondence to: seyedreza.omranian@uantwerpen.be

References

[1] Jenny, R. CO2 reduction on asphalt mixing plants potential and practical solutions. 2009. Transportation Research Board.

[2] NAPA. Asphalt Paving Industry- A Global Perspective: Productions, Use, Properties, and Occupation; 2011. Exposures Reduction Technologies and Trends. National Asphalt Pavement Association & European Asphalt Pavement Association.

[3] Keches, C. Reducing greenhouse gas emissions from asphalt materials. (Doctoral dissertation, 2007. WORCESTER POLYTECHNIC INSTITUTE).

[4] Hamzah, M.O., Omranian, S.R., Golchin, B. A review on the effects of aging on properties of asphalt binders and mixtures, Caspian J. Appl. Sci. Res. 4 (6), 2015. 15-34.

[5] Lu, X., and Isacsson, U. Artificial aging of polymer modified bitumens. Journal of applied polymer science 76, 2000. 1811-1824.

[6] Khodaii, A., H. Haghshenas, and H. Kazemi Tehrani. Effect of Grading and Lime Content on HMA Stripping Using Statistical Methodology. Construction and Building Materials. 2012. 34: 131-135.

[7] Jamshidi, A., M. O. Hamzah, and M. Zahed. Rheological Evaluation and Modeling of Sasobit®-Modified Asphalt Binder at High Temperatures. Petroleum Science And Technology. 2013. 31: 1574-1584.

[8] Kavussi, A., M. Qorbani, A. Khodaii, and H. F. Haghshenas. Moisture Susceptibility of Warm Mix Asphalt: A Statistical Analysis of The Laboratory Testing Results". Construction and Building Materials. 2014. 52: 511-517.

[9] Hamzah, M.O.; Omranian, S.R.; Golchin, B.; Hainin, M.R. Evaluation of effects of extended short-term aging on the rheological properties of asphalt binders at intermediate temperatures using respond surface method. J. Technol. 2015, 73, 133-139.

[10] Hamzah, M.O., and Omranian, S.R., Effects of extended short-term aging duration on asphalt binder behaviour at high temperatures. The Baltic Journal of Road and Bridge Engineering, 2016. 11 (4), 302-312.

[11] Saha, G., and Biligiri, K.P. Cracking performance analysis of asphalt mixtures using response surface methodology: experimental investigations and statistical optimization. Materials and Structures, 2017. 50 (1), 33.

[12] Omranian, S. R., Hamzah, M. O., Valentin, J., & Hasan, M. R. M. Determination of optimal mix from the standpoint of short term aging based on asphalt mixture fracture properties using response surface method. Construction and Building Materials, 2018. 179, 35-48.

[13] Omranian, S. R., Hamzah, M. O., Pipintakos, G., Vuye, C., & Hasan, M. R. M. Effects of Short-Term Aging on the Compactibility and Volumetric Properties of Asphalt Mixtures Using the Response Surface Method. Sustainability, 2020. 12(15), 6181.

[14] Li, W., Cai, L., Wu, Y., Liu, Q., Yu, H., & Zhang, C. Assessing recycled pavement concrete mechanical properties under joint action of freezing and fatigue via RSM. Construction and Building Materials, 2018. 164, 1-11.

[15] Long, X., Cai, L., & Li, W. RSM-based assessment of pavement concrete mechanical properties under joint action of corrosion, fatigue, and fiber content. Construction and Building Materials, 2019. 197, 406-420.

[16] Hou, D., Chen, D., Wang, X., Wu, D., Ma, H., Hu, X., & Yu, R.

RSM-based modelling and optimization of magnesium phosphate cement-based rapid-repair materials. Construction and Building Materials, 2020. 263, 120190.

[17] Bala, N., Napiah, M., & Kamaruddin, I. Nanosilica composite asphalt mixtures performance-based design and optimisation using response surface methodology. International Journal of Pavement Engineering, 2020. 21(1), 29-40.

[18] Lapian, F. E., Ramli, M. I., Pasra, M., Arsyad, A., & Yatmar, H. Application of Surface Response Methodology (RSM) for Improving the Marshall Quotient of AC-WC Mixtures Containing PET Plastic Waste. In Proceedings of the International Conference on Civil, Offshore and Environmental Engineering, 2021. (pp. 966-975). Springer, Singapore.

[19] Malaysian Public Works Department (PWD). Standard specification for road works, section 4, flexible pavement, Jabatan, Kerja Raya Malaysia, Kuala Lumpur, 2008.

[20] Wang, D., Falchetto, A. C., Riccardi, C., Poulikakos, L., Hofko, B., Porot, L., & Moon, K. H. Investigation on the combined effect of aging temperatures and cooling medium on rheological properties of asphalt binder based on DSR and BBR. Road Materials and Pavement Design, 20(sup1), 2019. S409-S433.

[21] Yan, C., Huang, W., & Tang, N. Evaluation of the temperature effect on Rolling Thin Film Oven aging for polymer modified asphalt. Construction and Building Materials, 2017. 137, 485-493.

[22] Omranian, S. R., Hamzah, M. O., Gungat, L., & Teh, S. Y. Evaluation of asphalt mixture behavior incorporating warm mix additives and reclaimed asphalt pavement. Journal of Traffic and Transportation Engineering (English Edition), 2018. 5(3), 181-196.

[23] Chen, X., & Solaimanian, M. Effect of long-term aging on fracture properties of virgin and recycled asphalt concrete. Advances in Civil Engineering Materials, 2019. 8(1), 527-543.

[24] Hamzah, M. O., Teh, S. Y., Golchin, B., & Voskuilen, J. Use of imaging technique and direct tensile test to evaluate moisture damage properties of warm mix asphalt using response surface method. Construction and Building Materials, 2017. 132, 323-334.

Chapter 8

Selection of Optimal Processing Condition during Removal of Methylene Blue Dye Using Treated Betel Nut Fibre Implementing Desirability Based RSM Approach

Amit Kumar Dey and Abhijit Dey

Abstract

Adsorption of Methylene Blue onto chemically (Na_2CO_3) treated ripe betel nut fibre (TRBNF) was studied using batch adsorption process for different concentrations of dye solutions (50, 100, 150 and 200 mg/L). Experiments were carried out as a function of contact time, initial solution pH (3 to11), adsorbent dose (10 gm/L – 18 gm/L) and temperature (293, 303 and 313 K). The adsorption was favoured at neutral pH and lower temperatures. Adsorption data were well described by the Langmuir isotherm and subsequently optimised using a second-order regression model by implementing face-centred CCD of Response Surface Methodology (RSM). The adsorption process followed the pseudo-second-order kinetic model. The maximum sorption capacity (q_{max}) was found to be 31.25 mg/g. Thermodynamic parameters suggest that the adsorption is a typical physical process, spontaneous, enthalpy driven and exothermic in nature. The maximum adsorption occurred at pH 7.0. The effect of adsorption was studied and optimum adsorption was obtained at a TRBNF dose of15 gm/L.

Keywords: Adsorption, Methylene Blue, betel nut fibre, RSM, Desirability

1. Introduction

Colour plays a significant importance in the human world as everybody likes colourful clothes, our food, medicine etc. is also having various colours. It is quite obvious that many researchers have carried out various studies on colour and its production. In this present situation, about ten thousand or more dyes are available commercially and the annual production of dye is about seven lakh tons [1]. There are numerous structural varieties available for dye like azo dye, acidic dye, basic dye, disperse, anthraquinone based and metal complex dyes. Dyes are mainly used in textile industries. A large portion of synthetic dyes do not bind during the process of colouration and it is then discharged to the waste streams [2]. The amount of dye that is discharged into the environment during the colouration process is about 10–15%. Those dyes discharge into waste streams are highly coloured and those are not pleasing aesthetically. Thus the textile industries cause the discharge of a large number of dyes and other additives into the environment, produced during the dying process [3]. The conventional water treatment process is not found to be effective in the case of removal of these dyes. Due to their high solubility in water, dyes are easily transported through a sewer and it finally reaches the natural water bodies. Carcinogenic and products having high toxicity are produced by the degradation of these dyes [4]. These dyes may cause hazardous effects to living organism too. Special concern should be there to prevent the contamination caused by these dyes and to do this the quantity estimate of dyes discharged into natural bodies should be done properly. It is well known that the use of activated carbon for the treatment of wastewater (removal of dyes from wastewater) is a very well established technique, but due to the high cost involved in the process, researchers are constantly working on finding other low-cost bio-sorbents which are effective in the removal of dyes from wastewater [5–7]. In this work, we have attempted to use an agricultural product, chemically treated Ripe Betel Nut fibre (TRBNF) for the removal of a textile dye namely Methylene Blue (MB) from an aqueous solution. Azo dyes form covalent bonds with the fibres they colour, e.g. cotton, rayon, wool silk and nylon. Methylene Blue is a commercial cationic dye with chemical formulaC16H18ClN3Sand Molar weight = 0319.9 g·mol^{-1}). The functional groups present in the dye molecule react with the -OH, -SH and -NH2 groups present in the fibre rich in cellulosic materials. Azo dyes are mostly preferred in the textile industries due to their fastness of the substrate. Understanding the process of kinetic and mass transfer is very essential for the design of an adsorption treatment system [8–14]. Several techniques and methodologies has been incorporated inorder to removal of dyes. Eventually along with the experimental analysis, researchers were focused to identify the approximate solution of these problems using different mathemathic modeling along with several multi criteria decision making approaches. Several studies have also been reported to the implementation of Response surface Methodology for improving the dye removal process by adjusting the process variables [15–17]. RSM is employed to remove ethylene blue dye using cheap adsorbent. The regression analysis has been used for the removal of colour of aqueous dye solution by using a novel adsorbent [18–23].

The approach of RSM can better predicts the impact of process variables on performance characteristics as well as it can be considered as a better option for optimization [24]. The CCD of RSM have been implemented for the design of experiments. In this study experimentation have been made to remove the methyene blue dye using treated betel nut fibre. Optimum adsorption capacities have been identified using the second order quadratic model of face centered RSM approach. The influence of each process variables and their percentage contribution on the developed quadratic model is explored with the help of ANOVA.

2. Materials and methods

It is known that components high in cellulose and hemicellulose composition are good in removing azo dyes from an aqueous solution (**Figure 1**). The composition of Ripe Betel Nut fibre consists of Cellulose (52.09%), Hemi-cellulose (12.04%), Lignin (22.34%), Fat and ash (5.94%) and Water-soluble matter (1.5%). Sun-dried ripe betel nut fibre was collected from the market and cut into sizes of 1 mm size and washed with distilled water and dried at 60°C thus raw betel nut fibre was obtained. The sample was then treated with 0.01 M Na_2CO_3 at room temperature for 4 hours, then distilled washed to remove excess chemicals in fibres and pH was reduced to 7, then dried for 4 hours at 100°C in a hot air oven and was kept in a container. Thus we get the treated ripe betel nut fibre (TRBNF). An azo dye Methylene

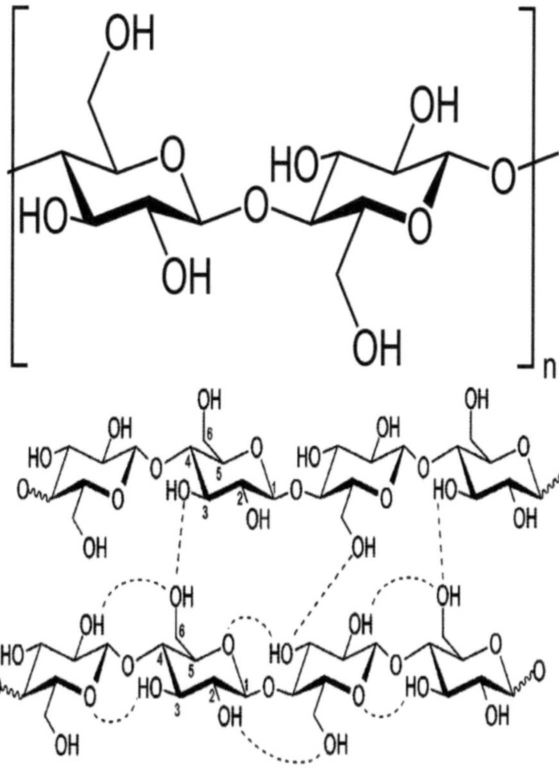

Figure 1.
(a) Cellulose structure; (b) Hydrogen bond structure of cellulose.

Blue, having a strong, though apparently non covalent, affinity to cellulose fibres, having molecular formula C16H18ClN3S was chosen as adsorbate. All the chemicals used were obtained from Himedia. A stock solution (1000 mg/L; pH 7) of dye was prepared using doubly distilled water.

3. Experiments and equilibrium studies

Batch adsorption studies were carried out and at first, the effect of pH variation on the removal of MB by TRBNF was studied and found that the maximum adsorption occurred at pH 7.0. From the stock solution of 1000 mg/L, different combinations of dye solutions were prepared for solutions of different initial concentrations viz. 50, 100, 150 and 200 mg/L at pH 7.0. Initial TRBNF dose was taken as 10 g/L and the same rate of the dose was mixed with each of the prepared solutions, agitated mechanically with the help of a rotary shaker at 303 K at 150 rpm until the equilibrium was reached. For time t = 0 minutes, 5 minutes, 10 minutes and so on, until equilibrium, the dye concentrations were measured by UV/VIS spectroscopy. The data were used to calculate the amount of dye adsorbed, q (mg/g). Effect of TRBNF dose was studied upon the absorption of MB dye by varying TRBNF dose at 10, 15 and 20 g/L. Experiments were carried out at different pH values ranging from 3 to 11. A fixed amount of TRBNF (1 gm) was added to the 100 ml of 50 mg/L of MB solution at different pH values (3–11) and agitated for 3 hours at 303 K to assess the influence of initial pH on MB concentration, by taking and measuring the samples after every five minutes of agitation. Experiments were also carried out to check for adsorption of MB by the container walls in the absence of betel nut fibre.

It was found that there was no degradation or adsorption of MB by container walls. Variation of temperature effect was evaluated for 293, 303 and 313 K. Experiments were carried out in duplicate and mean values were taken. The amount of dye adsorbed per unit adsorbent (mg dye per gm adsorbent) was calculated according to a mass balance on the dye concentration using the Eq. (1):

$$q_{\max} = \frac{\left(c_i - c_f\right)}{m} V \qquad (1)$$

Where,
Q_{max} = Maximum adsorption capacity (mg/g).
C_i = Initial concentration of dye in solution (mg/L).
C_f = Final concentration of dye in solution (mg/L).
V = Volume of solution (L).
m = adsorbent weight (g).

4. Adsorption studies by employing response surface methodology

Experiments were carried out batch-wise to obtain the maximum adsorption capacity (Qmax) as a function of pH, Temperature (K), TRBNF dose (g/L) and rotational speed (RPM). Keeping the other process parameters as constant, the value of Qmax was obtained once at a time by altering any one of the process parameters. Similarly, by using a combination of 30 different sets of a process parameter, the value of Qmax was obtained 30 times and the values were utilized to obtain the most desirable condition, using response surface methodology. Response surface methodology (RSM) is used for the modelling and optimization of response characteristics with quantitative independent variables.

A regression model is also known as a polynomial quadratic model of order two as shown in Eq. (2) shows the system quality characteristic. The software 'Design Expert 11.0' gives an approximation of the regression model coefficient [22–26].

$$Y = C_0 + \sum_{i=1}^{n} C_i X_n + \sum_{i=1}^{n} d_i X_i^2 \pm \varepsilon \qquad (2)$$

Face cantered central composite second order design (CCD) technique was mainly used for the design experiment of the present study. The "face-centred CCD" possesses 30 combinations of 4 different process variables with 3 level of each [22–24]. **Tables 1** and **2** delineated the layout of process variables and the combination of 30 experimental runs obtained by RSM CCD experimental design approach.

The model fit summery demonstrated that the developed regression model is fit significantly for Qmax on the selected experimental domain. The statistical analysis for the generated model has been demonstrated by the ANOVA analysis (**Table 3**). The model significance has been confined by the model F value. The probability of higher F value will be confirmed by value of model term less than 0.05 or in other words 95% confidence interval. It proves that the particular model terms are statistically significant for the developed model [20]. When the value of multiple coefficients of regression R_2 becomes unity, the response models fit better with actual data. The deviation becomes very less between the actual values and predicted values. The actual and predicted plots for Qmax have been delineated in **Figure 2** demonstrating the degree of proximity of the model terms. The errors are normally

Parameters	Labels	Levels		
		-1	0	+1
Temperature, (^{0}k)	A	293	303	313
TRBNF dose, (g/l)	B	10	14	18
pH.	C	3	7	11
Rotational speed, (RPM)	D	100	150	200

Table 1.
Operating variables and their levels.

Exp. no.	Factor1 A: pH	Factor 2 B: Temp	Factor 3 C: Jute Dose	Factor 4 D: RPM	Response Q_{max} (mg/g)
1	3	313	10	200	10.21
2	3	293	10	100	11.42
3	11	293	18	200	15.67
4	3	313	10	100	5.19
5	11	313	18	100	10.43
6	3	313	18	200	12.69
7	3	293	10	200	11.76
8	11	293	10	200	11.28
9	7	303	14	150	28.91
10	11	313	10	200	11.19
11	11	313	10	100	11.01
12	7	303	14	150	28.78
13	7	303	14	150	29.56
14	3	293	18	100	12.64
15	3	293	18	200	16.34
16	11	313	18	200	13.41
17	11	293	10	100	11.29
18	11	293	18	100	13.38
19	3	313	18	100	11.31
20	7	303	14	150	29.59
21	7	303	14	150	29.19
22	7	303	14	150	29.49
23	7	303	14	100	27.48
24	3	303	14	150	24.91
25	7	303	10	150	27.21
26	7	293	14	150	32.11
27	7	313	14	150	30.38
28	7	303	14	200	28.19
29	7	303	18	150	31.56
30	11	303	14	150	27.87

Table 2.
Experimental results obtained with the setting of processing variables.

Source	Sum of Squares	df	Mean Square	F-value	p-value	
Model	20334.01	14	1452.43	21.85	< 0.0001	Significant
A-Temperature	341.82	1	341.82	5.14	0.0385	
B-TRBNF dose	472.47	1	472.47	7.11	0.0176	
C-pH	80.69	1	80.69	1.21	0.2879	
D-Rotational speed	261.14	1	261.14	3.93	0.0661	
AB	6.33	1	6.33	0.0952	0.7620	
AC	77.70	1	77.70	1.17	0.2967	
AD	20.21	1	20.21	0.3040	0.5895	
BC	84.36	1	84.36	1.27	0.2776	
BD	0.9702	1	0.9702	0.0146	0.9054	
CD	59.68	1	59.68	0.8978	0.3584	
A^2	168.79	1	168.79	2.54	0.1319	
B^2	434.59	1	434.59	6.54	0.0219	
C^2	1079.96	1	1079.96	16.25	0.0011	
D^2	658.42	1	658.42	9.91	0.0066	
Residual	997.02	15	66.47			
Lack of Fit	994.19	10	99.42	175.51	< 0.0001	Significant
Pure Error	2.83	5	0.5665			
Cor Total	21331.03	29				

Table 3.
The ANOVA results for Qmax.

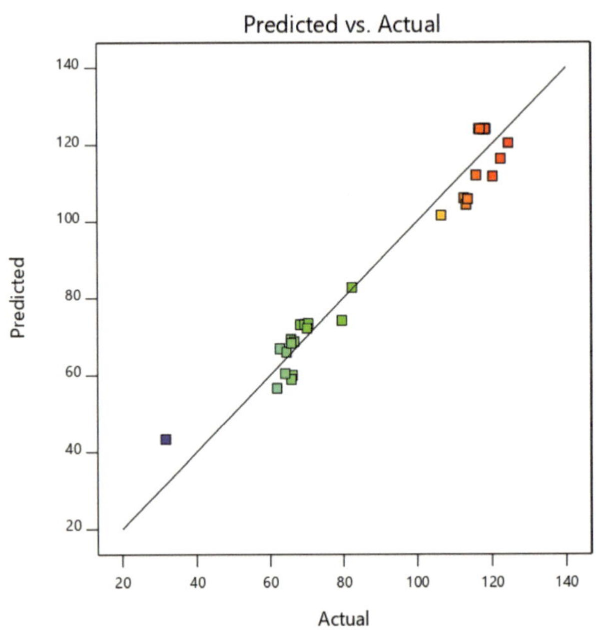

Figure 2.
Predicted vs. actual plot for Q_{max}.

distributed as a maximum of the values are close to a straight line. The standard normal distributions of the experimental data are obtained in the residual plots which validate the mathematical models [17, 19, 22]. The typical residual plots of the wear rates of composites are represented in **Figure 3**. The normal distribution of the data points in all the typical residual plots; normal probability plot (**Figure 3(a)**), Residual vs. run (**Figure 3(d)**); distribution of the predicted vs. Residual data points (**Figure 3(b)**); and the defit vs. Run (**Figure 3(c)**) suggested that the residual and the predicted model for all the responses of the composites are observed to be distributed normally. The residuals are observed to be distributed near to the straight line revealing the normal distributions of the random errors. There were no unpredictable patterns observed on the residual plots as most of the run residues lie in between the range. The AP value was found to be 3. The comparison to the mean predicted error with the predicted value span at the design space can be represented by AP values arresting the adequate model discrination [18, 26–30]. A larger values of AP (14.003) and coefficient of determination (R_2 = 0.953) have been predicted by the model for Qmax. Consequently, the insignificant lack of fit obtained for the developed model presumed that the generated model was best suited for selected operational domain for Qmax. It is possible to eliminate the insignificant model terms from the developed quadratic model and only the significant model terms would have been consider for the response surface for Qmax. A significant lack of fit was obtained due to retention of the insignificant model term in the developed model for Qmax (Eq. (3)). The developed surface model can be used to navigate

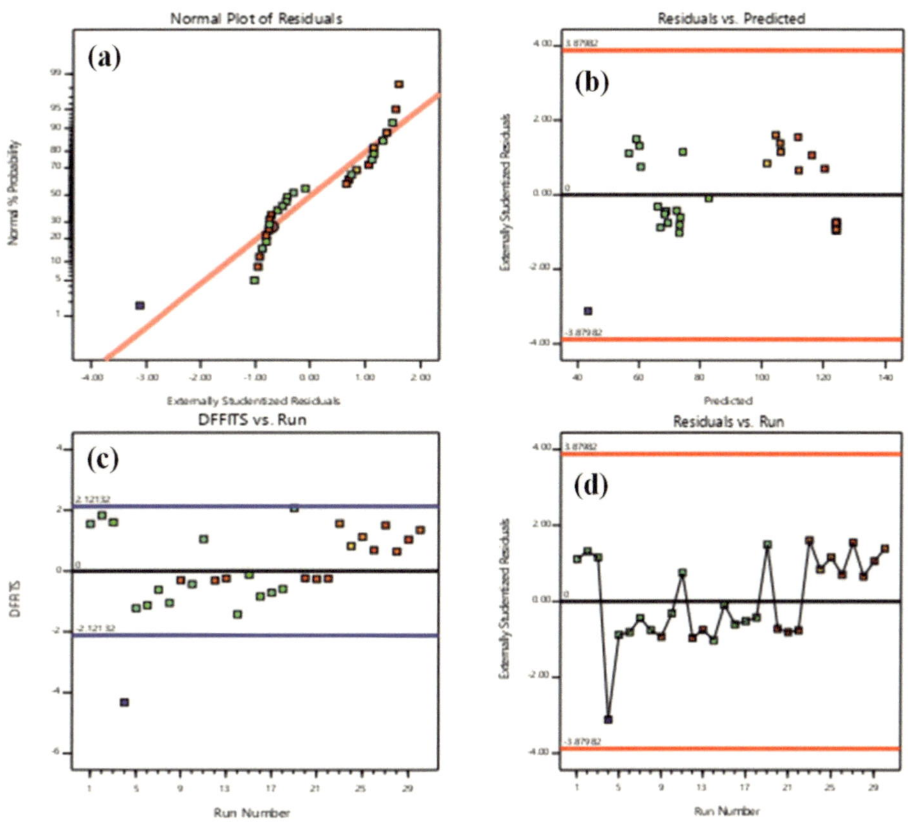

Figure 3.
Model summary statistics for Q_{max}, (a) Normal plots of Residuals; (b) Residuals Vs. Predicted; (c) Dffits Vs. Run; and (d) Residual Vs. Run.

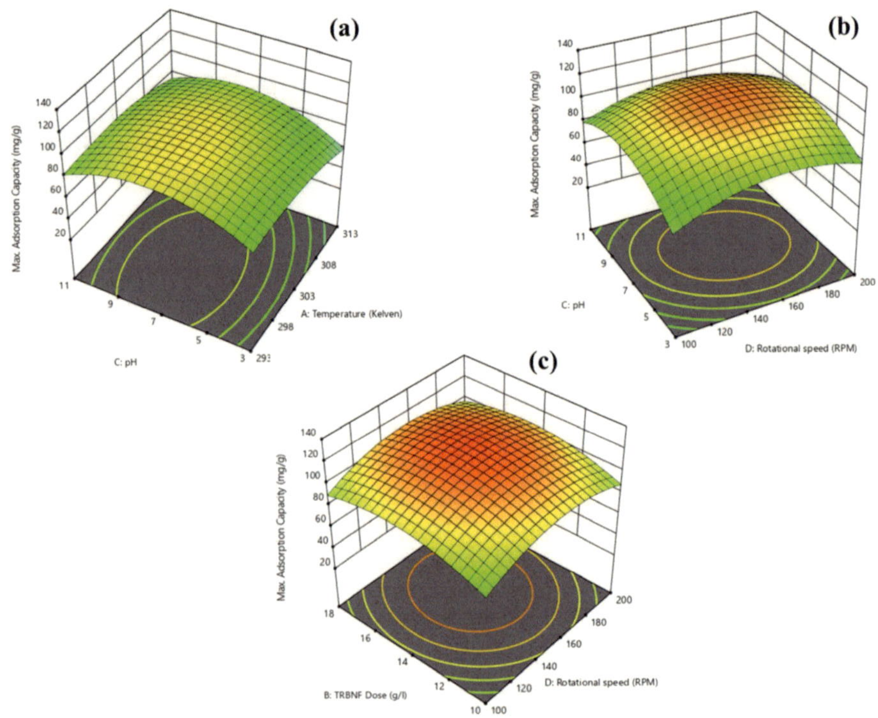

Figure 4.
Surfaceplots obtains for Q_{max}, (a) Temperature vs. pH; (b) Temperature vs. RPM; and (c) TRBNF dose Vs. RPM.

in the selected research domain by the adequate signal provided by the AP ratio. The developed response surface model shows the Maximum adsorption capacity as below,

$$\begin{aligned} \text{max.Adsorption Capacity} = &+124.10 - 4.36\,A + 5.12\,B + 2.12\,C + 3.81\,D \\ &+ 0.6288\,AB + 2.20\,AC + 1.12\,AD - 2.30\,BC \\ &+ 0.2462\,BD - 1.93\,CD - 8.07\,A^2 - 12.95\,B^2 \\ &- 20.42C^2 - 15.94\,D^2 \end{aligned} \quad (3)$$

Where, A, B, C and D are the coded factors(processing independent variables). The highest and the lowest levels of any particular coded factor are given as +1 and – 1 respectively.

Moreover, **Figure 4(a)-(c)** depicts the approximated response surface plot for Qmax concerning the process parameters of solution pH, TRBNF dose, Temperature and RPM.

5. Optimization of the process variables based on desirability function analysis

Table 4 shows the various goals and ranges of process variables viz. pH, Temperature, TRBNF dose and rotational speed (RPM) and the response characteristics viz. Qmax. The aim of using the RSM desirability was to obtain the optimum processing condition by maximizing the desirability as 1. The range of desirability

Name	Goal	Lower Limit	Upper Limit	Lower Weight	Upper Weight	Importance
A:Temperature	is in range	293	313	1	1	3
B: pH.	is in range	3	11	1	1	3
C:TRBNF dose	is in range	10	18	1	1	3
D:RPM	is in range	100	200	1	1	3
Qmax	maximize	31.56	32.11	1	1	3

Table 4.
Limits of Input and output process parameters for DFA.

would be in between 0 to 1. 0 desirability value is practically impossible to obtained as it purely reliant on how the real optimum value are differ from the upper and lower point [22].

30 sets of the optimal solution are acquired for the specific design space constraints for Qmax using statistical Design Expert software11.0. The set of parametric conditions consisting of the maximum value of desirability is preferred as the optimal processing condition for the performance characteristics that are desired [22]. The **Table 5** depicts the highest desirability obtained along with the optimum desirability. After identifying the optimal processing condition, the subsequent step would be analysis the variation of performance measure obtains using optimal processing condition. Experimental measures have been taken place so as to ensure the verification of the predicted optimal setting of the input variables (pH, Temperature, TRBNF dose and rotational speed (RPM)). The deviation observed within the results obtained from the predicted optimal parameter settings and the experimental validation have been delineated in **Table 6**. It was found that the deviation was very minimal.

Figure 5(a) and **(b)** demonstrate the desirability ramp function and the bar graph respectively. The dot point on the ramp can be the measure of a particular variable within the specified experimental domain and the elevation would be responsible for how much desirable it is. The linear graph of ramp function obtained is demonstrating the weightage that how far the goal or target are from the high values and accordingly the weight factor is distributed as 1 [27].

Parameter	Goal	Optimum value
pH	in range	7.5
Temperature, (^{0}k)	in range	303
TRBNF dose, (g/l)	in range	15.1
Rotational speed, (RPM)	in range	158.5

Table 5.
Predicted optimum levels of process variables.

Responses	Goal	Predicted value	Observed value	Error (%)
Q_{max}(mg/g)	Maximize	32.11	31.56	1.71

Table 6.
Predicted and observed values of responses of Q_{max}.

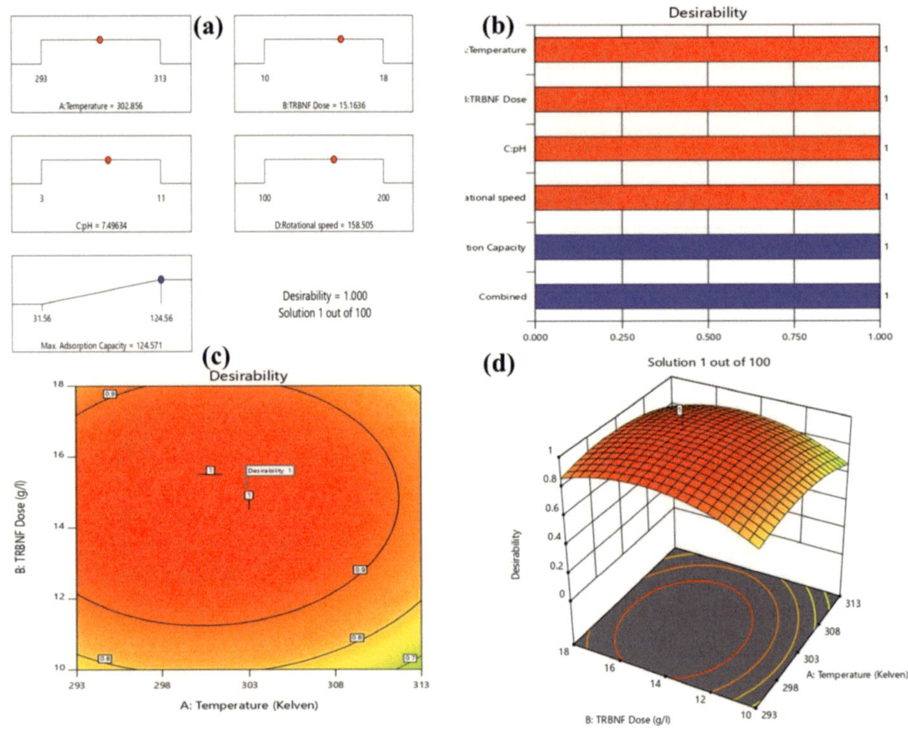

Figure 5.
Ramp function plot of Desirability (a); Bar graph of Desirability (b); 3D Surface plot of desirability (c) 2D view (d).

The overall desirability of the performance characteristics have been demonstrated by the bar graph of desirability. The value has been chosen in between 0 to 1 depends on the proximity of the output towards the target. The value of desirability close to 1 is considered as acceptable.

As it is single response, the maximum weightage have been given to it and similar weighage have been given to all the input processing variables and a 3D desirability plot were drawn. **Figure 5C** and **D** demonstrates the desirability function distribution for Q max during varying the input responses. It can be observed that the value of overall desirability is less at a higher pulse current and pulse on-time region. The region for optimal desirability was placed near the topmost area of the plot, which shows the overall desirability value '1' that slowly decreased while moving to the right side and backwards. Hence, the elucidated desirability value of '1' illustrates the proximity of the response towards the target [22–26].

6. Adsorption mechanism

Understanding adsorption is the most important part of any study of adsorption. Due to this reason, it is important to understand two essential points, (1) The adsorbate structure and (2) To know the functional groups present in an adsorbent responsible for adsorption. In the present study for the adsorption of Methylene Blue (MB), the presence of amino groups in the dye results to the formation of hydrogen bond in between the amino group and the hydroxyl groups in TRBNF. The silica content was observed to decrease and the crystallinity of cellulose fraction was increases due to the treatment of ripe betel nut fibre by Na_2CO_3 results in changed surface morphology

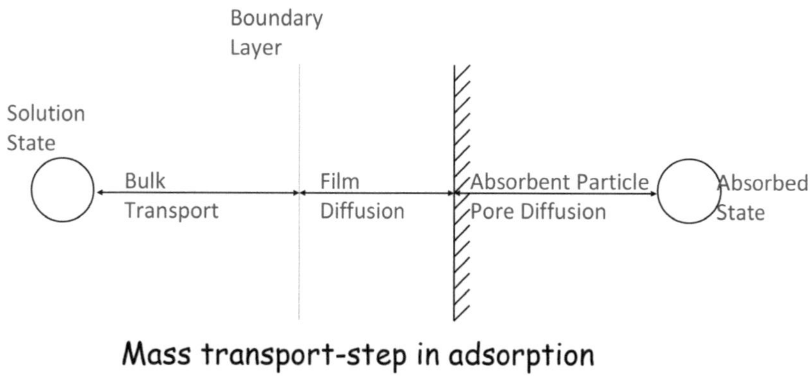

Figure 6.
General Mass transport step in adsorption.

of the betel nut fibre. The chances of chemical reaction to takes place between the hydroxyl exposed adsorbent and dye ions because of the changed surface morphology of the betel nut fibre and subsequently mechanical bonding takes place due to linkage of ions with the modified molecular structure of the absorbent.

As discussed in Section 5.6, during the first 60 minutes there was a very rapid rate of adsorption for all the cases, thereafter gradually slowed down and eventually, the equilibrium time reached 90 minutes revealing that the diffusion film has support the intra-particle diffusion. The maximum sorption was observed at pH value around 7.0. The primary adsorption mechanism were observed as below,

- The ripe betel nut fibre surface was absorbed the MB dye from the entire solution.

- The dye particles defused from aqueous solution to the adsorbent surface through the formation of boundary layer.

- The formation of Hydrogen bond between amino groups and the exposed hydroxyl group present in ripe betel nut fibres would be responsible for the successful adsorption by the of betel nut fibre. **Figure 6** delineated the general mass transfer phenomenon occurred in adsorption process.

7. Conclusions

In this study, ripe betel nut fibre treated with Na_2CO_3 was used as an adsorbent for the adsorption of Methylene Blue dye. For the design of the maximum adsorption capacity output, the usage of process parameters like pH, Temperature, TRBNF dose and RPM were successfully checked by using composite design face centred central response surface methodology by attending 30 experimental trials with repetition of three in each of the process parameters at three different levels. Results have shown that for optimum adsorption capacity, minimum to moderate temperature, moderate to high TRBNF dose and moderate RPM will be critical. The models are adequate which is proven by the obtained predicted value of R_2 for Qmax as 0.958. The result of Qmax was influenced by the two factors TRBNF dose and RPM. At moderate RPM and with an increase in TRBNF dose, the rate of adsorption increased. In this paper, the influence of all the process parameters is discussed. The combination of optimum parameter setting for Qmax obtained are pH 7.5,

Temperature, 303 K, TRBNF dose 15 gm/L and Rotational speed 158 RPM for maximizing the Qmax. The agreeable error percentage of 1.71 between the predicted and observed values for Qmax confirm the precision of the methodology.

Author details

Amit Kumar Dey[1*] and Abhijit Dey[2]

1 Department of Civil Engineering, Central Institute of Technology Kokrajhar, BTAD, Assam

2 Department of Mechanical Engineering, National Institute of Technology Srinagar, J&K, India

*Address all correspondence to: ak.dey@cit.ac.in

References

[1] H. Zollinger. Colour Chemistry – Synthesis, Properties of Organic Dyes and Pigments. VCH Publishers, New York. (1987)92-100.

[2] E.J. Weber, R.L. Adams, Chemical and sediment mediated reduction of the azo dye Disperse Blue 79. Environmental Science & Technology, 29 (1995) 1163-1170.

[3] C. Wang, A. Yediler, D. Linert, Z. Wang, A. Kettrup, Toxicity evaluation of reactive dye stuff, auxiliaries and selected effluents in textile finishing industry to luminescent bacteria vibrio fisheri. Chemosphere, 46 (2002) 339-344.

[4] E. Rindle, W.J. Troll, Metabolic reduction of benzidine azo dyes to benzidine in the Rhesus monkey. Journal of National Cancer Institute, 55 (1975)181.

[5] Ali, R.; Zarei, M.; Moradkhannejhad, L. (2010) Application of response surface methodology for optimization of azo dye removal by oxalate catalyzed photoelectro-Fenton process using carbon nanotube-PTFE cathode. *Desalination,* 258: 112.

[6] Chatterjee S.; kumarA.; Basu, S.; Dutta, S. (2012) Application of Response Surface Methodology for Methylene Blue dye removal from aqueous solution using low cost adsorbent. *Chem. Eng. J,* (181-182): 289.

[7] Chen, Y.; Zhang, D. (2014) Adsorption kinetics, isotherm and thermodynamics studies of flavones from VacciniumbracteatumThunb leaves on NKA-2 resin. *Chem. Eng. J,* 254: 579.

[8] Chiang, K. T. (2008)Modeling and analysis of the effects of machining parameters on the performance characteristics in EDM process of Al_2O_3 + TiC mixed ceramic.*Int. J. Adv. Manuf. Technol,* 37: 523.

[9] Cho, Il-H.; Zoh, K. D. (2007) Photocatalytic degradation of azo dye (*Reactive Red 120*) in TiO_2/UV system: Optimization and modeling using a response surface methodology (RSM) based on the central composite design. *Dye.Pigm,* 75: 533.

[10] Chong, M.N.; Jin, B.; Chow, C.W.K.; Saint, C.P. (2009) A new approach to optimize an annular slurry photoreactor system for the degradation of congo red: statistical analysis and modelling.*Chem. Eng. J,* 152: 158.

[11] Chowdhury, S.; Mishra, R.; Saha, P.; Kuskwaha, P. (2011) Adsorption thermodynamics, Kinetics and isosteric heat of adsorption of malachite green onto chemically modified rice husk. *Desalination,* 265: 159.

[12] Chowdhury, S.; Saha, P. (2010) Sea shell powder as a new adsorbent to remove Basic Green 4 (Malachite Green) from aqueous solutions: Equilibrium, kinetic and thermodynamic studies. *Chem. Eng. J,* 164: 168.

[13] Reddy, D.H.K.; Yun, Y.S. (2016) Spinel ferrite magnetic adsorbents: alternative future mate-rials for water purification. *Coord. Chem. Rev.,* 315: 90.

[14] Crini, G.; Peindy, H. N.; Gimbert, F.; Robert, C. (2007) Removal of C.I. Basic Green 4 (Malachite Green) from aqueous solutions by adsorption using cyclodextrin-based adsorbent: Kinetic and equilibrium studies. *Separ.Purif. Technol,* 53: 97.

[15] Agarwal, A.; Singh, H.; Kumar, P.; Singh, M. (2008) Optimisation of power consumption for CNC turned parts using response surface methodology and Taguchi's technique – a comparative study.*J. Mater. Proc. Technol,* 200: 373.

[16] Ali, R.; Zarei, M.;Moradkhannejhad, L. (2010) Application of response

surface methodology for optimization of azo dye removal by oxalate catalyzed photoelectro-Fenton process using carbon nanotube-PTFE cathode. *Desalination,* 258: 112.

[17] Chatterjee S.; kumarA.; Basu, S.; Dutta, S. (2012) Application of Response Surface Methodology for Methylene Blue dye removal from aqueous solution using low cost adsorbent. *Chem. Eng. J,* (181-182): 289.

[18] Cho, Il-H.; Zoh, K. D. (2007) Photocatalytic degradation of azo dye (*Reactive Red 120*) in TiO_2/UV system: Optimization and modeling using a response surface methodology (RSM) based on the central composite design. *Dye.Pigm,* 75: 533.

[19] Chong, M.N.; Jin, B.; Chow, C.W.K.; Saint, C.P. (2009) A new approach to optimize an annular slurry photoreactor system for the degradation of congo red: statistical analysis and modelling. *Chem. Eng. J,* 152: 158.

[20] Ravikumar, K.; Krishnan, S., Ramalingam, S. (2007) Optimization of process variables by the application of response surface methodology for dye removal using a novel adsorbent. *Dye. Pigm.* 72: 66.

[21] Ravikumar, K.; Deebika, B.; Balu, K. (2005) Decolourization of aqueous dye solutions by a novel adsorbent: Application of statistical designs and surface plots for the optimization and regression analysis. *J. Hazar.Mater.,* 122: 75.

[22] Dey, A. K.; Dey, A. (2021) Selection of optimal processing condition during removal of Reactive Red 195 by NaOH treated jute fibre using adsorption, *Groundwater for Sustainable Development,* 12: 100522.

[23] Dey, A. K.; Kumar, U.; Dey, A.; (2018) Use of response surface methodology for the optimization of process parameters for the removal of Congo Red by NaOH treated jute fibre, .*Desalination and Water Treatment*, 115: 300.

[24] Dey, A. K.; Kumar, U.; (2017) Adsorption of anionic azo dye Congo red from aqueous solution onto NaOH-modified jute fibre, *Desalination and Water Treatment*, 92:, 301.

[25] Dey, A. K.; Kumar, U. (2017) Adsorption of reactive red 195 from polluted water upon Na2CO3 modified jute fibre, *Int J Eng Technol,* 9: 53.

[26] Dey, A.; Pandey, K. M. (2018) Selection of optimal processing condition during WEDM of compocasted AA6061/cenosphere AMCs based on grey-based hybrid approach, *Mater. Manuf. Proc.,* 33(14): 1549.

[27] Rahman, M, ; Dey, A.; Pandey, K. M.; (2018) Machinability of cenosphere particulate–reinforced AA6061 aluminium alloy prepared by compo-casting, *Proc. Inst. Mech. Eng. Part B: J. Eng. Manuf.,* 232(14): 2499.

[28] Dey, A.; Debnath, M.; Pandey, K. M. (2017) Analysis of effect of machining parameters during electrical discharge machining using Taguchi-based multi-objective PSO, *Int. J. Comp. Intel. App.,* 16 (02): 1750010.

[29] Niraj, N.; Pandey, K. M.; Dey, A. (2018) Tribological behaviour of Magnesium Metal Matrix Composites reinforced with fly ash cenosphere, *Mater. Today: Proc.,* 5(9): 20138.

[30] Dey, A.; Pandey, K. M. (2018) Wire electrical discharge machining characteristics of AA6061/cenosphere as-cast aluminum matrix composites, *Mater. Manuf. Proc.,* 33(12):1346.

Chapter 9

Uses of the Response Surface Methodology for the Optimization of Agro-Industrial Processes

José Manuel Pais-Chanfrau, Jimmy Núñez-Pérez, Rosario del Carmen Espin-Valladares, Marco Vinicio Lara-Fiallos and Luis Enrique Trujillo-Toledo

Abstract

Response surface methodology is a tool for the design of experiments, widely used today to optimize industrial processes, including agro-industrial ones. Since its appearance in the last century's fifties, hundreds of articles, chapters of books, and books attest to this. In this work, a general overview of this tool's general practical aspects is made. This statistical tool's usefulness and popularity, used in the optimization of agro-industrial processes and in making them more efficient and sustainable, is described through multiple examples.

Keywords: response surface methodology, agro-industry, central composite design, independent variables, uncontrolled variables, response variables, optimization

1. Introduction

The response surfaces methodology (RSM) is a set of statistical tools for the design of experiments aimed at finding the value or values of the independent variables, which allow developed, improved and optimization (i.e., finding the maximum, minimum, or equal to a certain convenient value) one or more dependent variables or responses [1, 2].

Since the first works reported by Box and coworkers [3–5], RSM has been gaining popularity among researchers, developers, and engineers, and today it has become one of the preferred tools for increasing the productivity and efficiency of R&D processes and the production of goods and services.

The agro-industry, on the other hand, comprises a set of process industries that use agricultural and livestock resources to transform them into products of higher added value. Processed and improved foods, nutraceutical foods and beverages, chemical products and bioactive substances for the chemical, pharmaceutical and cosmetic industries, industrial enzymes and above all, vast and abundant quantities of plant and animal biomass, which could be the primary renewable raw materials with which that will count the industry of the future, are some of the main "outputs" of the agro-Industry.

By their nature, the sources of the raw materials of the agro-industry are renewable and could be a strategical industrial sector for the sustainable development of the whole of humankind, given the constant growth of the human population, Ambiental deterioration, and the evident depletion of the natural sources of raw materials. It is for this reason, that it is required to have well-designed processes that generate a minimum negative impact on the already deteriorated ecosystems, and in which yields, and productivity are maximized.

The objective of this work is to show, through a group of examples, the utility of RSM for the design of efficient, productive agro-industrial processes with a minimum negative impact on agro-ecosystems.

2. Agro-industries: the pillar in sustainable development that the world needs

Agroindustry can be defined as the process industries that use agricultural, livestock and aquaculture products as raw materials, transforming them into valuable, more elaborate products with greater added value. Among the products emerging from agroindustries are processed foods and beverages, dry and canned foods with greater durability, fermented foods and beverages with nutraceutical properties, as well as basic chemicals, precursors of other chemical compounds, biofuels, industrial enzymes, bioactive products, such as antibiotics, probiotics, prebiotics and synbiotics substances, vitamins, organic acids, phytohormones, antioxidant agents, growth factors, etc. (**Figure 1**).

Agro-industries can be subdivided into primary, secondary and tertiary transformation agroindustries, depending on the set of predominant operations carried out in them and the degree of complexity of their output products (**Figure 1**).

In primary transformation, the selection, crushing, separation, isolation, concentration, or drying of the product or products of interest usually predominate. The sugar factories made from sugar cane or sugar beet [6], the traditional dairy industry were powdered, evaporated, or condensed milk is produced (whole, defatted or lactose-free) [7], or the slaughter of cattle meat [8] or industries that produce concentrated juices or condiments and canned foods are examples of industries where these operations of physicochemical transformation of raw materials from agriculture, aquaculture and livestock, into products prevail.

On the other hand, in secondary transformation agro-industries, the products, by-products and residuals of the first transformation are usually used as starting

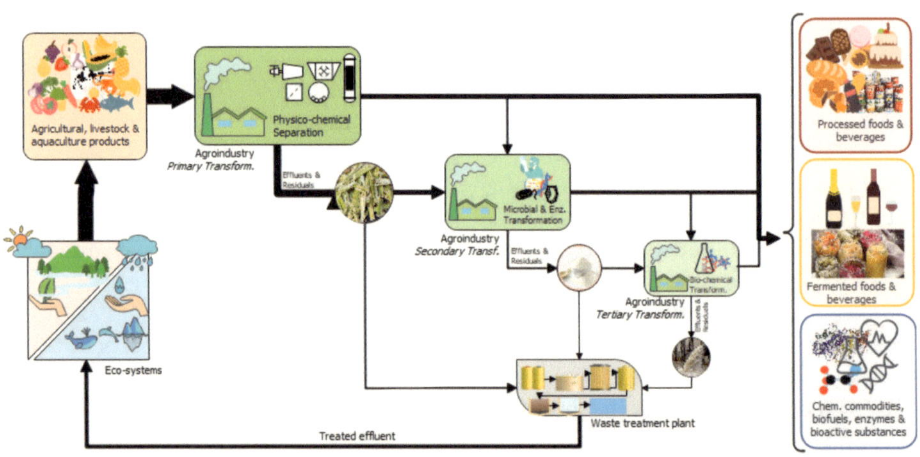

Figure 1.
Agroindustry: Its source of raw materials and main productions.

raw materials, and in their transformation performed by microorganisms, perfectly adapted to grow and develop using these raw materials, obtaining, as a result of their microbial activity, fermented products and beverages, with beneficial nutritional and food properties. Fermented food and beverage production industries [9, 10], such as yoghurt [11, 12], kefir [13, 14] and the manufacturing of beer [15] and wine [16], fermented sauces and condiments, such as soy sauce, as well as the production of bio-ethanol [17, 18], vinegar [19, 20] and some organic acids [21, 22], such as citric acid [23] and lactic acid [24], are practical examples of these agro-industries, where microorganisms and enzymes carry out the transformation of raw materials to product.

Finally, in the tertiary transformation agro-industries, the products, by-products or residues of the primary and secondary transformation agro-industries continue to be transformed chemically and/or biochemically into new chemical compounds, like bioactive compounds, enzymes, polysaccharides, gums, phytohormones, growth factors, etc. These, as a rule, are the products derived from agro-industries that have the highest added value. Some industrial enzymes such as cellulases [25], lipases [26] amylases [27], fructosyltransferases and invertases [28]; macromolecules like fructo- and galactooligosaccharides (FOS and GOS) [29–31], etc., are examples of agro-industries of tertiary transformation.

Currently, the production volumes of tertiary transformation agro-industries are significantly lower than the previous two. However, they should increase in the future, stimulated by the high prices of these products and the depletion of oil, the main raw material from which the traditional chemical industry's precursors are obtained [32].

An agro-industrial process can be considered as a set of operations that allow the gradual transformation of the process inputs (for example, raw materials, material, and energy resources) into the outputs (such as the main product (s), by-products, disposable materials and waste) (**Figure 2**).

As a rule, the added value of the product or products is significantly higher than the value of the inputs and other elements of the outputs.

An agro-industrial process, like any other, is made up of a series of stages of transformation processes. Each stage can be made up of one or more unit operations. In each of the stages of the transformation process of raw materials or intermediate products, a set of factors or variables can influence the efficiency and speed of said transformation. These factors can be subdivided into controllable and non-controllable factors or variables (**Figure 3**). The first ones are all those intensive variables of the process (such as temperature, pH, the concentration of certain analyte, etc.), whose values must be kept within a certain range on any scale and which, besides, are the ones that have the real possibility of being controlled within pre-established ranges in

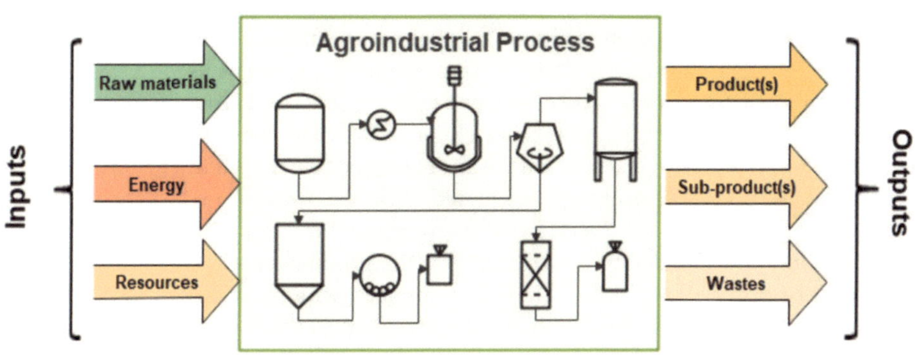

Figure 2.
General scheme of an agro-industrial process.

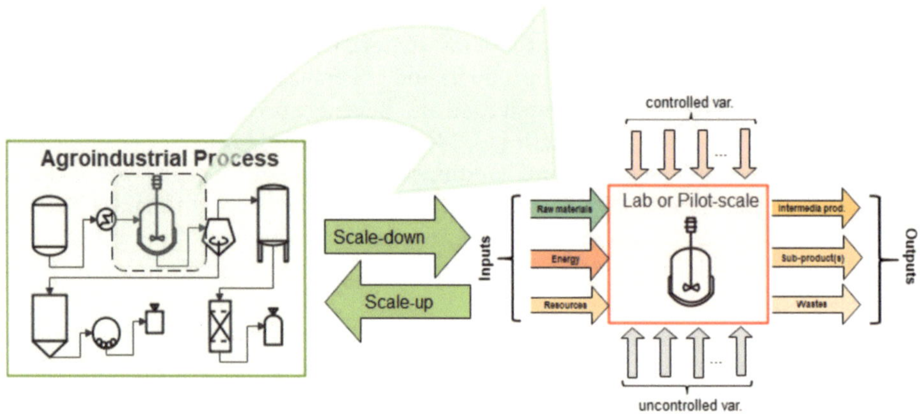

Figure 3.
The two ways to optimise agro-industrial processes: Scale-down of established processes and the Scale-up of new processes.

the different scales. The second, for their part, are all those variables, including those whose existence is still unknown, that may influence the transformation process but are not within reach of the processes and technologies to be controlled. By focusing attention on the intensive controllable variables, not only will it be possible to find the combination of these that allow the development of an optimal transformation process, but it will also allow knowing the values that must be achieved in the productive scale of a certain variable or response factor, commonly associated with some quality attribute of the final or intermediate product within the process.

An efficient and sustainable agro-industrial process will *maximize* the efficiency of the transformation of raw materials to finished products, *minimizing*, at the same time, the use of energy resources and the generation of by-products disposable and residual materials. The latter can be achieved by optimizing each of the stages of the process.

To do this, normally, you can proceed in two ways. If it is intended to optimize an already established large-scale non-optimal process, the established process could be scale-down to a smaller scale, a pilot-scale for example, or later to a laboratory, where all the necessary optimization experiments could be developed (**Figure 3**).

If it is intended to optimize the design of new processes, these can be optimized on a laboratory scale, first and later, these optimized processes would be scaled-up to pilot and further to industrial scale.

Due to the wide range of useful products that emerge from the agro-industry, ranging from products that improve the durability, texture and nutritional composition of natural foods, through simple chemical substances, precursors of other more complex and elaborated, to complex substances like antibiotics, prebiotics, hormones, enzymes, polysaccharides, etc. (**Figure 1**). There are numerous niches where modern techniques of experiment design and process optimization can be used [33, 34].

Among the most popular and effective tools to know the optimal parameters of a process is the response surface methodology (RSM), frequently associated with searching for an extreme value of one or more objective functions. The objective function or response is frequently associated with one or more of the product's attributes of a stage or unit operation of the process (for example, the concentration, the yield, the efficiency, the conversion, the productivity, etc.). Additionally, there may be other objective functions or responses, which can also integrate into the same optimization process, which may be related to other needs of the unit process or stage, such as the reduction of some by-product or residue, the decrease in consumption of energy, cleaning agents, shortening of the processing time, etc. In such cases, we would be in the presence of multi-objective optimization.

3. A short overview of response surface methodology (RSM)

The idea of the RSM is, through the design of experiments, to find the relationship between a certain response variable, commonly associated with one of the attributes of the experimental unit's output product with a few controllable variables. Strictly speaking, any response variable depends on both controllable and non-controllable variables. However, it is necessary to try to find a certain objective function, dependent only on a few controllable variables (usually from two to six), which allows navigating its surface in search of the combination of controlled variables with which an extreme value of the objective function is reached.

The response variable, as mentioned before, will also depend on the contribution of a certain "noise" function that depends on the rest of the controllable variables not taken into account in the objective function, as well as on the non-controllable factors. Still, it is must seek that noise's influence on the response is low enough or not significant. The determining influence on the response can be exerted mainly by the contribution of the objective function (**Figure 4A**).

To find the relationship between the variable response and the independent variables or controllable factors requires careful design of the experiments. These experiments must be carried out randomly and making the necessary replications to have the necessary certainty of their results [35]. In this sense, first of all, there must be solid evidence that the independent variables to be evaluated significantly influence the response variable or variables under study. This is based on our own experiments previously carried out or abundant reports published in the scientific literature. And secondly, must choose a suitable range for the independent variable analyzed, neither too narrow nor too long, so that the different values obtained from the response variable are notable.

The experimental runs must be carried out so that they cover the entire possible range of the independent variables that are being evaluated in the best way. Thus, the influence that each one, separately and combined with other independent variables, exerts on the response variables being analyzed can be evaluated. This can be ensured when the total sum of the products of all the coded independent variables of each run is equal to zero. The latter is known as the *orthogonal* design of experiments.

In addition to randomization and to minimize the influence that uncontrolled factors or variables may exert on the response variables, the different treatments are usually grouped into experimental blocks [36]. The latter can be beneficial, especially when various factors are evaluated, and the experiments must be performed over several days (**Figure 4B**).

In this way, the experiments must be *random*, with *replications* and designed in an *orthogonal* and *block-based* manner [37].

The most common experiment designs used in response surface methodology are the central composite design (CCD) and the Box–Behnken design (BBD). As a rule, in the BBDs, there are fewer experimental runs than in the CCDs (three levels for each factor in BBD, against five levels in CCD, for example). Therefore, it may be the preferred choice when the experiments are costly or when you need to have resulted in a shorter time. However, more robust and reliable models are obtained in CCDs, and they better support the loss or mismeasured response of the runs. The latter makes CCDs the "workhorse" and the first choice of researchers trying to optimize agro-industrial processes [37].

Fortunately, statistical packages accurately plan these experiment designs. Commercial statistical software such as Design-Expert®, JMP®, and Minitab® stand out, which are very useful and popular among scientific researchers and engineers communities. Additionally, there are becoming more popular every day; some free tools, such as the R and Python languages, are somewhat more

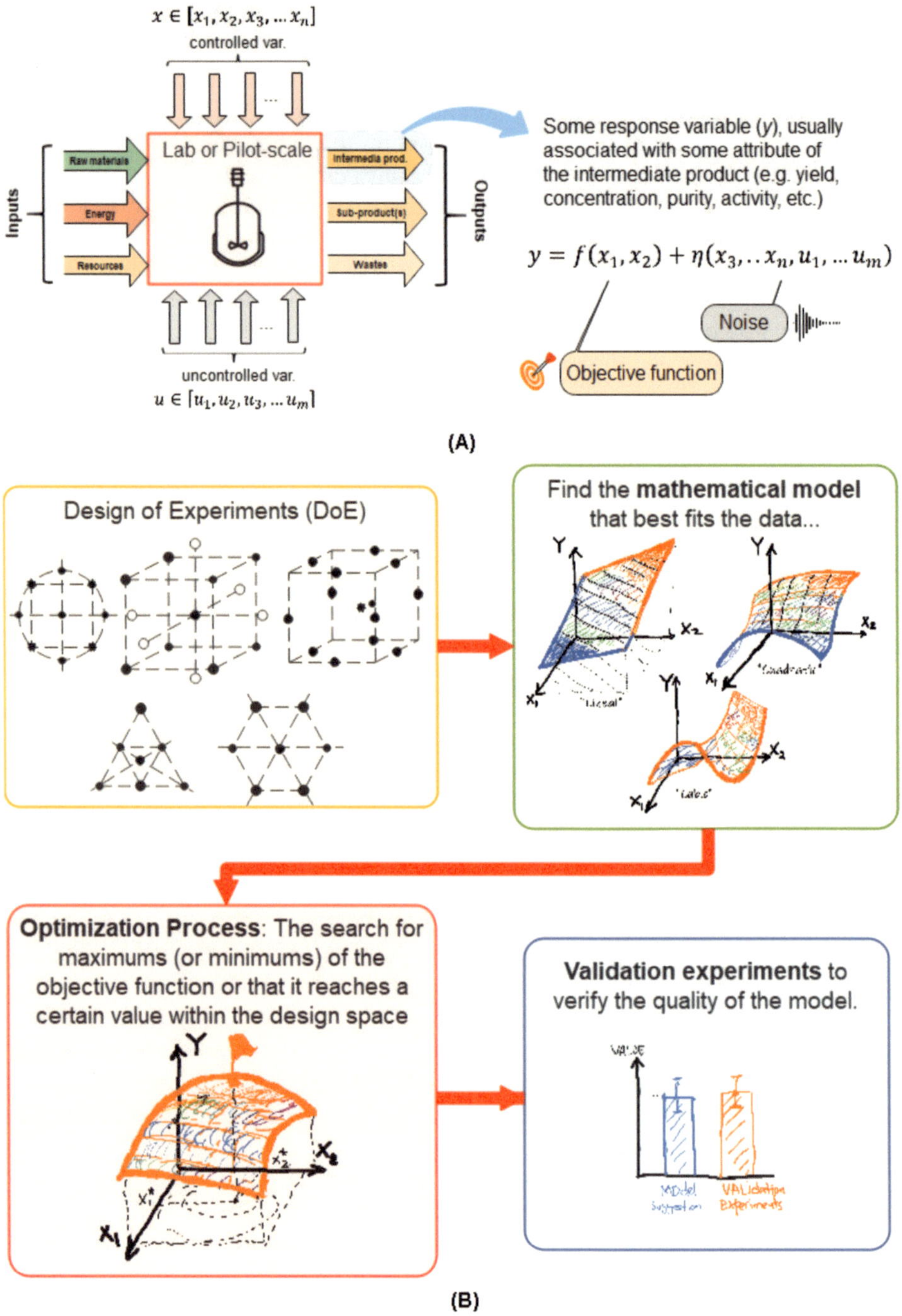

Figure 4.
(A) The response variable and its links with objective functions and noise, (B) general workflow for implementing response surface methodology.

complex to use and require greater statistical and programming knowledge for their successful use.

Once the experiments have been designed and executed, it is necessary to adjust the experimental data to a certain function that represents a close approximation to the response variable or variables obtained to each treatment, depending on the experiments' independent variables or factors (**Figure 4B**).

Depending on the values obtained in the response variable's experiments and if the difference between the maximum and minimum values obtained is huge, it will be necessary to transform the response variable before finding the adjustment function. Frequently the response variable (s), transformed or not, are adjusted to a certain polynomial, depending on the chosen design.

Subsequently, the polynomial that best fits the experimental data will be chosen. An analysis of variance will be carried out to the chosen polynomial. Those coefficients that are not significant for the model will be discarded as long as they do not sacrifice different runs' orthogonality. The quality of the model is evidenced through different statistical parameters, such as the adjusted quadratic coefficient of regression, adequate precision, the graphs of the predicted value versus the real value or through the graphs of the distribution of the residuals; the model may or may not be used, to find, within the design space of the independent variables, the optimal values of the function that adjusts the response variable(s) with the independent variables.

Once the appropriate model has been chosen, it is proceeding to explore the extreme or optimal values of the response function within the design space, that is, to find the combination of independent variables or factors that make the response objective function reach its maximum value, minimum value or equal to a specific value, depending on the response function. During the response function's optimisation process, and depending on the model's precision and variance, one or more extreme values may appear. If the extreme values found are related to variables or independent factors that are not related or are very far from each other, new experiments may be necessary around these points to improve the model's precision (**Figure 4B**).

On the contrary, if within the explored design space there is a single extreme value or only a few within a close region of the independent variables, can choose this solitary extreme value or can select one representative of the set of comparable response values can be chosen to proceed with a group of model confirmation or validation experiments.

There is no hard and fast rule about how many of these validation experiments are necessary, but they are usually between three and five. The model is validated when all the response function values are located within the range predicted by the model. The average weight of these does not differ significantly ($p < 0.05$) from the value predicted by the model. Otherwise, it would be suggesting that the model does not have the necessary accuracy. It is essential to continue exploring the search for extreme values within or outside of the original design space. In the latter case, an additional set of RSM-experiments would need to achieve (**Figure 4B**).

4. Agroindustry: a suitable receptor for the use of the response surface methodology (RSM)

Since Box and Wilson in 1951 [3] proposed this methodology, hundreds of scientific articles have been published [38–41].

Due to the wide range of useful products that emerge from the agro-industry, ranging from products that improve the durability, texture and nutritional composition of natural foods, through simple chemical substances, precursors of other more complex and elaborated, to complex substances like antibiotics, hormones, enzymes, polysaccharides, etc. (**Figure 2**). There are numerous niches where modern techniques of experiment design and process optimization can be used [42, 43].

Among the most popular and effective tools to know the optimal parameters of a process is the response surface methodology, frequently associated with development of new products [44–46] and processes [47, 48], the maximization of the productivity or yield of the process products [49–52], the reduction of their

Title	Clase[1]	Source/Product	RSM DOE	Optimal values	Ref.
Control of selected fermentation indices by statistically designed experiments in industrial scale beer fermentation.	AI PT	Optimization of beer fermentation	Box–Behnken	Pitching rate $6{\cdot}10^6$ cells/mL; fermentation temp. 11.2 °C; aeration level 10.5 mg/L; and CCTs filling time 13.5 h.	[57]
Effect of ohmic heating on quality and storability of sugarcane juice.	AI PT	Sugarcane juice/ sugarcane juice	Box–Behnken	Ohmic heating of sugarcane juice at 70°C for 3 min holding time.	[58]
Extraction of steviol glycosides from dried *Stevia rebaudiana* by pressurized hot water extraction.	AI PT	*Stevia rebaudiana* Bertoni leaves/Stevioside	CCD	2 bars of pressure, 20 min reaction time, and 20% dry leaves to water ratio	[59]
Optimization of spray-drying parameters for the production of 'Cempedak' (*Artocarpus integer*) fruit powder.	AI PT	Fruit juice/fruit powder	CCD	Air temperature of 160°C and maltodextrin conc. of 15% (w/w)	[60]
Optimizing the extraction of bioactive compounds from pu-erh tea (*Camellia sinensis var. assamica*) and evaluation of antioxidant, cytotoxic, antimicrobial, antihemolytic, and inhibition of α-amylase and α-glucosidase activities.	AI PT	Pu-erh tea/Antioxidants	CCD	Temperature of 85.4°C and time of 3 min	[61]
Maize stover as a feedstock for enhanced laccase production by two gammaproteobacteria: A solution to agroindustrial waste stockpiling.	AI PT	Maize stover/laccase	Box–Behnken	pH 5, 0.50 g biomaterial, 100 rpm and 0.10 $NaNO_3$	[62]
Evaluation of textural properties of corn based extruded products.	AI PT	Three corn varieties/ Extruded product	Box–Behnken	Temperature: 127.66°C, 18.96% feed moisture and 92:4:4 feed composition	[63]
Response Surface Methodology approach for optimization of endoglucanase from alkaliphilic *Fusarium oxysporum* VSTPDK and its potential application in pulp and paper industry.	AI PT	Rice straw/CMCase	CCD	pH 8.5, temperature 45°C, ammonium sulphate concentration 3% and 8 day incubation.	[64]
Antioxidant and prebiotic effects of a beverage composed by tropical fruits and yacon in alloxan-induced diabetic rats.	AI PT	Yacon extract + fruit juice/ fructo-oligosaccharides	CCD	Yacon extract: 50% and sweetener: 0.07%	[65]
Optimization of concentrating process using rotary vacuum evaporation for pineapple juice.	AI PT	Pineapple juice/ concentrated juice	CCD	Temp. 60°C and pressure 200 mBar for 75 min.	[66]

Title	Clase[1]	Source/Product	RSM DOE	Optimal values	Ref.
Pre-treatment optimization of barley straw as agro-industrial waste via alkaline peroxide and ultrasound for soluble sugar production and degradation.	AI PT	Barley straw/Sugars	Box–Behnken	US Power: 20 kj/kg DM Particle size: 0.6 mm	[67]
Techno-economic feasibility of bioethanol production via biorefinery of olive tree prunings (OTP): Optimization of the pretreatment stage.	AI PT	OTP/bioethanol	Box–Behnken	Minimum of Total Capital Cost: Temp. 199.98°C, 8 g H_2SO_4/100 g; 35% (w/v)	[68]
Design of experiments for enhanced production of bioactive exopolysaccharides from indigenous probiotic lactic acid bacteria.	AI ST	Lactose/ Exo-polysacharides	Box–Behnken	*Enterococcus faecium* K1: Lactose: 10.07g/L, Ammonium citrate 2.48 g/L, pH 5.4	[69]
Response surface methodology to optimize a bioprocess for kefiran production.	AI ST	WP/kefiran	CCD	Temp.: 25°C and 44.1% (w/w) of WP	[70]
Microwave-assisted extraction of pectin from "Saba" banana peel waste: Optimization, characterization, and rheology study.	AI ST	Banana peel waste/pectin	CCD	195° C, 8% solid–liquid ratio, and pH 3 HCl	[71]
Hydrolysis of orange peel with cellulase and pectinase to produce bacterial cellulose using *Gluconacetobacter xylinus*.	AI ST	Orange peel/cellulose	Box–Behnken	cellulase of 1589.41 U/g, pectinase of 31.75 U/g and a reaction time of 5.28 h	[72]
Valorization of sugarcane bagasse to high value-added xylooligosaccharides and evaluation of their prebiotic function in a synbiotic pomegranate juice	AI ST	Sugarcane bagasse/xylan	CCD	5.63% H_2O_2, 12.91% NaOH, and extraction time of 17.51 h	[50]
Playing with the senses: application of Box–Behnken design to optimize the bukayo formulation.	AI ST	Coconut meat and juice+sugar/bukayo acceptability	Box–Behnken	430 g young coconut meat, 400 g sinakob, and 340 g coconut juice	[73]
Utilization of Atlantic salmon by-product oil for omega-3 fatty acids rich 2-monoacylglycerol production: Optimization of enzymatic reaction parameters.	AI ST	Salmon By-product Oil/ Omega-3	Box–Behnken	Reaction temp. 42.5°C, time 4.15 h, enzyme load 42.81%, & ethanol: oil mol. Ratio 49.82	[74]
Bioconversion of cheese whey permeate into fungal oil by *Mucor circinelloides*.	AI ST	Whey permeate/fungal oil	CCD	Fermentation temp. 33.6°C and pH 4.5	[75]
Olive mill and winery wastes as viable sources of bioactive compounds: A study on polyphenols recovery.	AI ST	olive pomace residues/ polyphenols	Box–Behnken	Olive pomace microwave-extraction: ethanol:water 50:50 (v/v), 90°C, 5 min	[76]

Title	Clase[1]	Source/Product	RSM DOE	Optimal values	Ref.
Development of a low-temperature and high-performance green extraction process for the recovery of polyphenolic phytochemicals from waste potato peels using hydroxypropyl β-cyclodextrin.	AI ST	Potato peel/polyphenols	Box–Behnken	pH 5.0, ratio solvent-to-dry weigth 80 mL g^{-1} and agitation speed 800 rpm	[77]
Optimized preparation of activated carbon from coconut shell and municipal sludge.	AI ST	Coconut shell/activated carbon	Box–Behnken	Temp.: 800°C, activation time: 60 min, activator concentration: 2.5 mol/L, a 50% coconut shell.	[78]
Response surface methodology as a tool for modeling galacto-oligosaccharide (GOS) production.	AI ST	DWP/GOS	CCD	DWP: 18 g/ml, 0.20 g/L of β-galactosidase	[79]
Optimization of β-galactosidase production by batch cultures of *Lactobacillus leichmannii* 313 (ATCC 7830™).	AI ST	Lactose/β-galactosidase	CCD	pH 7.06 and 15.3 g/L lactose	[80]
An eco-friendly pressure liquid extraction method to recover anthocyanins from broken black bean hulls.	AI ST	Broken black bean hulls/anthocyanins	Box–Behnken	Ratio ethanol and citric acid sol.n 0.1 mol/L of 30:70 (v/v), flow rate: 4 mL/min, 60°C.	[81]
Canola meal as a promising source of fermentable sugars: Potential of the *Penicillium glabrum* crude extract for biomass hydrolysis.	AI ST	Canola meal/β-glucosidase	CCD	Fermentation time: 6.5 days, pH adjusted to 6.0, and substrate concentration of 2%	[82]
Optimization of galacto-oligosacharides (GOS) synthesis using response surface methodology.	AI ST	Lactose/GOS	CCD	Lactose conc. of 400 g/l, enzyme conc. of 13.5 g/l and reaction time of 13 min	[83]
Pre-treatment of sugarcane bagasse with aqueous ammonia–glycerol mixtures to enhance enzymatic saccharification and recovery of ammonia.	AI ST	Sugarcane bagasse/Sugars	Box–Behnken	Conc. of ammonia: 9.25%, pre-treatment time: 1.86 h, pre-treatment temp.: 180°C	[84]
Low-cost production of PHA using cashew apple (*Anacardium occidentale* L.) juice as potential substrate: optimization and characterization.	AI ST	Cashew apple juice/PHA	CCD	Conc. of total reducing sugar of 50 g/L, inoculum of 50 mL/L, and urea of 3 g/L.	[85]
A facile noncatalytic methyl ester production from waste chicken tallow (WCT) using single step subcritical methanol: Optimization study.	AI ST	WCT/biodiesel (FAME)	CCD	167°C, 36.8 min., and 42.7:1 (methanol/WCT, mol/mol)	[86]

DOI: http://dx.doi.org/10.5772/intechopen.98283

Title	Clase[1]	Source/Product	RSM DOE	Optimal values	Ref.
Recovery and bio-potentialities of astaxanthin-rich oil from shrimp (*Peneanus monodon*) waste and mackerel (*Scomberomous niphonius*) skin using concurrent supercritical CO_2 extraction.	AI ST	Astaxanthin-rich oil/shrimp waste & fish skin	CCD	Extraction temp. 45.7°C; pressure 264.09 bar, and shrimp waste-to-fish skin mixing ratio 79.63:20.37.	[87]
Zero-waste biorefinery of oleaginous microalgae as promising sources of biofuels and biochemicals through direct transesterification and acid hydrolysis.	AI ST/ TT	Microalgae (marine *Chlorella* sp.)/biofuels (FAME) & sugars.	Box–Behnken	FAME yield: Temp. 70 °C, ratio of chloroform:methanol 1.35:1 and reaction time 120 min. Sugar yield: 7.5% H_2SO_4, 60 min hydrolysis time, 3% biomass loading, and 100°C hydrolysis temp.	[88]
Sequential production of lignin, fatty acid methyl esters and biogas from spent coffee grounds (SCG) via an integrated physicochemical and biological process.	AI TT	SCG/Lignin/FAME & biogas	CCD	Temp. 161.0°C, sulfuric acid: 3.6% and methanol:SCG ratio: 4.7 mL/g	[89]
Heterogeneous catalytic conversion of rapeseed oil to methyl esters: Optimization and kinetic study.	AI TT	Rapeseed oil/FAME	CCD	Catalyst ratio (bentonite/NaOH): 1:20; catalyst amount: 6%wt.; reaction time: 3.5 h.	[90]

Abbreviations: DWP: demineralised whey powder; WP: whey powder; CMCase: carboxy-methyl-celulase; PHA: poly-hydroxy-alkanoate; FAME: fatty-acid methyl-ester.
[1]AI PT, ST, TT: Agro-industry (AI) of primary, secondary, or tertiary transformations.

Table 1.
Some of the recent work on the response surface methodology related to the agro-industry.

production costs [53], the minimization of risks for the human health [54] or its negative impacts on ecosystems [55, 56].

A summarized sample of some of the recent work in agro-industry related to the response surface methodology is shown in **Table 1**.

Table 1 shows more than thirty selected examples chosen from the last five years (2017–2021) from the multiple reports in the specialized bibliography, using RSM in all areas of agro-industry. In the selected cases, it is confirmed that the BBD and CCD designs are the most widely used and the utility that these experimental design and optimisation tools provide to researchers and engineers working in the agro-industry is demonstrated.

On the other hand, RSM is conveniently intertwined with the concepts of the "circular economy" [91] applied to agro-industry toward a broader framework of a sustainable bioeconomy [92, 93], where it is intended to maximize the efficiency and productivity of the transformation processes of raw materials into products, with a minimal negative impact on the environment and to minimize generation of by-products, wastes, and residuals of the agro-industrial processes, and, at the same time, reuse the latter as sources of raw materials for other products, valuing them.

On some occasions the by-products and wastes of some agro-industrial processes become sources of raw materials for obtaining other valuable products through chemical, enzymatic or biological transformation. Some examples such as whey, a by-product of the cheese industry [94–96], and molasses and bagasse by-products of the sugar cane industry, from which some valuable products are obtained [97–100].

This fact gives rise to the concept of biorefineries and circular economies [91, 93], applicable in certain economic agricultural crops exploited on a large scale, where a group of valuable products could be obtained from abundant and renewable raw materials, in addition to those that have traditionally been obtained previously.

5. Conclusion

Despite an appreciable decrease in publications related to the response surface methodology in agro-industry, in 2020 and so far in 2021, due to the effects of the impact of the SARS-CoV-2 pandemic in the world, everything indicates that RSM is a prevalent tool among researchers and engineers to improve agro-industrial processes, as demonstrated in this work, and that it will continue to be very useful and necessary to achieve efficient, sustainable and friendly agro-industrial processes with the environment. For this reason, once the effects of the pandemic have passed, new reports of applications of the use of this statistical tool will surely continue to appear.

Acknowledgements

We wish to express our gratitude to the authorities of the **Universidad Técnica del Norte** (UTN, Ibarra, Imbabura, Ecuador) for their unconditional support, and to **Dr. Bolívar Batallas** and **Dr. Hernán Cadenas**, Dean and Vice-Dean of our Faculty, respectively, for their support for this publication.

Author details

José Manuel Pais-Chanfrau[1*], Jimmy Núñez-Pérez[1], Rosario del Carmen Espin-Valladares[1], Marco Vinicio Lara-Fiallos[1] and Luis Enrique Trujillo-Toledo[2]

1 North-Technical University (Universidad Técnica del Norte, UTN), FICAYA, Ibarra, Imbabura, Ecuador

2 University of the Armed Forces (Universidad de las Fuerzas Armadas, ESPE), Quito, Pichincha, Ecuador

*Address all correspondence to: jmpais@utn.edu.ec

References

[1] Myers R, Montgomery D, Anderson-Cook C. Response Surface Methodology: Process and Product Optimization Using Designed Experiments. 4th ed. Wiley Series in Probability and Statistics. Hoboken, New Jersey: John Wiley & Sons, Inc; 2016. 854 p. DOI: 10.1017/CBO9781107415324.004

[2] Khuri AI. Response Surface Methodology and Its Applications In Agricultural and Food Sciences. Biometrics Biostat Int J. 2017;5(5): 155-63. Available from: https://medcraveonline.com/BBIJ/BBIJ-05-00141.pdf

[3] Box GEP, Wilson KB. On the Experimental Attainment of Optimum Conditions. J R Stat Soc Ser B. 1951;13: 1-38. DOI: 10.1111/j.2517-6161.1951.tb00067.x

[4] Box GEP, Hunter JS. Multi-Factor Experimental Designs for Exploring Response Surfaces. Ann Math Stat. 1957;28:195-241. DOI: 10.1214/aoms/1177707047

[5] Box GEP, Draper NR. A Basis for the Selection of a Response Surface Design. J Am Stat Assoc. 1959;54:622-54. DOI: 10.1080/01621459.1959.10501525

[6] van der Poel P, Schiweck H, Schwartz T. Sugar Technology, Beet and Cane Sugar Manufacture. 1st ed. Berlin: Verlag; 1998. 1118 p. DOI: 10.36961/st

[7] van Asselt AJ, Weeks MG. Dairy processing. In: Sustainable Dairy Production. 1st ed. Chichester: Wiley; 2013. p. 87-118. DOI: 10.1002/9781118489451.ch5

[8] Anonymous. Meat Processing. In: Meat Industry Guide. London: UK Food Std. Agency; 2015. p. 1-36. Available from: https://webarchive.nationalarchives.gov.uk/20191202184215/https://www.food.gov.uk/sites/default/files/media/document/chapter15-meat-processing-final_verion-2_0.pdf

[9] Corbo MR, Bevilacqua A, Petruzzi L, Casanova FP, Sinigaglia M. Functional Beverages: The Emerging Side of Functional Foods: Commercial Trends, Research, and Health Implications. Compr Rev Food Sci Food Saf. 2014;13:1192-206. DOI: 10.1111/1541-4337.12109

[10] Wu Y, Yan B, Zhou J, Lian H, Yu X, Zhao J, et al. Effects of sourdough on improving the textural characteristics of microwave-steamed cake: A perspective from dielectric properties and water distribution. J Food Sci. 2020;85:3282-3292. DOI: 10.1111/1750-3841.15424

[11] Bamforth CW, Cook DJ. Yoghurt and Other Fermented Milk Products. In: Food, Fermentation, and Micro-organisms. 2nd ed. Hoboken: Wiley; 2019. p. 183-186. DOI: 10.1002/9781119557456.ch11

[12] Behare P, Kumar H, Mandal S. Yogurt: Yogurt Based Products. In: Encyclopedia of Food and Health. 1st ed. London: Elsevier; 2015. p. 625-631. DOI: 10.1016/B978-0-12-384947-2.00767-4

[13] Kesenkaş H, Gürsoy O, Özbaş H. Kefir. In: Fermented Foods in Health and Disease Prevention. 1st ed. London: Elsevier; 2017. p. 339-361. DOI: 10.1016/B978-0-12-802309-9.00014-5

[14] Prado MR, Blandón LM, Vandenberghe LPS, Rodrigues C, Castro GR, Thomaz-Soccol V, et al. Milk kefir: Composition, microbial cultures, biological activities, and related products. Front Microbiol. 2015;6:1-10. DOI: 10.3389/fmicb.2015.01177

[15] Lager T. The Development and Manufacturing of Beer — An

Introduction and Background for the Process-Industrial Illustrative Case. In: Contemporary Quality Function Deployment for Product and Process Innovation. 1st ed. London: World Scientific Publishing Company; 2019. p. 23-36. DOI: 10.1142/9789813279889_0002

[16] Pozo-Bayón MA, Moreno-Arribas M V. Sherry Wines: Manufacture, Composition and Analysis. In: Encyclopedia of Food and Health. 2015. 1st ed. London: Elsevier; 2015. p. 779-784. DOI: 10.1016/B978-0-12-384947-2.00626-7

[17] Gupta A, Verma JP. Sustainable bio-ethanol production from agro-residues: A review. Renew Sustain Energy Rev. 2015;41:550-567. DOI: 10.1016/j.rser.2014.08.032

[18] O'Hara IM, Mundree SG, editors. Sugarcane-based biofuels and bioproducts. Sugarcane-based Biofuels and Bioproducts. 1st ed. Chichester: Wiley; 2016. 386 p. DOI:10.1002/9781118719862

[19] Garcia-Parrilla MC, Torija MJ, Mas A, Cerezo AB, Troncoso AM. Vinegars and Other Fermented Condiments. In: Frias J, Martinez-Villaluenga C, Peñas E, editors. Fermented Foods in Health and Disease Prevention. 1st ed. London: Elsevier; 2017. p. 577-591. DOI: 10.1016/B978-0-12-802309-9.00025-X

[20] Bekatorou A. Advances in Vinegar Production. 1st ed. Boca Raton: CRC Press. 2019. 522 p. DOI: 10.1201/9781351208475

[21] Rogers P, Chen JS, Zidwick MJ. Organic acid and solvent production: Acetic, lactic, gluconic, succinic, and polyhydroxyalkanoic acids. In: Dworkin M, Falkow S, Rosenberg E, Schleifer KH, Stackebrandt E, editors. The Prokaryotes: Applied Bacteriology and Biotechnology. New York: Springer; 2014. p. 3-75. DOI: 10.1007/0-387-30741-9_19

[22] Singh R, Mittal A, Kumar M, Mehta PK. Organic acids: An overview on microbial production. Int J Adv Biotechnol Res. 2017;8(1):104-111.

[23] Ciriminna R, Meneguzzo F, Delisi R, Pagliaro M. Citric acid: Emerging applications of key biotechnology industrial product. Chem Cent J. 2017;11(22):1-9. DOI: 10.1186/s13065-017-0251-y

[24] Gasca-González R, Prado-Rubio OA, Gómez-Castro FI, Fontalvo-Alzate J, Pérez-Cisneros ES, Morales-Rodriguez R. Techno-economic analysis of alternative reactive purification technologies in the lactic acid production process. In: Kiss A, Zondervan E, Lakerveld R, Özkan L, editors. 29th European Symposium on Computer Aided Process Engineering. Computer Aided Chemical Engineering. vol. 46. London: Elsevier; 2019. p. 457-462. DOI: 10.1016/B978-0-12-818634-3.50077-1

[25] Srivastava N, Mishra PK, Upadhyay SN. Microbial cellulase production. In: Srivastava N, Mishra PK, Upadhyay SN, editors. Industrial Enzymes for Biofuels Production. 1st ed. London: Elsevier; 2020. p. 19-35. DOI: 10.1016/b978-0-12-821010-9.00002-4

[26] Vargas M, Niehus X, Casas-Godoy L, Sandoval G. Lipases as biocatalyst for biodiesel production. Methods Mol Biol. 2018; 1835: 377-390. DOI: 10.1007/978-1-4939-8672-9_21

[27] Mohanan N, Satyanarayana T. Amylases. In: Schmidt, T, editor. Encyclopedia of Microbiology. 4th ed. London: Academic Press; 2019. p. 107-126. DOI: 10.1016/B978-0-12-809633-8.13003-1

[28] Trujillo-Toledo LE, García DM, Cruz EP, Intriago LMR, Pérez JN,

Pais-Chanfrau JM. Fructosyltransferases and invertases: Useful enzymes in the food and feed industries. In: Kuddus M, editor. Enzymes in Food Biotechnology: Production, Applications, and Future Prospects. 1st ed. London: Academic Press; 2018. p. 451-469. DOI: 10.1016/B978-0-12-813280-7.00026-8

[29] Flores-Maltos DA, Mussatto SI, Contreras-Esquivel JC, Rodríguez-Herrera R, Teixeira JA, Aguilar CN. Biotechnological production and application of fructooligosaccharides. Crit Rev Biotechnol. 2016;36:259-267. DOI: 10.3109/07388551.2014.953443

[30] Pérez Cruz ER, Hernández García L, Martínez García D, Trujillo Toledo LE, Menéndez Rodríguez, Carmen, et al. Method for obtaining 1-kestose. Cuba; WO2014044230 A1, 2014. p. 12. Available from: https://www.google.com/patents/WO2014044230A1?cl=en

[31] Fischer C, Kleinschmidt T. Synthesis of galactooligosaccharides using sweet and acid whey as a substrate. Int Dairy J. 2015;48:15-22. DOI: 10.1016/j.idairyj.2015.01.003

[32] Pais-Chanfrau JM, Núñez-Pérez J, Espin-Valladares R del C, Lara-Fiallos MV, Trujillo-Toledo LE. Bioconversion of Lactose from Cheese Whey to Organic Acids. In: Gutiérrez-Méndez N, editor. Lactose and Lactose Derivatives. 1st ed. London: IntechOpen; 2020. p. 53-74. DOI: 10.5772/intechopen.92766

[33] Lipták BG. Optimization of Industrial Unit Processes. 2nd ed. Boca Raton: CRC Press; 2020. 432 p. DOI: 10.1201/9780138744847

[34] Lahiri SK. Optimization of Industrial Processes and Process Equipment. In: Lahiri SK, editor. Profit Maximization Techniques for Operating Chemical Plants. 1st ed. NJ: Wiley; 2020. p. 131-158. DOI: 10.1002/9781119532231.ch8

[35] Jones B, Hunter JS, Montgomery DC. Partial replication of definitive screening designs. Qual Eng. 2020;32:1-6. DOI: 10.1080/08982112.2019.1680847

[36] Montgomery DC, Runger GC. Design of experiment with several factors. In: Montgomery DC, Runger GC, editors. Applied Statistics and Probabilities for Engineers. 7th ed. Hoboken: Wiley; 2018. p. 375-433.

[37] Montgomery DC. Design and analysis of experiments. 9th ed. Hoboken: Wiley; 2017. 748 p.

[38] Myers RH, Montgomery DC, Geoffrey Vining G, Borror CM, Kowalski SM. Response Surface Methodology: A Retrospective and Literature Survey. Journal of Quality Technology. 2003;36:53-78. DOI: 10.1080/00224065.2004.11980252

[39] Khuri AI, Mukhopadhyay S. Response surface methodology. Wiley Interdisciplinary Reviews: Computational Statistics. 2010;2:128-149. DOI: 10.1002/wics.73

[40] Bezerra MA, Santelli RE, Oliveira EP, Villar LS, Escaleira LA. Response surface methodology (RSM) as a tool for optimization in analytical chemistry. Talanta. 2008;76: 965-977. DOI: 10.1016/j.talanta.2008.05.019

[41] Manojkumar N, Muthukumaran C, Sharmila G. A comprehensive review on the application of response surface methodology for optimization of biodiesel production using different oil sources. Journal of King Saud University - Engineering Sciences. DOI: 10.1016/j.jksues.2020.09.012

[42] Yolmeh M, Jafari SM. Applications of Response Surface Methodology in the Food Industry Processes. Food and Bioprocess Technology. 2017;10:413-433. DOI: 10.1007/s11947-016-1855-2

[43] Granato D, de Araújo Calado VÔM, Jarvis B. Observations on the use of statistical methods in Food Science and Technology. Food Res. Inter. 2014;55:137-149. DOI: 10.1016/j.foodres.2013.10.024

[44] O'Neill CM, Cruz-Romero MC, Duffy G, Kerry JP. The application of response surface methodology for the development of sensory accepted low-salt cooked ham using high pressure processing and a mix of organic acids. Innov Food Sci Emerg Technol. 2018;45:401-411. DOI: 10.1016/j.ifset.2017.12.009

[45] Dastras M, Soltanzadeh M, Peighambardoust SH. Production of cereal-based probiotic beverage optimized by response surface methodology and investigation of its properties. Iran J Nutr Sci Food Technol. 2019;14: 47-56.

[46] Yih Hui B, Mohamad Zain NN, Mohamad S, Varanusupakul P, Osman H, Raoov M. Poly (cyclodextrin-ionic liquid) based ferrofluid: A new class of magnetic colloid for dispersive liquid phase microextraction of polycyclic aromatic hydrocarbons from food samples prior to GC-FID analysis. Food Chem. 2020;314:126214. DOI: 10.1016/j.foodchem.2020.126214

[47] Hesam F, Tarzi BG, Honarvar M, Jahadi M. Pistachio (*Pistacia vera*) shell as a new candidate for enzymatic production of xylooligosaccharides. J Food Meas Charact. 2021;15:33-41. DOI: 10.1007/s11694-020-00594-y

[48] Nadeem F, Mehmood T, Naveed M, Shamas S, Saman T, Anwar Z. Protease Production from *Cheotomium globusum* Through Central Composite Design Using Agricultural Wastes and Its Immobilization for Industrial Exploitation. Waste and Biomass Valorization. 2020;11:6529-6539. DOI: 10.1007/s12649-019-00890-9

[49] Zamri AI, Latiff NF, Abdullah QH, Ahmad F. Extraction and optimization of chitosan from razor clam (*Ensis arcuatus*) shells by using response surface methodology (RSM). Food Res. 2020;4:674-678. DOI: 10.26656/fr.2017.4(3).308

[50] Hesam F, Tarzi BG, Honarvar M, Jahadi M. Valorization of sugarcane bagasse to high value-added xylooligosaccharides and evaluation of their prebiotic function in a synbiotic pomegranate juice. Biomass Convers Biorefinery. DOI: 10.1007/s13399-020-01095-0

[51] Kalathinathan P, Kodiveri Muthukaliannan G. A statistical approach for enhanced production of β-galactosidase from *Paracoccus sp*. and synthesis of galacto-oligosaccharides. Folia Microbiol (Praha). 2020;65:811-822. DOI: 10.1007/s12223-020-00791-8

[52] Israni N, Venkatachalam P, Gajaraj B, Varalakshmi KN, Shivakumar S. Whey valorization for sustainable polyhydroxyalkanoate production by *Bacillus megaterium*: Production, characterization and in vitro biocompatibility evaluation. J Environ Manage. DOI: 10.1016/j.jenvman.2019.109884

[53] Moyo LB, Iyuke SE, Muvhiiwa RF, Simate GS, Hlabangana N. Application of response surface methodology for optimization of biodiesel production parameters from waste cooking oil using a membrane reactor. South African J Chem Eng. 2021;35:1-7. DOI: 10.1016/j.sajce.2020.10.002

[54] Kalagatur NK, Kamasani JR, Siddaiah C, Gupta VK, Krishna K, Mudili V. Combinational Inhibitory Action of *Hedychium spicatum* L. Essential Oil and γ-Radiation on Growth Rate and Mycotoxins Content of *Fusarium graminearum* in Maize: Response Surface Methodology. Front

Microbiol. 2018;9:1511. DOI: 10.3389/fmicb.2018.01511

[55] Kainthola J, Kalamdhad AS, Goud V V. Optimization of process parameters for accelerated methane yield from anaerobic co-digestion of rice straw and food waste. Renew Energy. 2020;149: 1352-1359. DOI: 10.1016/j.renene.2019.10.124

[56] Parra-Orobio BA, Torres-López WA, Torres-Lozada P. Response Surface Methodology as an Optimization Tool for Anaerobic Digestion of Food Waste. Water Air Soil Pollut. 2020;231:385. DOI: 10.1007/s11270-020-04764-y

[57] Kucharczyk K, Zyła K, Tuszyński T. Control of selected fermentation indices by statistically designed experiments in industrial scale beer fermentation. Czech J Food Sci. 2020;38:330-336. DOI: 10.17221/291/2019-CJFS

[58] Abhilasha P, Pal US. Effect of Ohmic Heating on Quality and Storability of Sugarcane Juice. Int J Curr Microbiol Appl Sci. 2018;7:2856-2868. DOI: 10.20546/ijcmas.2018.701.340

[59] Németh, Jánosi SZ. Extraction of steviol glycosides from dried *Stevia rebaudiana* by pressurized hot water extraction. Acta Aliment. 2019;48:241-252. DOI: 10.1556/066.2019.48.2.12

[60] Pui LP, Karim R, Yusof YA, Wong CW, Ghazali HM. Optimization of spray-drying parameters for the production of 'Cempedak' (*Artocarpus integer*) fruit powder. J Food Meas Charact. 2020;14:1-12. DOI: 10.1007/s11694-020-00565-3

[61] Armstrong L, Araújo Vieira do Carmo M, Wu Y, Antônio Esmerino L, Azevedo L, Zhang L, et al. Optimizing the extraction of bioactive compounds from pu-erh tea (*Camellia sinensis var. assamica*) and evaluation of antioxidant, cytotoxic, antimicrobial, antihemolytic, and inhibition of α-amylase and α-glucosidase activities. Food Res Int. 2020;137(109430). DOI: 10.1016/j.foodres.2020.109430

[62] Unuofin JO, Okoh AI, Nwodo UU. Maize stover as a feedstock for enhanced laccase production by two gammaproteobacteria: A solution to agroindustrial waste stockpiling. Ind Crops Prod. 2019;129: 611-623. DOI: 10.1016/j.indcrop.2018.12.043

[63] Shruthi VH, Hiregoudar S, Nidoni U. Evaluation of textural properties of corn based extruded products. Plant Arch. 2019;19:2405-2410. Available from: http://plantarchives.org/19-2/2405-2410 (5310).pdf

[64] Abdulhadi Y, Ashish V. Response Surface Methodology approach for optimization of endoglucanase from alkaliphilic *Fusarium oxysporum* VSTPDK and its potential application in pulp and paper industry. Res J Biotechnol. 2021;16(1):62-67.

[65] Dionisio AP, de Carvalho-Silva LB, Vieira NM, Wurlitzer NJ, Pereira AC da S, Borges M de F, et al. Antioxidant and prebiotic effects of a beverage composed by tropical fruits and yacon in alloxan-induced diabetic rats. Food Sci Technol. 2020;40:202-208. DOI: 10.1590/fst.34518

[66] Leong CY, Chua LS. Optimization of concentrating process using rotary vacuum evaporation for pineapple juice. Chem Eng Trans. 2020;78:7-12. Available from: https://www.aidic.it/cet/20/78/002.pdf

[67] Deveci EÜ, Gönen Ç, Akarsu C. Pre-treatment optimization of barley straw as agro-industrial waste via alkaline peroxide and ultrasound for soluble sugar production and degradation. Biomass Convers Biorefinery. 2020;10:1-9. DOI: 10.1007/s13399-020-00879-8

[68] Solarte-Toro JC, Romero-García JM, Susmozas A, Ruiz E, Castro E, Cardona-Alzate CA. Techno-economic feasibility of bioethanol production via biorefinery of olive tree prunings (OTP): Optimization of the pretreatment stage. Holzforschung. 2019;73:3-13. DOI: 10.1515/hf-2018-0096

[69] Bhat B, Vaid S, Habib B, Bajaj BK. Design of experiments for enhanced production of bioactive exopolysaccharides from indigenous probiotic lactic acid bacteria. Indian J Biochem Biophys. 2020;57(5):539-551.

[70] Pais-Chanfrau JM, Toledo LET, Cóndor PMA, Guerrero MJC, Pérez JN, Intriago LMR. Response surface methodology to optimize a bioprocess for kefiran production. La Prensa Medica Argentina. 2018;104:1-5. DOI: 10.4172/0032-745X.1000285

[71] Rivadeneira JP, Wu T, Ybanez Q, Dorado AA, Migo VP, Nayve FRP, et al. Microwave-assisted extraction of pectin from "Saba" banana peel waste: Optimization, characterization, and rheology study. Int J Food Sci. 2020;2020:1-9. DOI: 10.1155/2020/8879425

[72] Kuo CH, Huang CY, Shieh CJ, Wang HMD, Tseng CY. Hydrolysis of Orange Peel with Cellulase and Pectinase to Produce Bacterial Cellulose using *Gluconacetobacter xylinus*. Waste and Biomass Valorization. 2019;10:85-93. DOI: 10.1007/s12649-017-0034-7

[73] Domingo CJA, De Vera WM, Pambid RC, Austria VC. Playing with the senses: application of Box-Behnken design to optimize the bukayo formulation. Food Res. 2019;3(6):833-839. DOI: 10.26656/fr.2017.3(6).190

[74] Haq M, Pendleton P, Chun BS. Utilization of Atlantic Salmon By-product Oil for Omega-3 Fatty Acids Rich 2-Monoacylglycerol Production: Optimization of Enzymatic Reaction Parameters. Waste and Biomass Valorization. 2020;11:1-11. DOI: 10.1007/s12649-018-0392-9

[75] Chan LG, Cohen JL, Ozturk G, Hennebelle M, Taha AY, De Moura Bell JMLN. Bioconversion of cheese whey permeate into fungal oil by *Mucor circinelloides*. J Biol Eng. 2018;12(25):1-14. DOI: 10.1186/s13036-018-0116-5

[76] Tapia-Quirós P, Montenegro-Landívar MF, Reig M, Vecino X, Alvarino T, Cortina JL, et al. Olive mill and winery wastes as viable sources of bioactive compounds: A study on polyphenols recovery. Antioxidants. 2020;9(11):1-15. DOI: 10.3390/antiox9111074

[77] Lakka A, Lalas S, Makris DP. Development of a low-temperature and high-performance green extraction process for the recovery of polyphenolic phytochemicals from waste potato peels using hydroxypropyl β-cyclodextrin. Appl Sci. 2020;10:1-12. DOI: 10.3390/app10103611

[78] Liang Q, Liu Y, Chen M, Ma L, Yang B, Li L, et al. Optimized preparation of activated carbon from coconut shell and municipal sludge. Mater Chem Phys. 2020;241(2):122327. DOI: 10.1016/j.matchemphys.2019.122327

[79] Vénica CI, Bergamini C V., Perotti MC. Response surface methodology as a tool for modelling galacto-oligosaccharide production. J Dairy Res. 2017;84:464-470. DOI: 10.1017/S0022029917000541

[80] Deng Y, Xu M, Ji D, Agyei D. Optimization of β-galactosidase Production by Batch Cultures of *Lactobacillus leichmannii* 313 (ATCC 7830™). Fermentation. 2020;6:1-17. DOI: 10.3390/fermentation6010027

[81] Teixeira RF, Benvenutti L, Burin VM, Gomes TM, Ferreira SRS,

Zielinski AAF. An eco-friendly pressure liquid extraction method to recover anthocyanins from broken black bean hulls. Innov Food Sci Emerg Technol. 2021;67:1-8. DOI: 10.1016/j.ifset.2020.102587

[82] Martins EH, Ratuchne A, de Oliveira Machado G, Knob A. Canola meal as a promising source of fermentable sugars: Potential of the *Penicillium glabrum* crude extract for biomass hydrolysis. Biocatal Agric Biotechnol. 2020;27:1-10. DOI: 10.1016/j.bcab.2020.101713

[83] Carevic M, Banjanac K, Milivojevic A, Corovic M, Bezbradica D. Optimization of galacto-oligosacharides synthesis using response surface methodology. Food Feed Res. 2017;44:1-10. DOI: 10.5937/ffr1701001c

[84] Shi T, Lin J, Li J, Zhang Y, Jiang C, Lv X, et al. Pre-treatment of sugarcane bagasse with aqueous ammonia–glycerol mixtures to enhance enzymatic saccharification and recovery of ammonia. Bioresour Technol. 2019;289:1-9. DOI: 10.1016/j.biortech.2019.121628

[85] Arumugam A, Anudakshaini TS, Shruthi R, Jeyavishnu K, Sundarra Harini S, Sharad JS. Low-cost production of PHA using cashew apple (*Anacardium occidentale* L.) juice as potential substrate: optimization and characterization. Biomass Convers Biorefinery. 2020;10:1-12. DOI: 10.1007/s13399-019-00502-5

[86] Santosa FH, Laysandra L, Soetaredjo FE, Santoso SP, Ismadji S, Yuliana M. A facile noncatalytic methyl ester production from waste chicken tallow using single step subcritical methanol: Optimization study. Int J Energy Res. 2019;43:1-12. DOI: 10.1002/er.4844

[87] Roy VC, Getachew AT, Cho YJ, Park JS, Chun BS. Recovery and bio-potentialities of astaxanthin-rich oil from shrimp (*Peneanus monodon*) waste and mackerel (*Scomberomous niphonius*) skin using concurrent supercritical CO_2 extraction. J Supercrit Fluids. 2020;159:1-10. DOI: 10.1016/j.supflu.2020.104773

[88] Mandik YI, Cheirsilp B, Srinuanpan S, Maneechote W, Boonsawang P, Prasertsan P, et al. Zero-waste biorefinery of oleaginous microalgae as promising sources of biofuels and biochemicals through direct transesterification and acid hydrolysis. Process Biochem. 2020;95:214-22. DOI: 10.1016/j.procbio.2020.02.011

[89] Lee M, Yang M, Choi S, Shin J, Park C, Cho SK, et al. Sequential production of lignin, fatty acid methyl esters and biogas from spent coffee grounds via an integrated physicochemical and biological process. Energies. 2019;12:1-13. DOI: 10.3390/en12122360

[90] Ali B, Yusup S, Quitain AT, Bokhari A, Kida T, Chuah LF. Heterogeneous catalytic conversion of rapeseed oil to methyl esters: Optimization and kinetic study. In: Hosseini M, editor. Advances in Feedstock Conversion Technologies for Alternative Fuels and Bioproducts: New Technologies, Challenges and Opportunities. 1st ed. London: Elsevier; 2019. p. 221-38. DOI: 10.1016/B978-0-12-817937-6.00012-6

[91] Ubando AT, Felix CB, Chen WH. Biorefineries in circular bioeconomy: A comprehensive review. Bioresource Technology. 2020; 299:122585. DOI: 10.1016/j.biortech.2019.122585

[92] Imbert E. Food waste valorization options: Opportunities from the bioeconomy. Open Agric. 2017;2:195-204. DOI: 10.1515/opag-2017-0020

[93] Dahiya S, Kumar AN, Shanthi Sravan J, Chatterjee S, Sarkar O,

Mohan SV. Food waste biorefinery: Sustainable strategy for circular bioeconomy. Bioresource Technology. 2018;248:2-12. DOI: 10.1016/j.biortech.2017.07.176

[94] Pais-Chanfrau J, Núñez-Pérez J, Lara-Fiallos M, Rivera-Intriago L, Abril V, Cuaran-Guerrero M, et al. Milk Whey- From a Problematic Byproduct to a Source of Valuable Products for Health and Industry: An Overview from Biotechnology. La Prensa Medica Argentina. 2017;103(4):1-11. DOI: 10.4172/lpma.1000254

[95] Bosco F, Carletto RA, Marmo L. An integrated cheese whey valorization process. Chem Eng Trans. 2018;64:379-384. DOI: 10.3303/CET1864064

[96] Lappa IK, Papadaki A, Kachrimanidou V, Terpou A, Koulougliotis D, Eriotou E, et al. Cheese whey processing: Integrated biorefinery concepts and emerging food applications. Foods. 2019;8:1-37. DOI: 10.3390/foods8080347

[97] Martinez-Hernandez E, Amezcua-Allieri MA, Sadhukhan J, Anell JA. Sugarcane Bagasse Valorization Strategies for Bioethanol and Energy Production. In: de Oliveira AB, editor. Sugarcane - Technology and Research. 1st ed. London: IntechOpen; 2018. p. 71-83. DOI: 10.5772/intechopen.72237

[98] Guna V, Ilangovan M, Hu C, Venkatesh K, Reddy N. Valorization of sugarcane bagasse by developing completely biodegradable composites for industrial applications. Ind Crops Prod. 2019;131:25-31. DOI: 10.1016/j.indcrop.2019.01.011

[99] Ozdal M, Kurbanoglu EB. Citric Acid Production by Aspergillus niger from Agro-Industrial By-Products: Molasses and Chicken Feather Peptone. Waste and Biomass Valorization. 2019;10:631-640. DOI: 10.1007/s12649-018-0240-y

[100] Katakojwala R, Naresh Kumar A, Chakraborty D, Mohan SV. Valorization of sugarcane waste: Prospects of a biorefinery. In: Vara Prasad N, de Campos Fava PJ, Vithanage M, Mohan SV, editors. Industrial and Municipal Sludge: Emerging Concerns and Scope for Resource Recovery. 1st ed. London: Elsevier; 2019. p. 47-60. DOI: 10.1016/B978-0-12-815907-1.00003-9

Chapter 10

In Search of Optimal Laser Settings for Lithotripsy by Numerical Response Surfaces of Ablation and Retropulsion

Jian J. Zhang

Abstract

Even though ureteroscopic laser lithotripsy (URSL) has become the preferred treatment option for urolithiasis due to shorter operation time and a better stone-free rate, the optimum laser pulse settings for URSL with the shortest operative times remain unknown. In this chapter, two sets of design of experiments (DOE) were conducted with response surface methodology: 1) the quantitative responses of calculus ablation and retropulsion in terms of the pulse energy, pulse width, and the number of pulses of a prototype Chromium (Cr3+), Thulium (Tm3+), Holmium (Ho3+) triple doped yttrium aluminum garnet (CTH:YAG) laser system. The ablation or retropulsion is inversely proportional to the pulse width, and the pulse width has a higher impact coefficient for the ablation than for the retropulsion. The quadratic fit of the response surface for the volume of ablation has a nonlinear relationship with the pulse width and number of pulses. 2) the laser setting optimization of laser lithotripsy of a commercially available CTH: YAG laser system. The experimental setup is based on a benchtop model first introduced by Sroka's group. Comparing to frequency, the laser pulse energy or peak power has a higher impact coefficient to stone retropulsion as compared to stone ablation in CTH: YAG laser lithotripsy. The most efficient way to curtail stone retropulsion during laser lithotripsy is to lower the laser pulse peak power.

Keywords: stone, ablation, lithotripsy, retropulsion, response surface, design of experiment, pulse energy, peak power

1. Introduction

Urolithiasis, which is hard tissue (stone) formation in the urinary tract due to supersaturated body fluids, has risen steadily in recent decades. The leading causes of stone formation are the reduction of urine volume (or water intake), an increased calcium oxalate/calcium phosphate secretion, urine pH alteration, or urinary tract infections (urease forming bacteria) [1–4]. The prevalence in western countries is estimated at 10%-15%, and the recurrence rate is averaging up to 50% [5–7]. And according to Charles D. Scales [8], the prevalence of kidney stones nearly doubled in about 17 years from ~1995 to 2012. The prevalence of urolithiasis has been rising internationally over recent decades because of population growth, predicted obesity

IntechOpen

trends, and estimated increases in diabetes, just to name a few. The annual treatment cost of stone disease could reach >$5 billion/yr. (in 2014 prices) in the United States by the year 2030 [9, 10].

Shockwave lithotripsy (SWL) and ureteroscopic laser lithotripsy (URSL) are the most commonly performed procedures in the United States for the treatment of patients with urinary calculi [11, 12]. URSL is now the preferred treatment option for urolithiasis due to relatively shorter operative time and a better stone-free rate [1].

The first laser device was invented in 1960 by Maiman [13] based on the theoretical work by Townes and Schawlow. And in 1968, Mulvaney et al. [14] reported the first fragmentation of kidney stones with a pulsed ruby laser (λ: 694 nm) in an in vitro experiment by using quartz rods to deliver the laser light to the treatment site. Since then, a few laser lithotripters were clinically available, including the pulsed-dye laser, the frequency-doubled pulsed Neodymium ($Nd3+$) doped Yttrium aluminum garnet (Nd:YAG) laser (FREDDY), and the Ho: YAG laser [15–17]. The Ho: YAG laser with relatively long-pulse is the most effective and adaptable tool for lithotripsy among all the lasers comparing to nanosecond Nd: YAG lasers. It can disintegrate all kinds of calculus and provoke less calculus retropulsion during procedure than the short-pulsed lasers [18–22]. Since soon after its debut in the 1990s, the Ho: YAG laser has been the preferred lithotripter for the therapy of urinary calculus. It is a solid-state pulsed laser at a wavelength of 2.13 μm. This wavelength is easily absorbed by water (~26 cm^{-1} [23]), providing a wide safety margin for lithotripsy in the urinary tract [24–26]. Aside from treating calculi, it can be used for soft tissue applications such as treating urinary strictures and ablating urothelial tumors. The high-powered variant can also be used for the enucleation of the prostate (HoLEP). Recently, another technology has been explored for the next generation laser lithotripsy: the Thulium fiber laser [27, 28]. This promising technology offers several advantages that may expand the boundaries of laser lithotripsy [28, 29].

The dominant mechanism in Ho: YAG laser lithotripsy is photothermal along with minor effects of acoustic emission [29]. Because the Ho: YAG laser's thermal diffusion time in the water over the optical penetration depth is 286 ms [30], which is well above the laser pulse width (less than a few milliseconds, mostly in the 100 s of μs), in other words, it is photo-thermally confined. And the laser is not photo-mechanical or stress confined in water since the acoustic diffusion time over the optical penetration depth is 0.267 μs, much less than the Ho: YAG laser pulse width. The water has a strong absorption at the Ho: YAG 2.1 μm wavelength and the calculus ablation was dependent on the water content in calculus phantom [31, 32]. The temperature of the illuminated area of the urinary calculi due to the direct laser photon absorption raises above the ablation threshold, subsequently creating the expulsion of fractured crack-up pieces. Furthermore, the absorption of the laser photon by the liquid between the fiber end and calculus produces a vapor bubble, and the crashing of the bubble generates shockwave. The bubble should not be called the "cavitation" bubble as it does in [30, 33] because cavitation is a phenomenon in which rapid changes of pressure in a liquid lead to the formation of small vapor-filled cavities in places where the pressure is relatively low, while during laser lithotripsy, the bubble is generated by heated water vapor with relatively high pressure. This vapor bubble usually has a minimal mechanical effect on hard tissues but rather parts the water (the "Moses effect" [34]) for direct delivery of the remaining part of the laser light onto the stone [35]. The term "Moses effect" technology is also used by breaking one laser pulse into two, where the first pulse generates a bubble between the fiber tip and the stone to let the second pulse through this bubble to the surface of the stone [36]. The shock-wave image can be captured by a high-speed camera with ~1 μs frame interval [37]; it is a disturbance wave that is faster than the sound wave, which can quickly damp down to sound wave speed [38].

During laser-calculus interaction, the urinary calculus is subject to retropulsion forces caused by the combined effects of ablated particle ejection, interstitial water vaporization, and bubble expansion/collapse [39–41]. And an asymmetric collapse of the bubble near a solid boundary can generate a water jet in the time scale of milliseconds [30]. Therefore, because of the recoil momentum, the calculus is moved away from the end of the laser fiber. The calculus motion prolongs the procedure time because of the burdensome procedure needed to reorient the laser fiber to the new calculus locality. Earlier retropulsion studies quantified calculus retropulsion distance by altering laser pulse energy, pulse frequency, and fiber core size [42–44]. Retropulsion boosted with the laser pulse energy and the laser fiber core size. Moreover, Charles D. Scales et al. reported that a longer pulse width reduced calculus retropulsion distance during a procedure without diminishing ablation efficiency significantly [45].

Although laser lithotripsy is now the preferred treatment option for urolithiasis because it is capable of fragmenting calculus of all known composition, including hard calcium oxalate monohydrate, brushite, and cystine calculus [22, 24, 25, 29], and the rising prevalence of calculus disease has led to similarly increasing efforts to optimize ureteroscopic treatment [43, 45–52], the operative time for the stone procedure can be well above the one hour mark. According to Levi A. Deters et al. [53], URSL management of renal stones and ureteral stones were markedly different, with a significant increase in operative time (60% more) for renal stones and a significant lower stone-free rate (27% lower). And of the 213 cases, the average operative time for the renal group (98 cases) is 112 min and range up to 245 min, and the average operative time for the ureteral group (115 cases) is 70 min and range up to 185 min.

The response surface methodology (RSM) is a powerful statistical tool that can generate the numerical relationship between some key performance variables (responses) and device control parameters (control inputs). Although the model is only an approximation most of the time because of limited knowledge of the process, the RSM plus design of experiments (DOE) can produce analytical models (equations) that can depict 1) the relative impact of the control inputs on the responses by comparing their coefficients in the coded equations; 2) optimization of the responses with proper control inputs.

In this chapter, two sets of DOE experiments were conducted with response surface methodology: 1) the quantitative responses of calculus ablation and retropulsion in terms of the pulse energy, pulse width, and the number of pulses of a prototype CTH: YAG laser system. This step is to understand the dominant laser parameters that control the lithotripsy outcome, so that preferred laser settings can be derived for the next generation of laser lithotripter; 2) the laser setting optimization of laser lithotripsy of a commercially available CTH: YAG laser system. This experiment is to identify a series of laser settings for relatively efficient laser lithotripsy in terms of laser pulse energy and peak power.

2. Experimental method and setup

2.1 The quantitative responses of calculus ablation and retropulsion in terms of the pulse energy, pulse width, and the number of pulses of a prototype CTH: YAG laser system

In this study, the key components of the setup of the experimental materials are listed in **Table 1**.

A prototype CTH:YAG laser had pulse energy from 0.2 J to 3.0 J with variable pulse width from 150 μs to 1000 μs at 2.13 μm. This range of pulse duration is

Item	Description
Laser	Prototype laser
Fiber	365-μm Core (S-LLF365 SureFlex Fiber, American Medical Systems, San Jose, CA, USA)
Phantom	White gypsum cement UtralCal®30 (United States Gypsum Company, Chicago, IL) [54]
Camera	Photron Fastcam SA5 (Photron USA, Inc. San Diego, CA)
Digital microscope	VHX-900F (Keyence, Elmwood Park, NJ, USA)
Program	Design-Expert® (Stat-Ease, Inc., Minneapolis, MN 55413)

Table 1.
The list of components of the experimental setup.

known to generate a photothermal effect to fragment the calculus [51]. Each data point is the average of 10 sample measurements.

Figure 1 is the pictures of the test setup, (a) ablation test setup, and (b) retropulsion test setup. In the ablation test setup (a) submerged in the distilled water, the fiber was held vertically by a clamp with its tip in contact with the calculus phantom underneath the fiber inside a holder. The stone was held fixed during the ablation study.

The laser ablation crater volume in the phantom due to the laser pulse and calculus interaction was measured by a digital microscope (VHX-900F, Keyence, Elmwood Park, NJ, USA). In the retropulsion test setup (b), the fiber was held horizontally, pointing to an underwater pendulum phantom cube with a dimension of 10 × 10 × 10 mm3. The pendulum length was ~200 mm, and the phantom was held by 2 strings with a separation of ~10 mm in a clear plastic basket. The retropulsion motion of the calculus phantom was recorded and analyzed by a high-speed camera.

Figure 2 is a screenshot of DOE by Design-Expert®10. There are three categories of the laser parameter settings: energy, number of pulses, and electrical pump pulse widths (not the optical output pulse width). The ten-pulses range was selected since the typical retropulsion of a 10 × 10 × 10 mm3 will reach its maximum amplitude after ~1 s of 10 Hz 1 J pulse train from the fiber tip [52]. There are 5 × 3 × 3 = 45 data points with the combination of all the laser parameters.

2.2 The laser setting optimization of laser lithotripsy of a commercially available CTH:YAG laser system

It is challenging to characterizing the URSL performance (ablation and retropulsion) in one setup that can mimic the clinical situation, especially measuring retropulsion [55–61]. In this study, in vitro investigations of Ho:YAG laser-induced stone ablation and retropulsion were performed with a benchtop model first introduced by Sroka's group [55, 60]. It is a test that can be performed in a highly reproducible manner using a hands-free setup and measuring the effects of multiple pulses which are mimicking the clinical situation. The advantage of this setup has two folds: 1) No human factor, hands-free, independent repetitive experiments; 2) Providing measurement results for both ablation rate and retropulsion speed. Although the stone moves during the test, which means the distance between the fiber tip and the stone is not a constant, which will report a lower ablation rate, it is still an efficient way to generate meaningful data in terms of ablation and retropulsion for comparing different laser modes. **Table 2** is a description of the list of the key components of the setup.

Figure 1.
The pictures of the test setup, (a) ablation test setup with 10 mm phantom; (b) Retropulsion test setup with 10 mm phantom.

Select	Std	Run	Factor 1 A:Number of...	Factor 2 B:Energy J	Factor 3 C:Pulse width us
	13	1	3.00	3.00	1000.00
	14	2	10.00	3.00	1000.00
	12	3	1.00	1.00	1000.00
	1	4	1.00	0.20	333.00
	4	5	3.00	2.00	333.00
	11	6	10.00	0.20	1000.00
	6	7	10.00	3.00	333.00
	3	8	3.00	2.00	333.00
	10	9	1.00	3.00	667.00
	5	10	1.00	3.00	333.00
	9	11	10.00	2.00	667.00
	8	12	3.00	0.20	667.00
	2	13	10.00	0.20	333.00
	7	14	3.00	0.20	667.00

Figure 2.
A screenshot of DOE by DesignExpert-10.

The setup, an acrylic cylinder with a drill hole mimicking the ureter ending in a conical base, is illustrated in **Figure 3**. The diameter of the drill hole can be adapted to the clinical situation (e.g., ureter diameter) or stone size. The stone phantom is a 5 mm cubic shape Bego stone with a composition of 15:3 [62]. The setup is in an upright position filled with the saline at a designated flow speed. The optical fiber is attached through a borehole at the base of the acrylic cylinder. Therefore, laser energy can be delivered to the stone phantom to produce vertical displacement. The gravity and the viscosity of the water are the steady resistances to this motion.

The ablation is quantified by the stone phantom mass deficit after the laser stone interaction by a scale with a resolution of +/− 0.0001 g (Entris 224-1S Sartorius Lab Instruments GmbH & Co. KG, Goettingen, Germany).

The retropulsion is quantified by the vertical displacement velocity of the stone during the laser stone interaction. A high-speed camera, Sony RX100 IV (1000 fps), oriented perpendicular to the upright motion and aimed to the middle of the artificial ureter, registers the event for ~7 seconds. The video subsequently is analyzed in MATLAB, and a representative stone sample vertical displacement graph vs. video frame is illustrated in **Figure 3(B)**. Initial data assessment incorporates background rectification and color tracking algorithm, recognition of the

Item	Description
Laser	PowerSuite™ Ho:YAG 100 W (VersaPulse® 100 W, Lumenis Ltd., Yokneam, Israel)
Fiber	365-μm Core (S-LLF365 SureFlex Fiber, American Medical Systems, San Jose, CA, USA)
Phantom	BEGO Stone 15:3 (BEGO GmbH & Co. KG, Bremen, Germany)
Camera	Sony RX100 IV (Sony Corporation of America, NY, USA)
Balance	Sartorius Entris 224-1S (Sartorius Lab Instruments GmbH & Co. KG, Goettingen, Germany)
Program	Design-Expert® (Stat-Ease, Inc., Minneapolis, MN 55413)

Table 2.
The list of components of the experimental setup.

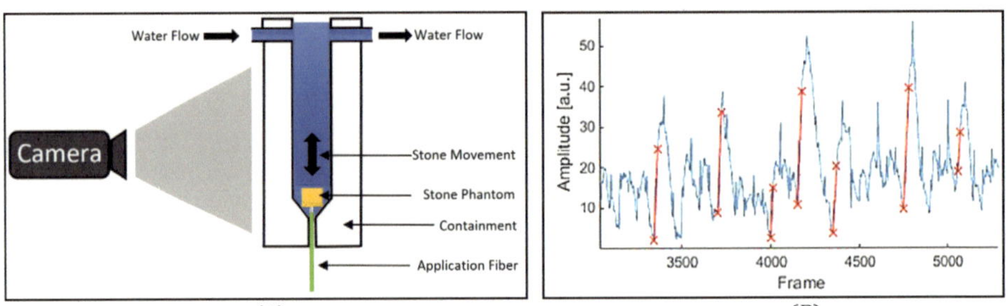

Figure 3.
(A) Test setup including the stone phantom containment vessel, the water flow inlet, and outlet tubes, application fiber, and a high-speed camera. (B) Stone phantom vertical movement vs. video frame.

Run	Factor 1 A:Frequency Hz	Factor 2 B:Energy J
1	40	0.4
2	20	1.5
3	30	1.2
4	5	1.2
5	30	0.8
6	5	0.4
7	40	1.5
8	30	0.8
9	30	0.8
10	20	0.4

(A)

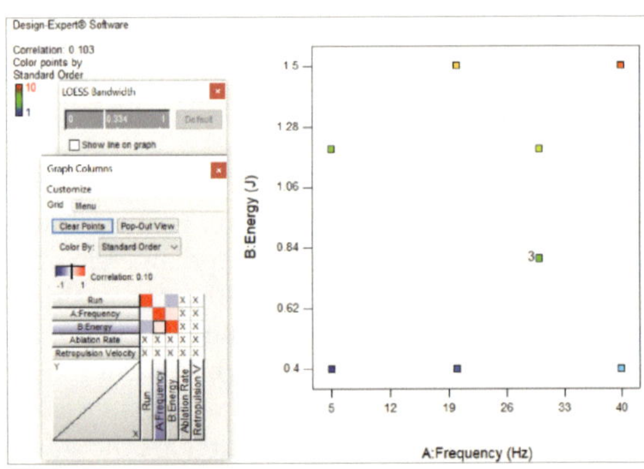

(B)

Figure 4.
A few screen shots of the DOE. (A) Design layout; (B) design graphic columns.

center of weight of the traced stone image in each frame and tracing the center of weight position as a function of time. Afterwards, each rising wing can be utilized to obtain the ascending vertical speed. The 1st Derivative of rising flanks displays the mean velocity of the stone $v = \partial x/\partial t$ (x-displacement, t-time), and the velocity of rising flanks is proportional to the applied momentum $p = m * \mathrm{v}$ (m-stone mass, v-stone velocity).

The experiment is designed by Design-Expert® with randomized optimal (custom), two replicate points, and two lack-of-fit points. **Figure 4** is a few screenshots of the DOE of laser pulse energy: 0.2, 0.4, 0.6, 0.8, 1.2, 1.5 J and frequency: 5, 10, 15, 20, 30, 40 Hz. A sample size of 14 is used for each data point, and each sample was applied with 15 seconds long laser dose.

3. Results

3.1 The quantitative responses of calculus ablation and retropulsion in terms of the pulse energy, pulse width, and the number of pulses of a prototype CTH:YAG laser system

3.1.1 Retropulsion amplitude data

The retropulsion videos taken by a high-speed camera at 10 k FPS were analyzed by a MATLAB program for the pendulum swing amplitude. **Figure 5(a)** is some sample curves of the retropulsion movement; each data point is the average of 10 measurements.

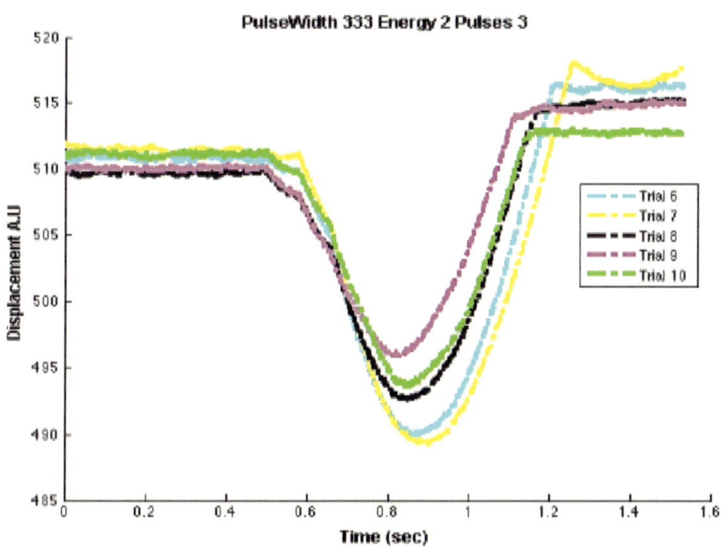

Figure 5.
Retropulsion amplitude measurement results.

4. Retropulsion amplitude response surface

Figure 6 is the screenshots of the response surface of retropulsion amplitude at the pulse energy level of (a) 1 J; (b) 2 J; (c) 3 J. The analytical formula of the response surface of retropulsion is shown in Formula (1). The ANOVA shows

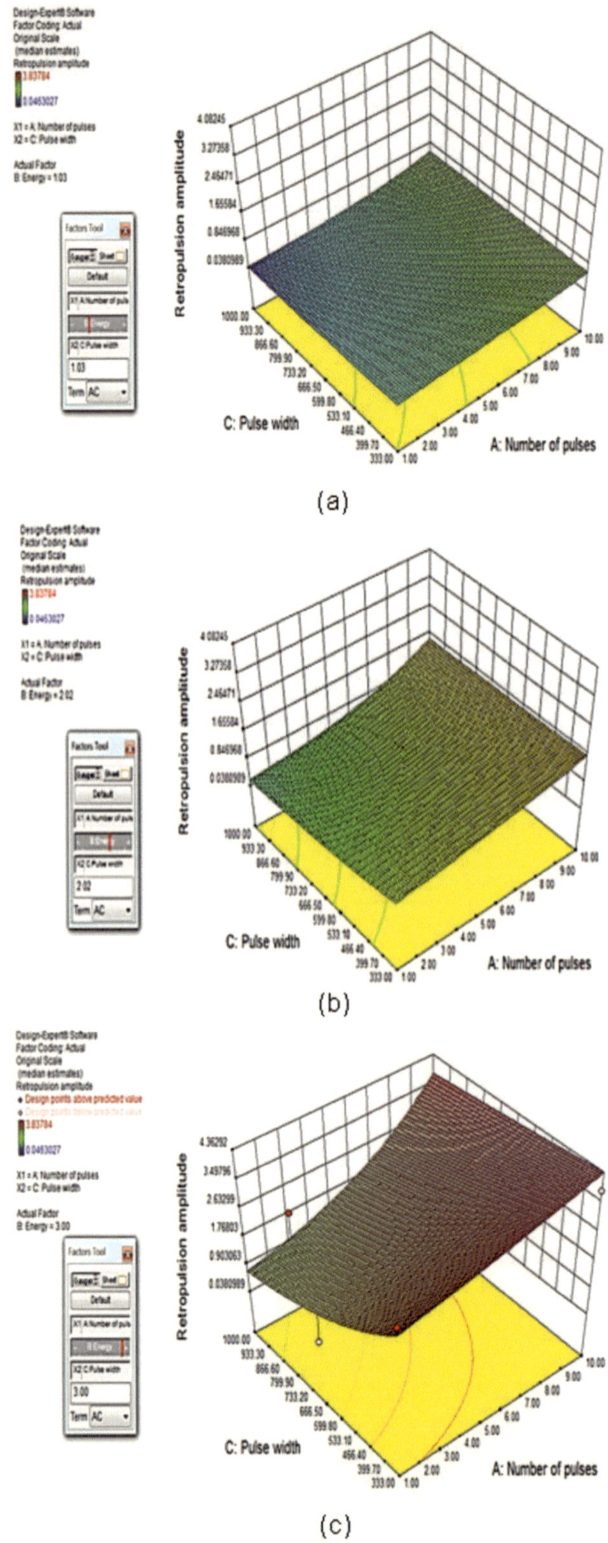

Figure 6.
The screen shots of the response surface of retropulsion amplitude against pulse width and number of pulses at pulse energy level of (a) 1 J; (b) 2 J; (c) 3 J.

an insignificant lack of fit, acceptable agreement of the Predicted and Adjusted R-Squares, and acceptable precision (8.324, > 4.0).

$$\mathrm{Ln}(A) = 0.56 + 0.08n + 1.42\varepsilon - 0.0021\tau - 0.039n\varepsilon + 0.00022n\tau - 0.00011\varepsilon\tau \quad (1)$$

Where A is the amplitude of retropulsion in mm, n is the pulse number, ε is the pulse energy in J, and τ is the pulse width in μs.

5. Volume of ablation data

The volume of the hole by laser ablation was quantified by a digital microscope. A representative picture is in **Figure 7**.

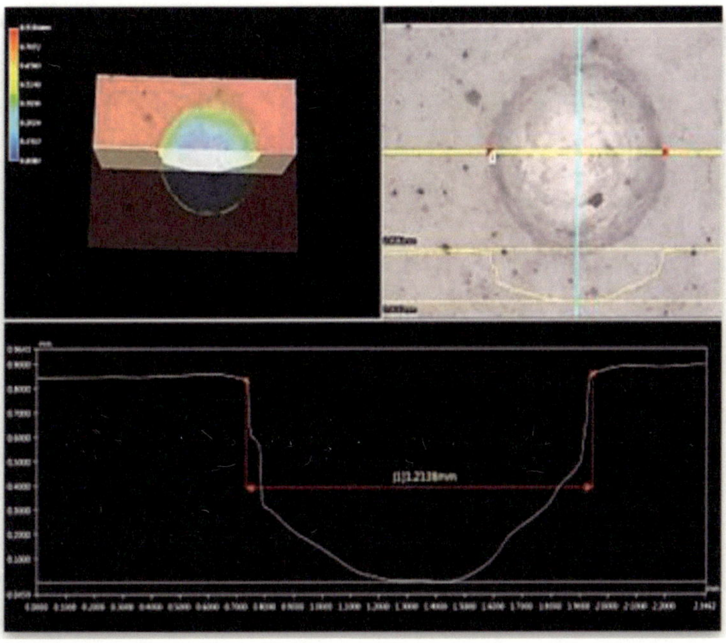

Figure 7.
Volume of ablation response measurement results. Screenshot of VHX-900F digital microscope;

6. Volume of ablation response surface

According to the response surface information from the above section, the Design-Expert® -10 app can produce a response surface and the analytical equation. **Figure 8** is the pictures of the response surface of volume of ablation versus the pulse width and the number of pulses at the pulse energy status of (a) 1 J; (b) 2 J; (c) 3 J. The analytical formula of the response surface of the volume of ablation, including the polynomial terms of two factors interactions, is shown in Formula (2). The ANOVA shows an insignificant lack of fit, acceptable agreement of the Predicted (0.8636) and Adjusted (0.9177) R-Squares, and acceptable precision (15.46, > 4.0).

$$\text{Ln}(V) = -0.27 + 0.023n + 1.11\varepsilon - 0.0083\tau + 0.011n\varepsilon + 0.00047n\tau + 0.0012\varepsilon\tau \quad (2)$$

Where V is the ablation volume in mm^3, n is the pulse number, ε is the pulse energy in J, and τ is the pulse width in μs.

Figure 9 is the pictures of the response surface of ablation volume under the quadratic fit with pulse width and the number of pulses at the pulse energy status of (a) 1 J; (b) 2 J; (c) 3 J. The analytical equation of the response surface of ablation volume, involving the polynomial terms of two factors interactions, is illustrated in Eq. (3). The ANOVA shows an insignificant lack of fit, acceptable agreement of the Predicted (0.9900) and Adjusted (0.9999) R-Squares, and acceptable precision (466.6, > 4.0).

$$\text{Ln}(V) = -1.16 + 0.94n + 3.46\varepsilon - 0.021\tau + 0.0031n\varepsilon + 0.00048n\tau + 0.0014\varepsilon\tau - 0.078n^2 - 0.77\varepsilon^2 + 0.0000093\tau^2 \quad (3)$$

Where V is the ablation volume in mm^3, n is the pulse number, ε is the pulse energy in J, and τ is the pulse width in μs.

Figure 10 is the ablation and retropulsion in percentages by 10 pulses of the 1000 μs pulses versus those of 333 μs. The variation of the volume of between long and short pulse is comparatively larger at 1 J and 2 J level contrasting to retropulsion. Namely, ablation declines more swiftly than retropulsion when expanding pulse width.

6.1 The laser setting optimization of laser lithotripsy of a commercially available CTH:YAG laser system

6.1.1 Response surface in terms of laser pulse energy

Figure 11(A) and **(B)** are the screenshots of the response surface of ablation rate and the retropulsion velocity; each data point is the average of 14 measurements. The ANOVA shows an insignificant lack of fit (except P = 0.04 for the ablation rate), acceptable agreement of the predicted and adjusted R-squares, and acceptable precision (> 4.0).

The coded equation can predict the response for each of the control parameters. By convention, the high elevations of the parameters are coded as +1, and the low elevations of the parameters are coded as −1. The coded equation is valuable at attaining the comparable impact parameters by the coefficients of each term.

Coded analytical equation of ablation and retropulsion by laser pulse energy is:

$$A_{P-100}{}^{0.5} = 0.50 + 0.34A + 0.49B + 0.18AB \quad (4)$$

$$R_{P-100} = 92.69 + 4.94A + 28.47B \quad (5)$$

$A_{P\text{-}100}$ – Ablation rate, $R_{P\text{-}100}$ -Retropulsion velocity, A-Frequency, B-Energy

From the coded equation, we can see the impact of the laser pulse energy is 1.4 times that of the frequency on the ablation rate, while for retropulsion velocity, the impact of the laser pulse energy is 5.8 times that of the frequency. This indicates the laser pulse energy setting has a vital impact on both ablation rate and retropulsion velocity.

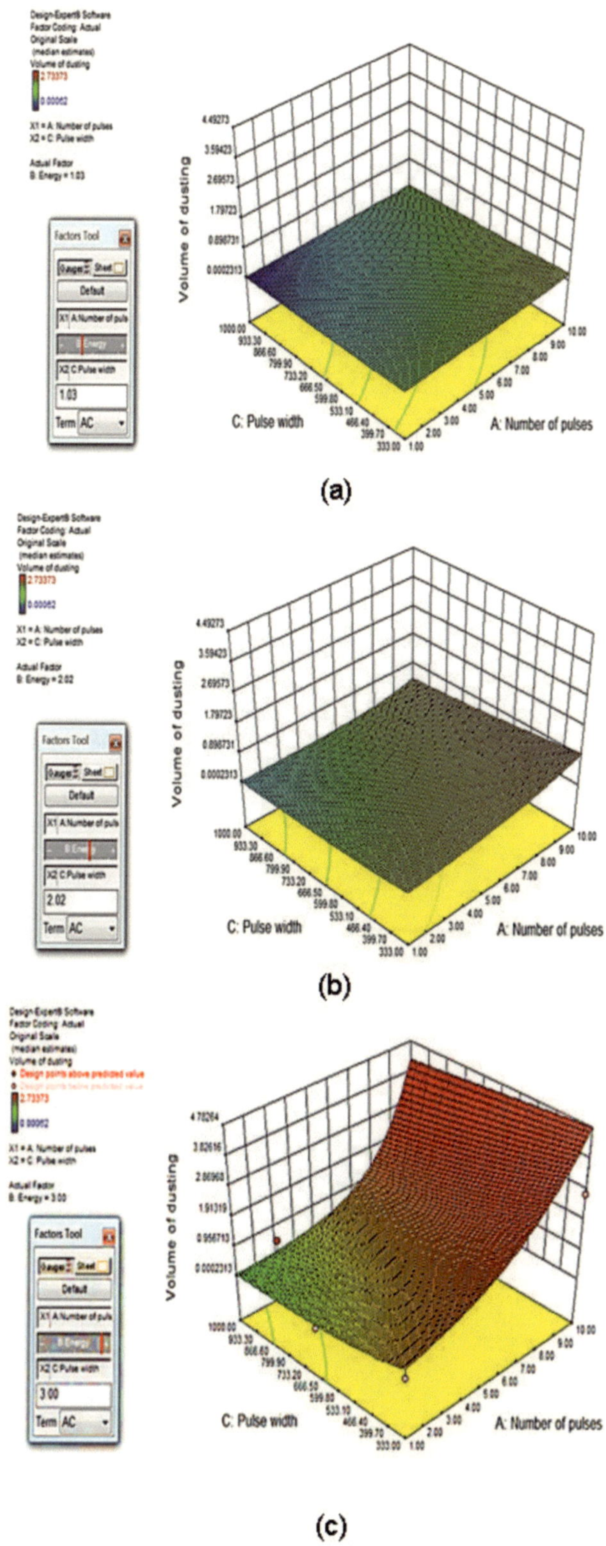

Figure 8.
The screenshots of the response surface of volume of ablation with two factors interactions fit against pulse width and the number of pulses at the pulse energy level of (a) 1 J; (b) 2 J; (c) 3 J.

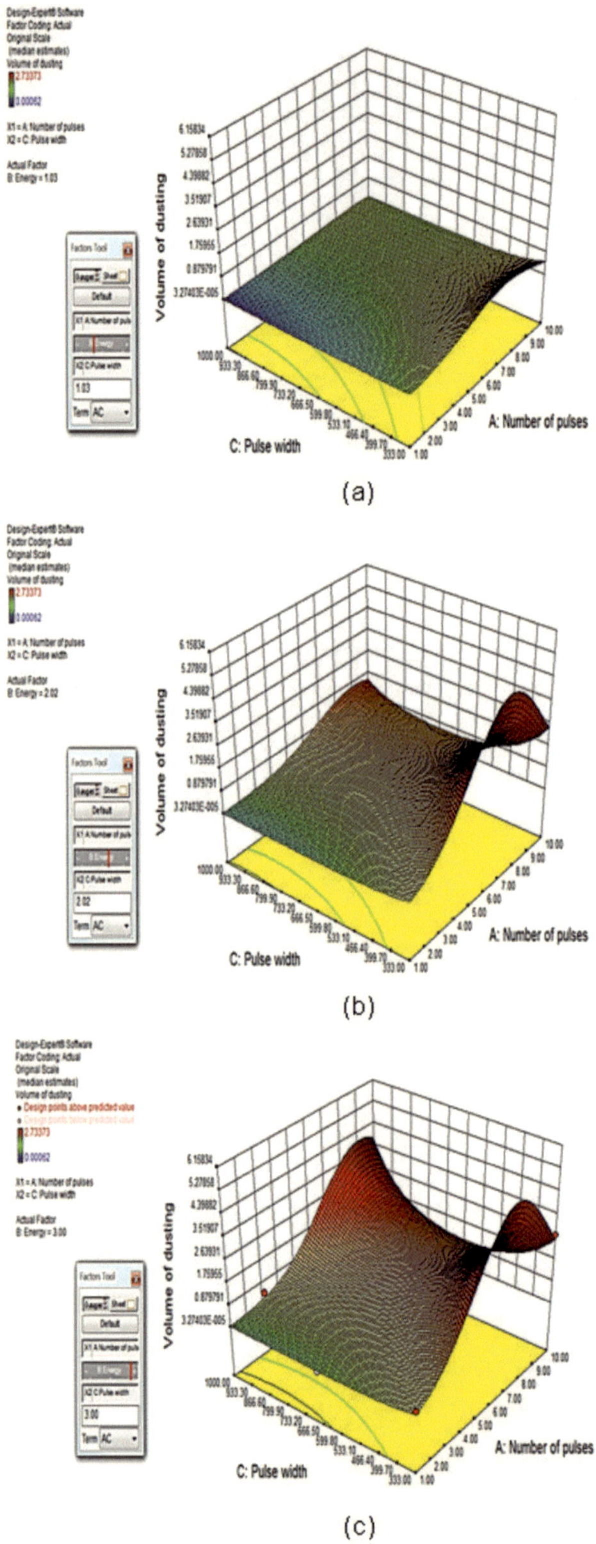

Figure 9.
The screenshots of the response surface of volume of ablation with quadratic fit against pulse width and the number of pulses at the pulse energy level of (a) 1 J; (b) 2 J; (c) 3 J.

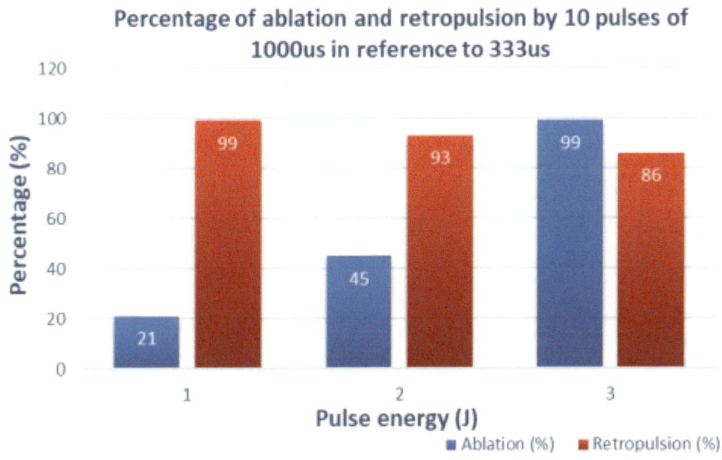

Figure 10.
The percentages of ablation and retropulsion by 10 pulses of the 1000 μs pulses in reference to those of 333 μs.

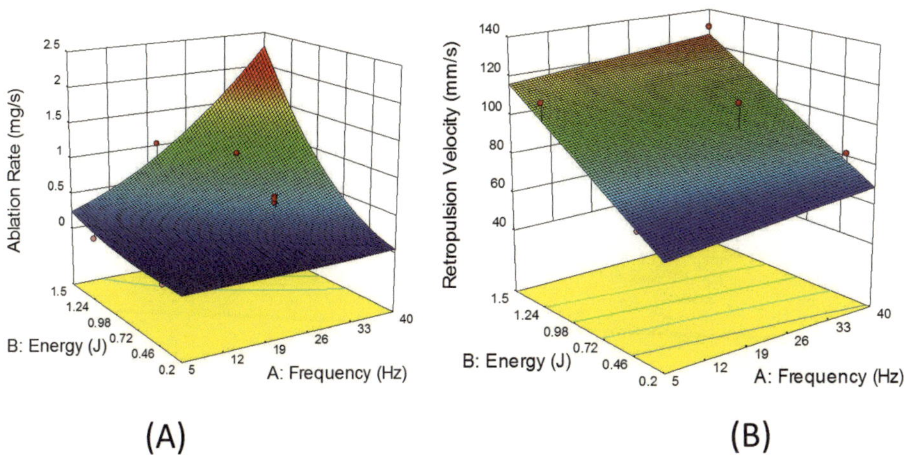

Figure 11.
(A) Response surface of ablation rate over laser pulse energy and frequency; (B) response surface of retropulsion velocity over laser pulse energy and frequency.

The actual analytical equation of ablation and retropulsion by laser pulse energy is:

$$A_{P-100}{}^{0.5} = -0.28 + 0.00064A + 0.41B + 0.015AB \quad (6)$$

$$R_{P-100} = 49.1 + 0.28A + 43.8B \quad (7)$$

Figure 12(A) and **(B)** are the optimized laser settings listed with equal weight ratio between ablation rate and retropulsion velocity. The 1st optimized laser setting is 1.2 J 40 Hz. IF we choose a 1:3 weight ratio between ablation rate and retropulsion velocity, the 1st optimized laser setting becomes 0.6 J 40 Hz. This means the "optimized" laser setting depends on the criterion used.

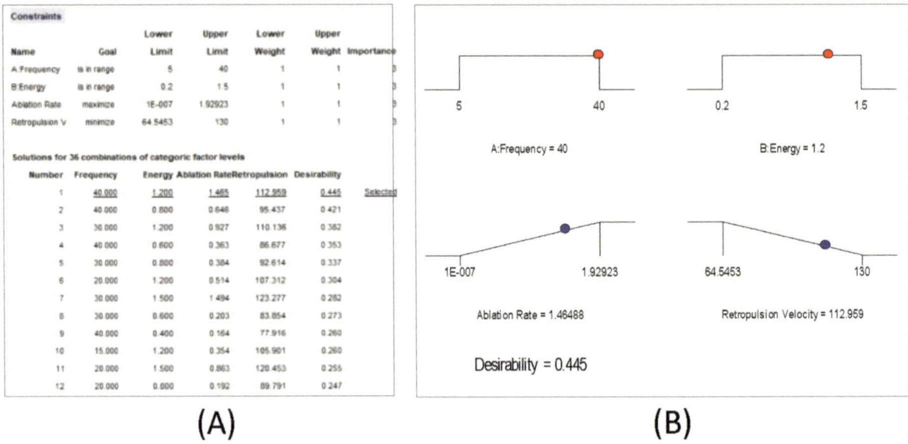

Constraints

Name	Goal	Lower Limit	Upper Limit	Lower Weight	Upper Weight	Importance
A:Frequency	is in range	5	40	1	1	3
B:Energy	is in range	0.2	1.5	1	1	3
Ablation Rate	maximize	1E-007	1.92923	1	1	3
Retropulsion V	minimize	64.5453	130	1	1	3

Solutions for 36 combinations of categoric factor levels

Number	Frequency	Energy	Ablation Rate	Retropulsion	Desirability	
1	40.000	1.200	1.465	112.959	0.445	Selected
2	40.000	0.800	0.646	95.437	0.421	
3	30.000	1.200	0.927	110.136	0.382	
4	40.000	0.600	0.363	86.677	0.353	
5	30.000	0.800	0.384	92.614	0.337	
6	20.000	1.200	0.514	107.312	0.304	
7	30.000	1.500	1.494	123.277	0.282	
8	30.000	0.600	0.203	83.854	0.273	
9	40.000	0.400	0.164	77.916	0.260	
10	15.000	1.200	0.354	105.901	0.260	
11	20.000	1.500	0.863	120.453	0.255	
12	20.000	0.800	0.192	89.791	0.247	

Figure 12.
(A) List of optimized laser parameters with even weight (50/50) of ablation and retropulsion speed; (B) ramp view of the chosen laser parameters with even weight (50/50) of ablation and retropulsion speed.

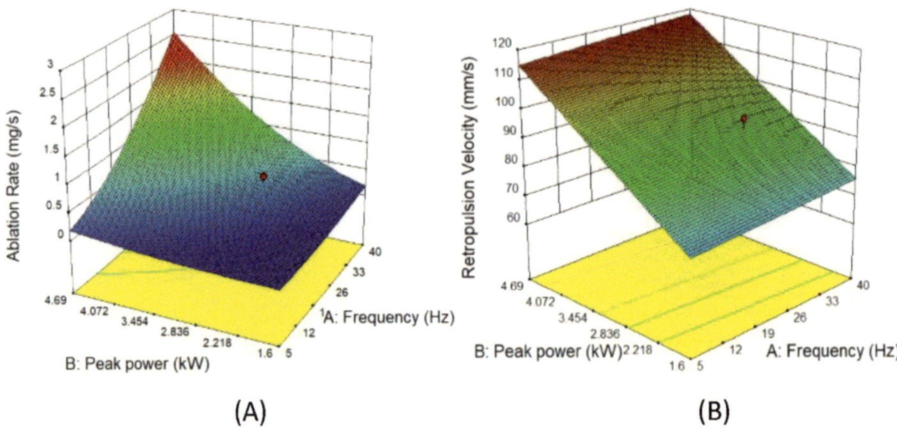

Figure 13.
(A) Response surface of ablation rate over laser pulse peak power and frequency; (B) response surface of retropulsion velocity over laser pulse peak power and frequency.

6.1.2 Response surface in terms of laser pulse peak power

The laser pulse peak power is another way of evaluating the laser damage to the stone [63]. The peak power value is defined by the laser pulse energy over the full pulse width (full width of pulse at ~10% of max amplitude). The ANOVA shows an insignificant lack of fit, acceptable agreement of the Predicted and Adjusted R-Squares, and acceptable precision (> 4.0).

Figure 13 (A) and **(B)** are the screenshots of the response surface of ablation rate and the retropulsion velocity in terms of laser pulse peak power.

Coded analytical equation of ablation and retropulsion by laser pulse Peak power is:

$$A_{P-100}{}^{0.5} = 0.60 + 0.41A + 0.40B + 0.19AB \tag{8}$$

$$R_{P-100} = 95.5 + 1.62A + 21.06B - 0.87AB \tag{9}$$

Constraints

Name	Goal	Lower Limit	Upper Limit	Lower Weight	Upper Weight	Importance
A:Frequency	is in range	5	40	1	1	3
B:Peak power	is in range	1.6	4.09	1	1	3
C:Pulse width	is in range	250	320	1	1	3
Ablation Rate	maximize	1E-005	2.2	1	1	3
Retropulsion V	minimize	62	118	1	1	3

Solutions

Number	Frequency	Peak power	Pulse width	Ablation Rate	Retropulsion	Desirability	
1	40.000	3.320	319.999	1.163	98.629	0.428	Selected
2	40.000	3.300	320.000	1.147	98.383	0.428	
3	40.000	3.337	319.977	1.177	98.863	0.428	
4	40.000	3.316	319.566	1.160	98.588	0.427	
5	40.000	3.290	319.587	1.139	98.234	0.427	
6	40.000	3.324	319.153	1.167	98.714	0.427	
7	39.990	3.272	320.000	1.124	97.979	0.427	
8	40.000	3.386	319.998	1.218	99.532	0.427	
9	40.000	3.245	319.999	1.102	97.607	0.427	
10	40.000	3.284	318.532	1.134	98.182	0.427	
11	39.925	3.350	319.998	1.184	99.034	0.427	
12	40.000	3.370	317.712	1.204	99.389	0.427	

(A)

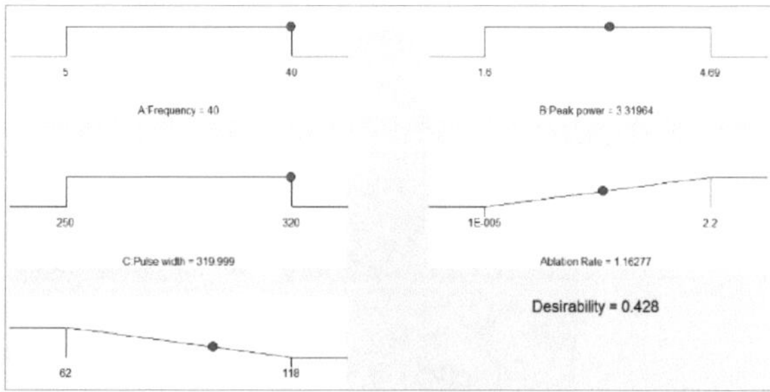

(B)

Figure 14.
(A) List of optimal laser settings with equal weight (50/50) of ablation rate and retropulsion velocity; (B) ramp view of the selected laser settings with equal weight (50/50) of ablation rate and retropulsion velocity.

$A_{P\text{-}100}$ is the Ablation rate, $R_{P\text{-}100}$ is the Retropulsion speed, A is the Frequency, B is the Peak power, and C is the Pulse width.

From the Eqs. 8 and 9, we can see the peak power has roughly the same influence as the frequency on the ablation rate; and for retropulsion speed, the peak power's influence is 13 folds that of the frequency. Namely, the peak power parameter is crucial to retropulsion speed.

The actual analytical equation of ablation and retropulsion by laser pulse peak power is:

$$A_{P-100}{}^{0.5} = -0.26 + 0.0016A + 0.11B + 0.0069AB \tag{10}$$

$$R_{P-100} = 57.6 + 0.093A + 13.6B - 0.025AB \tag{11}$$

Figure 14(A) and **(B)** are the optimized laser settings listed with an equal weight ratio of the ablation rate and retropulsion velocity. The 1st optimized laser setting is 3.3 kW 40 Hz 320 μs. IF we choose a 1:3 weight ratio of the ablation rate and retropulsion velocity, the 1st optimized laser setting becomes 2.2 kW 40 Hz 320 μs. This means the "optimized" laser setting depends on the criterion used.

7. Discussion

7.1 The quantitative responses of calculus ablation and retropulsion in terms of the pulse energy, pulse width, and the number of pulses of a prototype CTH:YAG laser system

In the coded formulas of the response surface (1) and (2), the pulse energy is the dominant control input factor for both the responses of retropulsion and ablation (1.42 and 1.11); while the control input pulse width has more than an order of magnitude less influence on the responses of ablation and retropulsion (−0.0083 versus −0.0021). And the two-factor terms have even lesser influence (a few times to an order of magnitude) than the first-order terms. The pulse number term seems to have some nonlinear effects between long and short pulses at pulse numbers ~7-8 from **Figure 9**. This effect could be due to the vapor bubble behavior of [47]. As it is shown in **Figure 15(b)** since the vapor bubble of the long laser pulse will have a much-elongated shape bubble which

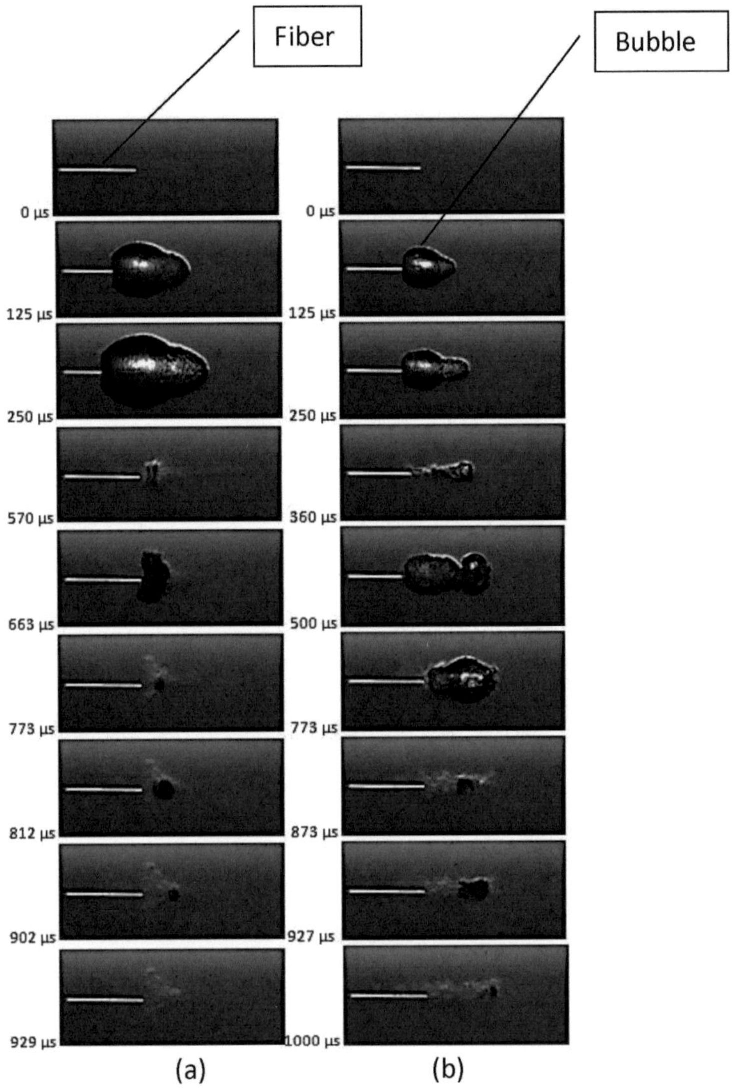

Figure 15.
Series of screenshots of vapor bubbles behavior of Ho lasers. (a) Ho at 1 J, 150 µs; (b) Ho at 1 J, 800 µs.

can be divided into two small bubbles which will collapse sequentially with the 2nd bubble collapses further away from the fiber tip as compared to the bubble of a short laser pulse. Therefore, the long laser pulse can make a deeper crater. This effect will be enhanced at higher pulse energy, and furthermore, since both fiber and calculus were fixed, the depth of the hole max out after ~7-8 pulses.

7.2 The laser setting optimization of laser lithotripsy of a commercially available CTH:YAG laser system

For the response of ablation rate, the coded formulas of the response surface reveal that the control input laser pulse energy is 1.4 times as impactful as that of the frequency, and the laser pulse peak power has the same impact as the frequency; while for the response of retropulsion, the control input laser pulse energy is 5.8 times as impactful as that of the frequency, and laser pulse peak power has 13 times as impactful as the frequency. The laser pulse peak power is the dominant control input factor for the response of stone retropulsion during laser lithotripsy.

As for the optimal lithotripsy laser dosimetry (setting), there are conflict interests to deliver more energy per time (more power or fluence) and achieve faster lithotripsy but at the cost of more retropulsion and larger fragments. As concluded by Sea J et al. in Ref [47], the optimal lithotripsy laser setting depends on the individual case condition (calculus type, size, location, etc.) and the desired outcome. The response surfaces are generated by analysis of variance (ANOVA) of the tested data points, and a ranked list of the optimized laser settings can be generated by the criteria the user selected. If the least retropulsion is the desired, the most effectual method to curtail stone retropulsion during laser lithotripsy is to decrease the peak power (which has the maximum influential coefficient in the coded response surface equations). Dongyul C et al. investigated the ablation thresholds of stone sample by peak power density [63], which presents a recommendation of the lowest peak power for Bego calculus phantom ablation.

8. Conclusions

In this chapter, the application of RSM were conducted by two sets of DOE experiments: 1) with a prototype CTH:YAG laser system, the RSM reveals that the dominant control input laser parameters that influence the responses of lithotripsy outcome: the ablation or retropulsion is Inversely proportional to the pulse width, and the pulse width has a higher impact coefficient to the ablation than that to the retropulsion. The quadratic fit of the response surface for the volume of ablation has a nonlinear relationship with the pulse width and number of the pulse. 2) the laser setting optimization of laser lithotripsy of a commercially available CTH:YAG laser system: a series of laser settings for relatively efficient laser lithotripsy (maximize the ablation rate while minimizing the retropulsion as well as to improve the discharge of fragments via the urinary tract) in terms of control input laser pulse energy and peak power. Comparing to the control input frequency, the laser pulse energy or peak power has a higher impact coefficient to the response of stone retropulsion as compared to the response of stone ablation in Ho:YAG laser lithotripsy. The most efficient way to curtail stone retropulsion during laser lithotripsy is to lower the laser pulse peak power.

More detailed investigation of the optimal conditions for the ablation of other kinds of calculus samples and the fiber size/burn back effects will be conducted as a future study.

Acknowledgements

The author thanks the colleagues of Boston Scientific Corporation: Jonathan Rutherford, Metasebya Solomon, and Dongyul Chai for experimental data collection; Sean Curran and Nicholas Nimchuk for their assistance with the Design-Expert® (DX10) software license, data analysis, fitting, and optimization; and Jason R Xuan, Thomas Hasenberg and Timothy Harrah for technical discussions on the results.

Disclaimer

The opinions expressed in this book chapter is solely those of the author and not necessarily those of Boston Scientific Corporation (BSC). BSC does not guarantee the accuracy or reliability of the information provided herein.

Author details

Jian J. Zhang
Boston Scientific Corporation, Marlborough, MA, USA

*Address all correspondence to: james.zhang@bsci.com

References

[1] Yang C, Li S, Cui Y; "Comparison of YAG Laser Lithotripsy and Extracorporeal Shock Wave Lithotripsy in Treatment of Ureteral Calculi: A Meta-Analysis." Urol Int; 98:373-381 (2017).

[2] Bader MJ, Pongratz T, Khoder W, Stief CG, Herrmann T, Nagele U, Sroka R. Impact of pulse duration on Ho:YAG laser lithotripsy: fragmentation and dusting performance. World J Urol. 2015; 33: 471-477.

[3] Graham A, Luber S, Wolfson AB. Urolithiasis in the Emergency Department. Emerg. Med. Clin. North Am. 2011;29(3):519-538.

[4] Türk C, Neisius A, Petřík A, Seitz C, Thomas K, Skolarikos A. EAU guidelines on urolithiasis 2018. in European Association of Urology Guidelines. Presented at the EAU Annual Congress, Copenhagen., 2018th ed., The European Association of Urology Guidelines Office, Arnhem, The Netherlands 2018.

[5] Matlaga BR, Jansen JP, Meckley LM, Byrne TW, Lingeman JE. Economic outcomes of treatment for ureteral and renal stones: a systematic literature review. J. Urol. 2012;188(8):449-454.

[6] Rizvi SAH, Naqvi SAA, Hussain Z, Hashmi A, Hussain M, Zafar MN, Mehdi H, Khalid R. The management of stone disease. BJU Int. 2002;89(Suppl. 1):62-68.

[7] Tiselius HG. Epidemiology and medical management of stone disease. BJU Int. 2003; 91: 758-767.

[8] Scales CD. Jr. Practice Patterns in the Management of Urinary Lithiasis. Curr Urol Rep 2013; 14: 154-157.

[9] Turney BW, Reynard JM. The Cost of Stone Surgery. EUROPEAN UROLOGY 2014; 66: 730-731.

[10] Antonelli JA, Maalouf NM, Pearle MS, Lotan Y. Use of the National Health and Nutrition Examination Survey to calculate the impact of obesity and diabetes on cost and prevalence of urolithiasis in 2030. Eur Urol 2014; 66:724-729.

[11] Pearle MS, Calhoun EA, Curhan GC, Urologic Diseases of America Project. Urologic diseases in America project: urolithiasis. The Journal of Urology. 2005; 173(3): 848-857.

[12] Seklehner S, Laudano MA, del Pizzo J, Chughtai B, Lee RK. Renal calculi: trends in the utilization of shock-wave lithotripsy and ureteroscopy. The Canadian Journal of Urology. 2015; 22(1): 7627-7634.

[13] Maiman TH (1967) Ruby laser systems. US Patent 3,353,115.

[14] Mulvaney WP, Beck CW. The Laser Beam in Urology. J. Urol. 1968; 99: 112-115.

[15] Watson G, Smith N. Comparison of the pulsed dye and holmium lasers for stone fragmentation: in-vitro studies and clinical experience. Proceedings of SPIE. 1993; 1879: 139-142.

[16] Tischer C, Koort HJ, Bazo A, Rasch R, Thiede C. Clinical experiences with a new frequency-doubled doublepulse Nd:YAG laser (FREDDY) for the treatment of urolithiasis. Proceedings of SPIE, 2002; 4609: 128-135.

[17] Sayer J, Johnson DE, Price RE, Cromeens DM. Endoscopic laser fragmentation of ureteral calculi using the holmium:YAG. Proceedings of SPIE. 1993; 1879: 143-148.

[18] Grasso M, Chalik Y. Principles and applications of laser lithotripsy: experience with the holmium laser

Lithotrite. Journal of Clinical Laser Medicine & Surgery, 1998; 16(1): 3-7.

[19] Marguet CG, Sung JC, Springhart WP, L'esperance JO, Zhou SL, Zhong P, Albala DM, Preminger GM. In vitro comparison of stone retropulsion and fragmentation of the frequency doubled, double pulse Nd:YAG laser and the holmium:YAG laser. The Journal of Urology, 2005; 173(5): 1797-1800.

[20] Marks AJ, Teichman JMH. Lasers in clinical urology: state of the art and new horizons. World Journal of Urology. 2007; 25(3): 227-233.

[21] Jansen ED, van Leeuwen TG, Motamedi M, Borst C, Welch AJ. Temperature dependence of the absorption coefficient of water for midinfrared laser radiation. Lasers in Surgery and Medicine. 1994; 14(3): 258-268.

[22] Teichman JMH, Vassar GJ, Glickman RD. Holmium: yttrium-aluminum-garnet lithotripsy efficiency varies with stone composition. Urology. 1998; 52(3): 392-397.

[23] Grasso M. Experience with the holmium laser as an endoscopic lithotrite. Urology. 1996; 48(2): 199-206.

[24] Chan KF, Vassar GJ, Pfefer TJ. et al. Holmium:YAG laser lithotripsy: a dominant photothermal ablative mechanism with chemical decomposition of urinary calculi. Lasers in Surgery and Medicine. 1999; 25(1): 22-37.

[25] Pierre S, Preminger GM. Holmium laser for stone management. World Journal of Urology. 2007; 25(3): 235– 239.

[26] Teichman JMH, Rogenes VJ, McIver BJ, Harris JM. Holmium:yttrium-aluminum-garnet laser cystolithotripsy of large bladder calculi. Urology. 1997; 50(1): 44-48.

[27] Fried NM, Irby PB. Advances in laser technology and fiberoptic delivery systems in lithotripsy. Nat Rev Urol. 2018; 15: 563-573.

[28] Traxer O, Keller EX. Thulium fiber laser: the new player for kidney stone treatment? A comparison with Holmium:YAG laser. World Journal of Urology. 2019; Feb 06

[29] Frenz M, Zweig AD, Romano V, Weber HP. Dynamics in laser cutting of soft media. Proceedings of SPIE. 1990; 1202: 22-33

[30] Niemz M. Laser- Tissue Interactions – Fundamentals and Applications. 2nd Edition, 2002; pp72, Springer ISSN 1618-7210 ISBN 3-540-42763-5.

[31] Rajabhandharaks D, Zhang JJ, Wang H, Xuan JR, Chia RWJ, Hasenberg T, Kang HW. Dependence of water content in calculus phantom during Q-switched Tm:YAG laser lithotripsy. Proc. SPIE 8565, Photonic Therapeutics and Diagnostics IX, 2013; 856519.

[32] Rajabhandharaks D, Zhang JJ, Wang H, Xuan JR, Chia RWJ, Hasenberg T, Kang HW. Water content contribution in calculus phantomablation during Q-switched Tm:YAG laser lithotripsy. Journal of Biomedical Optics 2015; 20(12): 128001.

[33] Zhang JJ, Xuan JR, Yu H, Devincentis D. Study of cavitation bubble dynamics during Ho:YAG laser lithotripsy by high-speed camera. Proc. SPIE 9689, Photonic Therapeutics and Diagnostics XII, 2016: 1-7.

[34] Isner JM, Clarke RH. Cardiovascular Laser Therapy. Raven Press, New York. 1989.

[35] Chan KF, Pfefer TJ, Teichman JMH, Welch AJ. A Perspective on Laser Lithotripsy: The Fragmentation Processes. JOURNAL OF ENDOUROLOGY. 2001; 15(3): 257-273.

[36] Strittmatter F, Eisel M, Brinkmann R, Cordes J, Lange B, Sroka R. Laser-induced lithotripsy: a review, insight into laboratory work, and lessons learned. Translational Biophotonics, 2020; e201900029.

[37] Zhang JJ, Getzan G, Xuan JR, Yu H. Study of fiber-tip damage mechanism during Ho:YAG laser lithotripsy by high-speed camera and the Schlieren method. Proc. of SPIE. 9303 Photonic Therapeutics and Diagnostics XI, 2015; 930311-1-10.

[38] Lauterborn W, Vogel A. ShockWave Emission by Laser Generated Bubbles. Springer-Verlag, C.F. Delale (Ed.): Bubble Dynamics & Shock Waves, SHOCKWAVES 8, 2013: 67-103.

[39] Kuznetsov LI. Recoil momentum at a solid surface during developed laser ablation. Quantum Electronics. 1993; 23(12): 1035-1038.

[40] Foth HJ, Meyer D, Stockel T. Side effects of laser-tissue interaction studied by laser Doppler vibrometry. Proceedings of SPIE, 2000; 4072: 392-400.

[41] White MD, Moran ME, Calvano CJ, Borhan-Manesh AL, Mehlhaff BA. Evaluation of retropulsion caused by holmium: YAG laser with various power settings and fibers. Journal of Endourology. 1998; 12(2): 183-186.

[42] Lee HO, Ryan RT, Teichman JMH, et al. Stone retropulsion during holmium:YAG lithotripsy. The Journal of Urology. 2003; 169(3): 881-885.

[43] Lee H, Ryan RT, Kim JH, et al. Dependence of calculus retropulsion dynamics on fiber size and radiant exposure during ho:YAG lithotripsy. Journal of Biomechanical Engineering. 2004; 126(4): 506-515.

[44] Finley DS, Petersen J, Abdelshehid C, et al. Effect of holmium: YAG laser pulse width on lithotripsy retropulsion in vitro. Journal of Endourology. 2005; 19(8): 1041-1044.

[45] Scales CD Jr, Smith AC, Hanley JM, Saigal CS, and Urologic Diseases in America Project, "Prevalence of kidney stones in the United States," European Urology, vol. 62, no. 1, pp. 160-165, 2012.

[46] Jian J. Zhang; Jonathan Rutherford; Metasebya Solomon; Brian Cheng; Jason R. Xuan; Jason Gong; Honggang Yu; Michael Xia; Xirong Yang; Thomas Hasenberg; Sean Curran, "The study of laser pulse width on efficiency of Ho:YAG laser lithotripsy", Proc. SPIE 10038, Photonic Therapeutics and Diagnostics XIII, 1-7 (2017)

[47] D. S. Finley, J. Petersen, C. Abdelshehid et al., "Effect of holmium:YAG laser pulse width on lithotripsy retropulsion in vitro," Journal of Endourology, vol. 19, no. 8, pp. 1041-1044, 2005.

[48] Spore SS, Teichman JMH, CorbinNS,ChampionPC,WilliamsonEA, Glickman RD. Holmium:YAG lithotripsy: optimal power settings. Journal of Endourology. 1999; 13(8): 559-566.

[49] Sea J, Jonat LM, Chew BH, et al. Optimal power settings for holmium:YAG lithotripsy. The Journal of Urology. 2012; 187(3): 914-919.

[50] Teichman JMH, Rao R, Glickman R, Harris J. Holmium:YAG percutaneous nephrolithotomy: the laser incident angle matters. The Journal of Urology. 1998; 159(3): 690-694.

[51] Vassar GJ, Teichman JMH, Glickman RD. Holmium:YAG lithotripsy efficiency varies with energy density. The Journal of Urology. 1998; 160(2): 471-476.

[52] Kang HW, Lee H, Teichman JMH, Oh J, Kim J, Welch AJ. Dependence of

calculus retropulsion on pulse duration during ho: YAG laser lithotripsy. Lasers in Surgery and Medicine. 2006; 38(8): 762-772.

[53] Deters LA, Pais VM Jr. Difference in Operative Time According to Stone Location for Endoscopic Management of Ureteral and Renal Stones. Urology. 2013; 81(3): 522-526.

[54] Carey RI, Kyle CC, Carey DL, and Leveillee RJ. "Preparation of artificial kidney stones of reproducible size, shape, and mass by precision injection molding," J Endourol 2008; 22(1), 127-131.

[55] Eisel M, Strobl S, Pongratz T, Strittmatter F, Sroka R. Holmium:yttrium-aluminum-garnet laser induced lithotripsy: in-vitro investigations on fragmentation, dusting, propulsion and fluorescence. Biomed. Opt. Express 2018; 9(11): 5115.

[56] Eisel M, Strobl S, Pongratz T, Strittmatter F, Sroka R. In vitro investigations of propulsion during laser lithotripsy using video tracking. Lasers Surg. Med. 2018; 50: 333-339.

[57] Zhang JJ, Rajabhandharaks D, Xuan RJ, Chia RWJ, Hasenberg TC. Characterization of Calculus Migration during Ho:YAG Laser Lithotripsy by High Speed Camera Using Suspended Pendulum Method. Proc. SPIE 8926, Photonic Therapeutics and Diagnostics X, 2014; 89261I-1-7.

[58] Kronenberg P, Traxer O. Update on lasers in urology 2014: current assessment on holmium:yttrium–aluminum–garnet (Ho:YAG) laser lithotripter settings and laser fibers. World J. Urol. 2015; 33: 463-469.

[59] T. C. Hutchens, D. A. Gonzalez, P. B. Irby, N. M. Fried, Fiber optic muzzle brake tip for reducing fiber burnback and stone retropulsion during thulium fiber laser lithotripsyJ. Biomed. Opt. 2017; 22(1): 18001.

[60] Sroka R, Haseke N, Pongratz T, Hecht V, Tilki D, Stief CG, Bader MJ. In Vitro Investigations of Repulsion During Laser Lithotripsy Using a Pendulum Setup. Lasers Med. Sci. 2012; 27: 637.

[61] Zhang JJ, Rajabhandharaks D, Xuan R J, Chia RWJ, Hasenberg TC. Calculus migration characterization during Ho:YAG laser lithotripsy by high-speed camera using suspended pendulum method. Lasers Med. Sci. 2017; 32: 1017-1021.

[62] Esch E, Simmons WN, Sankin G, Cocks HF, Preminger GM, Zhong P. A simple method for fabricating artificial kidney stones of different physical properties. Urol Res. 2010; 38(4): 315-319.

[63] Chai DY, Zhang JJ, Podana N, Xuan RJ, Hasenberg T, Harrah T. The study of Ho: YAG laser ablation thresholds of calculus phantom in terms of peak power density. Proc. SPIE 10852, Therapeutics and Diagnostics in Urology 2019; 108520D.

Chapter 11

Response Surface Methodology Applied to the Optimization of Phenolic Compound Extraction from *Brassica*

Valentin Reungoat, Morad Chadni and Irina Ioannou

Abstract

The response surface methodology (RSM) is a relevant mathematical and statistical tool for process optimization. A state of the art on the optimization of the extraction of phenolic compounds from *Brassica* has shown that this approach is not sufficiently used. The reason for this is certainly an apparent complexity in comparison with the implementation of a one-factor-at-a-time (OFAT) optimization. The objective of this chapter is to show how one implement the response surface methodology in a didactic way on a case study: the extraction of sinapine from mustard bran. Using this approach, prediction models have been developed and validated to predict the sinapine content extracted as well as the purity of the extract in sinapine. The methodology presented in this chapter can be reproduced on any other application in the field of process engineering.

Keywords: Response surface methodology, extraction, phenolic compounds, process engineering, biomass valorization

1. Introduction

Nowadays, bio-based molecules are more and more popular and used in everyday consumer products. Certain molecules such as phenolic compounds (PCs) are very appreciated for their biological activities which make it possible to fight against aging or to act as an antibacterial or anti-oxidant agent. Phenolic compounds are secondary metabolites of plants and are present in plant biomass as well as in agro-industrial by-products [1]. The latter are currently used in sectors with low added value such as methanization or animal feed. To provide additional value to its agro-industrial co-products, phenolic compounds could be extracted and concentrated [2]. For this, separation processes will have to be implemented and optimized. Thus, maximizing the extraction of phenolic compounds has become a topic of interest which would improve the profitability of crops and by-products resulting from their industrial transformation [3].

Many studies focus on maximizing extraction efficiency by optimizing using OFAT. This method, which seems simpler, is often either time consuming or leads to partial conclusions (e.g. no interpretation of the interactions between variables). Thus, to achieve such an optimization, it is recommended, if conditions allow it, to use the response surface methodology. RSM is a mathematical and statistical tool for exploring the relationships between several explanatory variables - called factors - and one or more variables to be optimized, called response(s). RSM is particularly relevant when the response is suspected to evolve in a curved way.

In this chapter, we will focus on the application of RSM for optimizing the extraction of phenolic compounds from *Brassica*. In the first part, we propose a state of the art of the studies on this topic with an analysis of the main tools used to determine the optimal operating conditions for the extraction of phenolic compounds. In a second part, a case study based on the work of Reungoat et *al.* (2020) is presented [4]. This study focuses on the optimization of a sustainable extraction process to improve the recovery (yield and purity) of sinapine from mustard bran. Sinapine has biological activities however, its first interest is the degradation product of its hydrolysis: the sinapic acid. It has been shown that providing bio-based sinapic acid is very relevant in various application fields [5]. Indeed, this platform molecules can be used for the chemo-enzymatic synthesis of various molecules such as an anti-UV agent [6, 7], a non-endocrine disruptive antiradical additive [8] and a bisphenol A substitute for polymer/resin synthesis [9]. The study will be detailed not from an application point of view but from a methodological point of view with the presentation of the different steps which led to obtaining optimum operating conditions of the extraction process.

2. State of the art on the optimization of PC extraction from *Brassica*

The studies reported in **Table 1** deal with the optimization of the extraction process of phenolic compounds from Brassica. These all relate to the use of a design of experiments (DoE). OFAT optimization has been excluded.

Twenty papers have been identified on various raw materials belonging to *Brassica* (rapeseed, mustard, cabbage, broccoli, cauliflower). The extraction processes implemented are the most popular ones: conventional solvent extraction (CSE), ultrasound-accelerated extraction (UAE), microwave-accelerated extraction (MAE). One study deals with an extraction assisted by pulsed electric field (PEF) [13] and another with accelerated solvent extraction (ASE) [5].

The operating conditions the most often optimized are the extraction temperature, the solvent concentration in water, the solid-to-matter ratio, and the extraction time. Some specific conditions can also be investigated such as ultrasonic or microwave power when UAE and MAE are carried out.

The predicted responses are diverse whether they are measurement of individual phenolic content obtained by HPLC, total phenolic compounds (TPC), or content of total flavonoid (TFC), or antioxidant activity (AA) which can be measured by different methods (**Table 2**).

Most studies have used RSM to model and/or predict responses. A mixture design was also used to determine the composition of an extraction solvent from three pure solvents; a simplex centroid mixture was carried out [3]. Some studies model responses using first order polynomial equations. These models are obtained from factorial design of experiments [5, 15, 16]. Concerning the implementation of the RSM, the experimental design carried out are mainly Box–Behnken (BB) [8, 10, 13, 14, 17] and Central Composite (CC) [1, 6, 7, 9, 18–20]. We also found a D-optimal [4] and a full factorial [2]. However, these DoEs are rarely associated with RSM. The predictions made by RSM are associated with second order polynomial models.

Compared to all the studies that exist in the literature on the extraction of phenolic compounds from *Brassica*, only a small proportion uses RSM.

Biomass	Extraction process	DOE	Factors	Response	Model	Ref
Mustard bran (*Brassica juncea*)	CSE	CCF	• T (C°): 45, 60, 75 • E (%): 45, 70, 95 • S/M (mL/g): 10, 20, 30	Sinapine yield, sinapine purity	Second-order polynomial	[4]
Rapeseed cake (*Brassica napus*)	CSE	2^3 full factorial	• T (C°) :20, 30, 45, 60, 70 • E (%): 0,17, 43, 68, 85 • S/M (g/100 mL): 5, 8, 12.5, 17, 20	TPC	Second-order polynomial	[10]
Mustard seeds (*Sinapis alba*, *Brassica nigra*)	CSE	Simplex centroid mixture	Mixture of solvents: • Water: 0,1/6, ½, 2/3,1 • Acetone: 0,1/6, ½, 2/3,1 • Methanol: 0,1/6, ½, 2/3,1	TPC, AA	Quadratic or cubic regression	[11]
Mustard bran (*Brassica juncea*)	CSE	D-optimal	• T (C°): 30, 42.5, 55 • Time (h): 0.5, 3.25, 6 • E (%): 30, 65, 100 • S/M (mL/g): 10 30 50 • Pretreatment: none, dried and defatted	TPC	Second-order polynomial	[12]
Canola meal (*Brassica napus*)	Accelerated Solvent Extraction	Factorial	• Particle size (mm): 0.5, 1. • T (C°): 140, 160, 180 • Type of solvent: water, ethanol, methanol • Percentage of solvent in water (%): 30, 40, 60, 70	TPC, TFC	First-order polynomial	[13]
Brassica oleracea leaves	CSE	CC	• Time (h): 12, 24, 36, 48, 60 • Polarity of extracting solvents in terms of dipole moment (D): 0.0 (hexane), 4.40 (diethyl ether), 2.80 (ethyl acetate), 5.10 (methanol) and 9.0 D (water)	TPA, TAA, AA	Second-order polynomial	[14]
Brassica oleracea seeds	CSE	CC	• T (C°): 50, 60, 70, 80, 90 • Power (W): 100, 125, 150, 175, 200 • Solvent concentration (methanol/water, v/v (%)): 50, 60, 70, 80, 90 • Time (min): 1, 7.5, 14, 20.5, 27	• Total extractable components • TPA • AA	Second order polynomial	[15]

Biomass	Extraction process	DOE	Factors	Response	Model	Ref
Red Cabbage (*Brassica oleracea*)	UAE	BB	• Ultrasonic time (min): 20, 40, 60 • Ultrasonic frequency (kHz): 0, 22.5, 45 • T (C°): 50, 60, 70	TPC	Second order polynomial	[16]
Broccoli leaves	CSE	CC	• Pressure (bar): 150, 225, 300 • T (°C): 35 ,47.5, 60 • E (%): 0, 10, 20 • Time (min): 10, 20, 30	AA	Second order polynomial	[17]
Rapeseed stems (*Cultiva Dkextorm*)	UAE	BB	• Ultrasound power (W): 0, 200, 400 • Treatment time (min): 5, 32.5, 60 • Sample length (cm): 0.5, 1.5, 2.5 • Agitation speed (rpm): 0, 300, 600	TPC	Second order polynomial	[18]
White cabbage *Brassica oleracea var. capitata*	UAE	CC rotational	• T (C°): 30, 36, 50, 64, 70 • E (%): 20, 32, 60, 88, 100	Phenolic compound content (HPLC)	-	[19]
Mustard seed (*Sinapis Alba L.*)	UAE	CC	• Solvent polarity: 32.6, 56.3, 78.4 • Ultasound power to sonication time ratio (W/min): 0,4,8	TPC, AA	Second order polynomial	[20]
Canola seed cake (*Brassica napus*)	MAE	BB	• Time (min): 1, 3, 5 • Microwave power (W): 440, 770, 1100 • S/M (mL/g): 4, 5, 6	TPC, TFC, AA	Second order polynomial	[21]
	UAE enzymatic extraction		• S/M (mL/g): 4:1, 6:1, 8:1 • Ultrasonic power (W): 300, 400, 500 • pH: 4.5, 5.5, 6.5 • T (C°): 50, 55, 60			
Broccoli (*Brassica oleracea*)	Alkaline Hydrolysis and UAE	BB	• T (C°): 40, 60, 80 • NaOH concentrations (M): 1, 2, 4 • Sonication time (min): 15, 30, 45	-	Second order polynomial	[22]
Cauliflower Waste (*Brassica oleracea*)	UAE	3^3 factorial	• Solvent volume (mL): 50, 75, 100 • Extraction time (min) :20, 30, 40 • T (C°): 50, 60, 70	TPC	First-order polynomial	[23]

Biomass	Extraction process	DOE	Factors	Response	Model	Ref
Canola seed cake (*Brassica napus*)	CSE	Orthogonal experiment	• T (C°): 30, 50, 70 • Liquid-to-solid ratio (mL/g): 10, 15, 20 • Time (min): 30, 60, 90 • E (%): 45, 70, 95	AA	First order polynomial	[24]
Maca flour (*Lepidium meyenii*)	CSE	BB	• T (C°): 30, 50, 70 • Liquid-to-solid ratio (mL/g): 10, 15, 20 • Time (min): 30, 60, 90 • E (%): 45, 70, 95	TPC, AA	Second order polynomial	[25]
Broccoli	MAE	CC	• T (C°): 50, 60, 70, 80, 90 • Power (W): 100, 125, 150, 175, 200 • Solvent concentration (methanol/water, v/v (%)): 50, 60, 70, 80, 90 • Time (min): 1, 7.5, 14, 20.5, 27	-	-	[26]
Purple Cabbage	MAE	CC	• Power (W): 100, 200, 300, 400, 500 • Liquid to solid ratio: 35, 47.5, 60 • E (%): 40, 50, 60, 70, 80 • Time (min): 25, 40, 55, 70, 85	Yield of proanthocyanidins	Second order polynomial	[27]
Fresh cabbage *Brassica oleracea var. capita*	CSE	CC	• S/M (mL/g): 10, 15, 20, 25, 30 • E (%): 0, 25, 50, 75, 100 • T (°C): 35, 50, 65, 80, 95	TPC, AA	Second order polynomial	[28]

Table 1.
Studies dealing with the extraction of PCs from Brassica.

3. Optimization of sinapine extraction from mustard bran

3.1 Context of the study

Mustard bran is one of the main by-products of the mustard seed industry whose production peaked at 710 thousand tonnes in 2018 [29]. By-products from their processing represent up to 60%w of seeds [30]. Mustard bran is rich in water with a content between 53 ± 1%. The dry matter is mainly composed of proteins (27 ± 7%), lipids (18 ± 1%), carbohydrates (34 ± 5%) and ash (12 ± 5%) [30–32].

Phenolic compounds represent between 1 and 4% of the wet matter of defatted mustard seeds [33]. They are mainly derivatives of sinapic acid, present at 90% as sinapine with relatively small amounts of sinapic acid. Sinapine can be used directly due to its many bioactivities [5, 34] or be hydrolyzed to sinapic acid by chemical or enzymatic means [35]. Thus, our work will focus on the extraction of sinapine from mustard bran. Moreover, bio-based sinapic acid is highly sought after thanks to its many applications, whether in cosmetics (anti-aging, anti-UV) or in the field of polymers [6, 8].

Thus, the implementation of a green extraction process to recover sinapine seems particularly relevant. The most widely used process in the various studies found in the literature is conventional solvent extraction (CSE). This is a solid/liquid extraction, the liquid being a solvent whose properties will define the sustainability of the process. Solvents such as acetone, methanol, ethanol or water, as well as a mixture, have been used [36]. To follow the principles of green extraction [37], the extraction process developed will use aqueous ethanol as solvent, the percentage of which will be determined during the optimization of the process.

3.2 Material and methods

Mustard bran, was supplied by Charbonneaux-Brabant (Reims, France). Mustard (*B. juncea*) grew in Canada and was processed in France. The treatment undergone by the seeds is cold mechanical pressure. The material has not been defatted, ground or dried. The raw mustard bran was stored in a cold room at 4°C until use.

A CSE using an ethanol/water mixture was implemented to remove sinapine from mustard bran. A fixed volume of 100 mL of solvent was used for each experiment. The extraction temperature was regulated with a digital thermometer in contact with the solvent and connected to the heating plate (IKA-RCT). Magnetic stirring was ensured throughout the duration of the extraction (2 h). Centrifugation was used (4713 g, 10 min) to separate the liquid extract from the solid residue. The sinapine content was measured by HPLC. More details on the materials and the methods can be found in Reugoat *et al.*, 2020 [1].

3.3 Implementation of RSM

RSM is the recommended approach to optimize process operating conditions, for example to maximize extraction yield or minimize impurity content. Indeed, the implementation of the RSM, and therefore of a design of experiments, makes it possible to minimize the number of experiments, to determine the quadratic effect of a factor or the interaction between several factors and to obtain a high precision on the prediction of an optimal value.

The implementation of RSM requires the identification of the factors that will be involved in the model. Thus, RSM is often used after a screening plan which allows the discrimination of the operating conditions leading to a significant variation in the response. Sometimes, prior knowledge of the process is sufficient to avoid the screening step and RSM can be applied after arbitrary choice of factors by the experimenter. RSM is a relevant approach if the response surface is suspected to be curved. Indeed, the equation of the model used includes quadratic terms which make it possible to translate the curvature of the response.

In order to apply RSM, it is necessary to follow a rigorous approach so as not to end up with wrong conclusions or an unusable data set for the prediction of an optimum. This approach is illustrated in **Figure 1**.

For each step, the reasoning adopted for our case study will be detailed, the choices will be explained so that the methodology can be easily implemented on other cases.

3.3.1 Definition of the objectives

The objective of the optimization study must be defined according to the overall objective of the application. In our case, the operating conditions of the extraction process leading to a maximum yield of sinapine are sought. However, the global objective of the application is to produce sinapine, that is, to obtain a high purity sinapine extract. Thus, a second variable to be optimized emerges in addition to the yield of sinapine: the purity of the sinapine extract. Under such considerations, the optimum operating conditions sought will be a compromise between those allowing

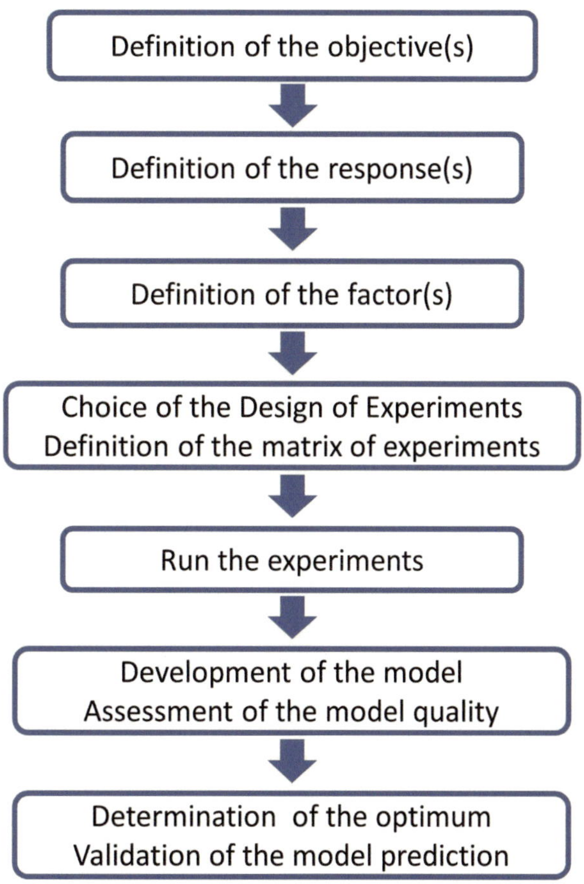

Figure 1.
Methodology for the implementation of RSM.

to maximize the yield of sinapine and the ones that maximize the purity of sinapine. Failure to correctly define the objective may lead to an incorrect definition of the responses, factors and their levels and thus induce a partial conclusion at the end of the study.

3.3.2 Definition of the responses

A response is defined as a variable to be explained. For the choice of responses, it is necessary to ensure that the measurement tools are sufficiently repeatable. Indeed, in statistics, it is common to say that the more the value is dispersed the more it will be difficult to highlight significant differences and therefore to obtain a valid prediction model. This is why the presence of a triplicate in the DOE is essential to quantify the repeatability of the measurement. If it is too large, the DOE will not be able to generate a valid model.

In our case study, the two responses to be optimized are the yield of sinapine in % (Y_1) and the purity of sinapine in % (Y_2) defined by Eq. (1,2)

$$Y_1 = \frac{C_{sinapine} * V_{solvent}}{m_{BDM}} \tag{1}$$

with $C_{sinapine}$ the sinapine content measured by HPLC in mg/L, $V_{solvent}$ the volume of solvent added during the extraction and m_{BDM} the mass of dry matter in mustard bran.

$$Y_2 = \frac{m_{sinapine}}{m_{EDM}} * 100 \tag{2}$$

with $m_{sinapine}$ (g) the mass of sinapine in the extract determined from $C_{sinapine}$ and m_{EDM} the mass of the dry matter in the extract (g).

3.3.3 Definition of the factors

A factor is defined as a variable that provides information to explain a response. Two strategies can be used to define the factors: to apply a screening plan (factorial or Plackett-Burman) or to use expertise on the process. In our case, the factors were chosen based on prior knowledge about the extraction process [4]. Note that the factors must be independent for the implementation of the experimental design. This should be checked before establishing the matrix of experiments.

According to theory, the liquid/solid extraction processes are influenced by a set of parameters which can modify their efficiency. These relate to: (i) the equipment used (stirring power, the configuration of the reactor), (ii) the operating conditions (extraction time, extraction temperature and pressure), and (iii) the biomass and the solvent (solvent-to-matter ratio, state of the biomass, nature of the solvent).

Some of these parameters are often fixed in the design of the experiments. Indeed, for laboratory experiments, the extraction reactor is always the same as well as the stirring system (type and power). Conventionally, the extraction time corresponds to the time required to reach the equilibrium. In our case, the biomass is wet and in the form of bran, so it cannot be crushed or sieved. This parameter cannot be taken as a factor. In addition, two constraints were imposed: to conduct the experiments at atmospheric pressure and working with ethanol (pure or aqueous) to design a sustainable process. Thus, the parameters that could be included as factors in the design of experiments are the solvent-to-matter ratio, the extraction temperature and the ethanol concentration. These parameters being independent, three factors will be used in models developed using RSM. The last point to be defined is the variation range of each factor.

3.3.3.1 Range of extraction temperature

Technological limits exist for the choice of extraction temperatures. The experimental domain cannot be extended above 75°C to avoid evaporation phenomena due to the boiling temperature of ethanol. Thus, the extraction temperature will be able to vary between room temperature and 75°C. However, according to the literature, it does not seem interesting to carry out experiments at temperatures close to room temperature. Indeed, it is known that an increase of temperature allows to improve the extraction of phenolic compounds. A range of values too large can adversely affect the quality and accuracy of the prediction model. We have, therefore, chosen to limit our temperature range between 45°C and 75°C.

3.3.3.2 Range of solvent-to-matter ratio (S/M)

There is also a technological limit for this factor. Indeed, it is not possible to extract with less than 10 mL per gram of mustard bran. In addition, the objective being not to consume too much solvent, no more than 30 mL per gram of mustard bran will be used. Thus, the range of the S/M factor will be between 10 and 30 mL/g.

3.3.3.3 Range of ethanol concentration

No technological limit was found for this factor. The use of extreme values (water or pure ethanol) is not interesting because the better extraction yields are obtained with aqueous ethanol. According to preliminary experiments, to maximize the yield of sinapine (Y1), the values to be studied should be between 40 and 80%. Considering the purity of sinapine (Y2), the values to be studied should be between 60 and 100% in order to limit the extraction of impurities such as sugars and proteins.

To define the range of variation of the ethanol concentration, we merged the two previous intervals by removing the extreme values so as not to widen the range of values to be studied too much. Thus, ethanol concentrations between 45 and 95% were studied in the design of experiments.

3.3.4 Choice and implementation of the design of experiments

The two most used design of experiments for the implementation of RSM are the composite center (CC) and Box–Behnken (BB) designs.

For a same number of factors and levels, a BB design generates fewer experiments than a CC design. However, BB designs have a certain rigidity in their implementation since the number of levels per factor is fixed. In addition, the BB designs do not include in the experiments the extreme values of the variation ranges of the factors. This can sometimes constitute a problem, when a precise knowledge of the interval is available and/or when the extreme values want to be tested.

In addition, CC design is to be able to integrate preliminary experiments. Thus, the results of certain experiments present in the screening plan carried out upstream can be used as experiments in the CC design. Thus, the number of new assays to realize will decrease.

The CCF design belongs to the category of the CC design. The experiments defined are located in the center of each face of the experimental domain.

In our application, a CCF design was used to optimize the extraction process for the recovery of sinapine. A total of 17 experiments including a repetition at the central point constitutes the set of the experiments to implement RSM. The different assays are presented **Table 2** in the form of coded and uncoded variables.

Assay	X_1	X_2	X_3	T(°C)	E ($\%_{v/v}$)	S/M (mL/g_{BDM})
1	−1	−1	−1	45	45	10
2	+1	−1	−1	75	45	10
3	−1	+1	−1	45	95	10
4	+1	+1	−1	75	95	10
5	−1	−1	+1	45	45	30
6	+1	−1	+1	75	45	30
7	−1	+1	+1	45	95	30
8	+1	+1	+1	75	95	30
9	−1	0	0	45	70	20
10	+1	0	0	75	70	20
11	0	−1	0	60	45	20
12	0	+1	0	60	95	20
13	0	0	−1	60	70	10
14	0	0	+1	60	70	30
15	0	0	0	60	70	20
16	0	0	0	60	70	20
17	0	0	0	60	70	20

Table 2.
CCF experimental design.

X_1 (extraction temperature; 45, 60, 75°C), X_2 (concentration of ethanol; 45-70-95$\%_{v/v}$) and X_3 (solvent-to-matter ratio; 10, 20, 30 mL/g_{BDM}) are the independent variables used to explain the responses Y_1 (sinapine yield on the mustard bran dry matter in g/g_{BDM}) and Y_2 (sinapine purity on the extract dry matter in $\%_{EDM}$).

The experimental data were fitted using a second-order polynomial (Eq. (3)):

$$Y_q = \beta_0 + \sum_{i=1}^{3} \beta_i . X_i + \sum_{i=1}^{3} \beta_{ii} . X_i^2 + \sum \sum_{i<j=1}^{3} \beta_{ij} . X_i . X_j + \varepsilon_q \tag{3}$$

where Y_q are the different responses (q = 1–2); β_0, β_i, β_{ij}, β_{ii} are the regression coefficients for the mean, linear, interaction and quadratic terms respectively. X_i and X_j are the independent variables. ε_q the residues between the observed and the predicted values.

3.3.5 Run the experiments

This step corresponds to data collection. Assays can be performed in random order. The material and methods of analysis were briefly introduced. More details can be found in Reungoat et al. (2020) [1].

3.3.6 Development and validation of the model

Once the data has been collected, they are processed by a software to generate a model and indicators that allow its quality to be assessed (fit to the data, ability to predict). The software used to carry out our case study is the commercial software MODDE v.12.0 (Umetrics AB, Sweden).

First, it is necessary to determine whether the model should be reduced. Reducing a model means removing variables whose coefficients are not significant. Significance tests are carried out for this purpose. The p-values obtained indicate whether the value of the coefficient can be considered equal to 0. In this case, the factors are considered to have no effect on the response. **Table 3** presents the scaled and centered coefficients of the model associated with each term as well as the results of the significance test for each coefficient.

The p-values in red in **Table 3** indicate the significant coefficients and factors to keep in the model.

3.3.6.1 Analysis of the prediction model of Y_1

The significant coefficients are β0 (constant), β1 (T) and β22 (E*E). Since the quadratic term E*E is significant, the variable E cannot be removed from the model. Thus, the factors to keep are temperature and ethanol. The S/M ratio has no effect on the sinapine yield. The data must be reprocessed by the software keeping the variables T, E and E*E. New values are found for the coefficients of the reduced model. Sinapine yield can be predicted according to Eq. (4) with unscaled coefficients.

$$\mathbf{Y_1 = -4.846 + 0.002\,T + 0.376\,E - 0.003\,(E * E)} + \varepsilon \tag{4}$$

3.3.6.2 Analysis of the prediction model of Y_2

The significant coefficients are β0 (constant), β1 (T), β2 (E), β3 (S/M), β23 (E*S/M) and β22 (E*E). Thus, all the factors should be kept. The data must be processed by the software by keeping the variables T, E, S/M, E*S/M and E*E. New values are found for the coefficients of the reduced model. Sinapine purity can be predicted according to Eq. (5) with unscaled coefficients.

$$\begin{aligned} Y_2 = &-6.452 + 0.010\,T + 0.174\,E + 0.297\,(\mathrm{S/M}) - 0.001\,(E * E) \\ &- 0.003(\mathrm{S/M} * \mathrm{S/M}) - 0.003(E * \mathrm{S/M}) + \varepsilon \end{aligned} \tag{5}$$

Variables	Coefficients scaled and centered	Y_1		Y_2	
		Value	p-value	Value	p-value
constant	β0	8.398	< 0.001	3.491	< 0.001
T	β1	0.622	<0.01	0.140	0.030
E	β2	−0.038	0.755	0.353	<0.01
S/M	β3	−0.021	0.863	−0.231	<0.01
T*E	β12	0.098	0.382	0.010	0.817
T*S/M	β13	0.227	0.070	−0.022	0.584
E*S/M	β23	−0.037	0.745	−0.417	<0.001
T*T	β11	−0.136	0.472	−0.061	0.473
E*E	β22	−0.745	<0.01	−0.388	0.007
S/M*S/M	β33	−0.258	0.194	−0.121	0.185

Table 3.
Values of the model coefficients and the p-values of their significance tests.

Secondly, the indicators calculated on each reduced model are interpreted to assess whether the correlation between the model and the experimental data is acceptable and whether these models can be considered as good prediction tools. These indicators are presented in **Table 4**.

Indicators	Reduced model (Y_1)	Reduced model (Y_2)
R^2	0.90	0.97
R^2 adj	0.86	0.94
Model regression (p value)	0.00003	0.00002
Reproducibility	0.98	0.99
Condition number	2.88	4.67

Table 4.
Indicators to assess the fit and quality of reduced models.

The coefficients of determination being close to 1, the reduced models have a good accuracy in their prediction. The values of the adjusted coefficients of determination are high enough to suggest a satisfactory correlation between the values predicted by the model and the values observed by the experiments. The p-values obtained by the ANOVA on the model regression are less than 0.01% which validates the models obtained. The condition number determines the correct orthogonality of the two models because it does not exceed 10. Each model reproducibility is also excellent with a value close to 1. All these statistical parameters indicate that the relationships between the variables and the responses are well described by the models.

3.3.7 Determination of the optima and validation of the model

In order to determine the optimal operating conditions for each response, the 3D response surfaces will be plotted. In a second time, the software optimizer tool based on the Nelder–Mead simplex method was implemented to obtain the optimal operating conditions. **Figure 2** presents the evolution of Y_1 according to the extraction temperature, the ethanol concentration for a solid-to-matter ratio of 10 mL/g.

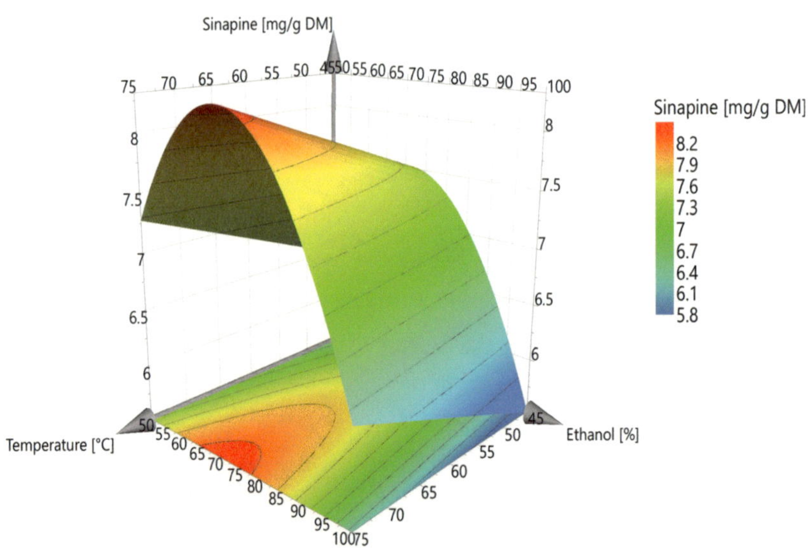

Figure 2.
3D response surface for a solid-to-matter ratio of 10 mL/g. for Y_1.

Variations of the sinapine yield from 5.3 to 8.9 mg/g_{BDM} were found among the 17 experiments of the CCF design.

As can be seen on **Figure 2**, the sinapine yield evolves in a parabolic shape. This can be explained by a strong influence of the quadratic term of the ethanol concentration. The maximum sinapine yield is achieved in the range 65–80% ethanol. The extraction temperature has a positive effect on the sinapine yield as observed in **Figure 2** with the inclination of the response surface towards the high temperature zone.

The optimal operating conditions determined for Y_1 by the software MODDE are 70% ethanol, 75°C.

An experimental sinapine yield of 8.8 ± 0.1 mg/g was achieved under these conditions.

Figure 3 presents the evolution of Y_2 according to the extraction temperature, the ethanol concentration for a solid-to-matter ratio of 10 mL/g_{BDM}.

Variations of the sinapine purity from 1.4%$_{EDM}$ and 4.4%$_{EDM}$ were found among the 17 experiments of the CCF design.

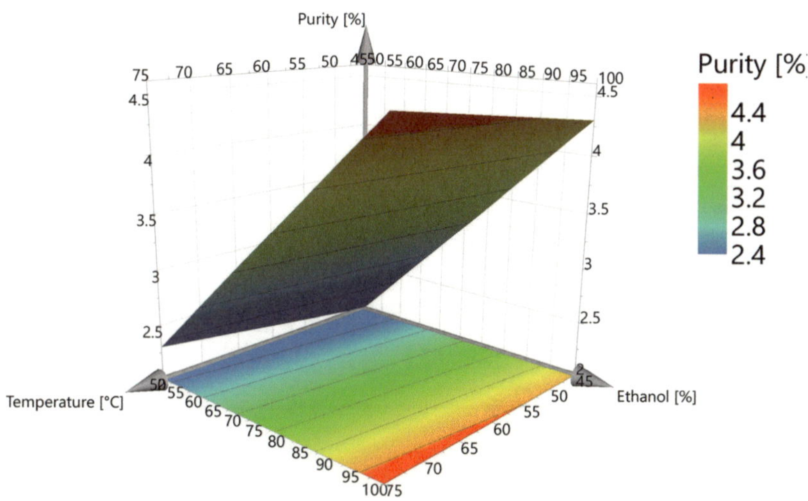

Figure 3.
3D response surface for a solid-to-matter ratio of 10 mL/g_{BDM}. For Y_2.

For a ratio of 20, the response surface is flat. Quadratic terms have little influence. The extract, containing the most sinapine compared to other extracted solutes, is obtained for a maximum temperature and ethanol concentration. This may be due to low solubility of proteins, sugars and minerals in ethanol compared to sinapine. However, an increase of the solvent-to-matter ratio increases the solubility of those impurities and decreases the sinapine purity in the extract.

The optimal operating conditions determined for Y_2 by the software MODDE are 100% ethanol, 75°C and, 10 mL/g_{BDM}. An experimental sinapine purity of 4.4 ± 0.1% $_{EDM}$ was achieved under these conditions.

Since the two optima are not the same, it will be necessary to find the operating conditions allowing to maximize the two responses at the same time. **Figure 4** presents the response surfaces for Y_1 and Y_2 on the same graph.

The MODDE software has determined that the optimal operating conditions that will provide the highest yield of sinapine while maintaining high purity, are 83% ethanol, 75°C, and 10 mL/g_{BDM}. An experimental sinapine yield of 8.0 ± 0.1 mg/g was obtained under these conditions with a purity of 4.2 ± 0.1% $_{EDM}$.

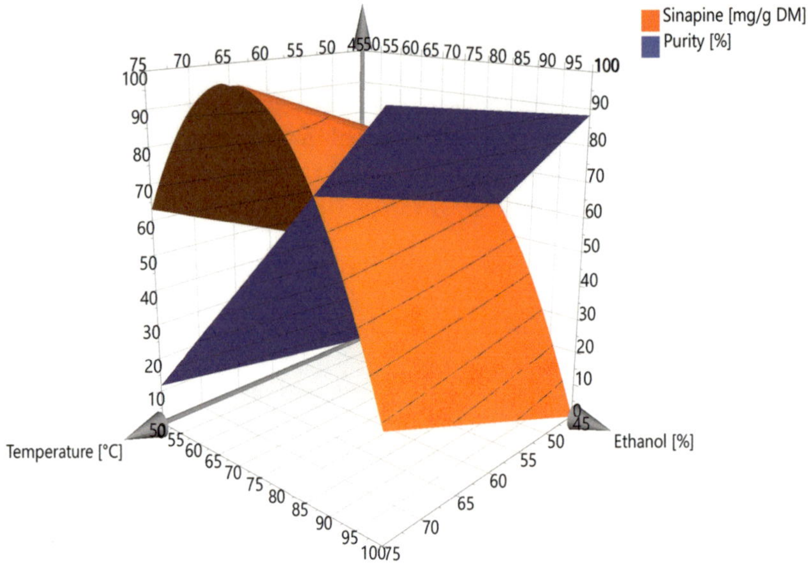

Figure 4.
3D response surface for a solid-to-matter ratio of 10 mL/g_{BDM} for Y_1 and Y_2.

S/M (10 mL/g_{BDM})	Predicted values	Observed values	p-value (Student-tests)
Sinapine yield (mg/g_{BDM})			
70% EtOH, 75°C	8.7 ± 0.2 mg/g	8.8 ± 0.1 mg/g	0.26
83% EtOH, 75°C	8.2 ± 0.2 mg/g	8.0 ± 0.1 mg/g	0.11
Sinapine purity (%$_{EDM}$)			
70% EtOH, 75°C	3.6 ± 0.7%	3.5 ± 0.2%	0.43
83% EtOH, 75°C	4.0 ± 0.7%	4.2 ± 0.1%	0.30

Table 5.
Validation of the models by performing student tests between predicted and observed values.

The last step to be carried out is the validation of the models on new experiments. For this, experiments were realized in duplicate under optimal conditions corresponding to the maximization of Y_1 and for the compromise between Y_1 and Y_2. Student's tests were performed to determine if the predicted values given by the models can be considered equivalent to the observed values. The results are shown in **Table 5**.

Experimental values correspond to predicted values since p-value>0.05. Thus, models developed by RSM are validated and can be used as prediction tool.

4. Conclusions

Concerning the extraction of sinapine from mustard bran, a CCF design was used to optimize the extraction process. A total of 17 experiments including a repetition at the central point constituted the set of the experiments to implement the RSM.

Two prediction models have thus been developed. These models have been validated, making it possible to predict the yield and the purity of sinapine from the operating conditions of the extraction process (extraction temperature, ethanol concentration and solvent-to-matter ratio).

An optimal sinapine content of 8.8 ± 0.1 mg/g was obtained at 75°C, 70% ethanol and 10 mL/g_{BDM} whereas an optimal purity of sinapine in the extract (4.2 ± 0.1% $_{EDM}$) was achieved under different operating conditions (75°C,100% ethanol and 10 mL/g_{BDM}).

Wishing to situate us as close as possible to the 2 optima, the MODDE software determined that the most appropriate operating conditions were 75°C, 83% ethanol and 10 mL/g_{BDM}. The loss in yield and purity remains low since the sinapine yield of 8.0 ± 0.1 mg/g and a purity of 4.0 ± 0.1% $_{EDM}$ are obtained.

The use of rigorous mathematical tools for optimization in process engineering remains under-exploited as we have shown for the extraction of phenolic compounds from *Brassica*. To remedy this, a generalization of the learning and use of experimental designs in universities and in the research community should be put in place. This is to encourage experimenters to optimize their process in a structured way rather than using OFAT approaches which seem easy to understand at first glance, but which may prevent the full exploitation of the information provided by the experiments. The case study, presented here, illustrated the potential in terms of process optimization using RSM.

Acknowledgements

The case study presented was supported by Extractis (Amiens, France). We would like to thank the Region Grand Est, the Conseil Départemental de la Marne and the Grand Reims for their financial support, as well as Charbonneaux-Brabant for providing the mustard bran.

Conflict of interest

There is no conflict of interest.

Nomenclature

AA	Antioxidant activity
ASE	Accelerated solvent extraction
BB	Box–Behnken
CC	Central Composite
CCF	Central Composite Face Centered
CSE	Conventional Solvent Extraction
E	Ethanol concentration
MAE	Microwave-accelerated extraction
PCs	Phenolic compounds
PEF	Pulsed electric field
S/M	Solvent to Matter ratio
T	Temperature
TAA	Total Antioxidant Activity
TPA	Total of Phenolic Acids
TPC	Total Phenol Content
UAE	Ultrasound-accelerated extraction

Author details

Valentin Reungoat, Morad Chadni and Irina Ioannou*
URD Industrial Agro-Biotechnologies (ABI), CEBB, AgroParisTech, Pomacle, France

*Address all correspondence to: irina.ioannou@agroparistech.fr

References

[1] Sharma P., Gaur V.K., Sirohi R., Varjani S., Kim S.H., Wong J.W.C. Sustainable processing of food waste for production of bio-based products for circular bioeconomy. Bioresource Technology. 2021; 325. DOI: 10.1016/j.biortech.2021.124684

[2] Lizárraga-Velázquez CE, Leyva-López N, Hernández C, Gutiérrez-Grijalva EP, Salazar-Leyva JA, Osuna-Ruíz I, Martínez-Montaño E, Arrizon J, Guerrero A, Benitez-Hernández A, Ávalos-Soriano A. Antioxidant Molecules from Plant Waste: Extraction Techniques and Biological Properties. *Processes*. 2020; 8(12):1566. https://doi.org/10.3390/pr8121566

[3] Laguna O., Guyot S., Yu X., Broudiscou L.P., Chapoutot P., Solé-Jamault V., Anton M., Quinsac A., Sicaire A.G., Fine F., Citeau M., Durand E., Barakat A., Villeneuve P., Lecomte J., Dauguet S. The PHENOLEO project or how to separate and add-value to phenolic compounds present in rapeseed and sunflower meals. OCL-Oilseeds and fats crops and lipids, 2020; 27(61). https://doi.org/10.1051/ocl/2020056

[4] Reungoat V, Gaudin M, Flourat AL, Isidore E, Mouterde LMM, Allais F, Ducatel H, Ioannou I. Optimization of an Ethanol/Water-Based Sinapine Extraction from Mustard Bran Using Response Surface Methodology. Food Bioproducts and Processing. 2020;122: 322–331. DOI:10.1016/j.fbp.2020.06.001

[5] Niciforovic N., Abramovic H. Sinapic acid and its derivatives: natural sources and bioactivity. Comprehensive reviews in food science and food safety. 2014;13 (1), 34–51. DOI: 10.1111/1541-4337.12041

[6] Baker L.A., Horbury M.D., Greenough S.E., Allais F., Walsh P.S., Habershon S., Stavros V.G. Ultrafast photoprotecting sunscreens in natural plants. Journal of Physical Chemical Letters, 2016;7, 56–61, http://dx.doi.org/10.1021/acs.jpclett.5b02474

[7] Dean J.C., Kusaka R., Walsh P.S., Allais F., Zwier T.S. Plant Sunscreens in the UV-B: ultraviolet spectroscopy of jet-cooled sinapoyl malate, sinapic acid, and sinapate ester derivatives. Journal of American Chemical Society, 2014;136, 14780–14795. http://dx.doi.org/10.1021/ja5059026.

[8] Janvier M., Hollande L., Jaufurally A. S., Pernes M., Ménard R., Grimaldi M., Beaugrand J., Balaguer P., Ducrot P.-H., Allais F. Syringaresinol: a renewable and safer alternative to bisphenol A for epoxy-amine resins. ChemSusChem, 2017;10, 738–746. http://dx.doi.org/10.1002/cssc.201601595.

[9] Jaufurally A.S., Teixeira A.R.S., Hollande L., Allais F., Ducrot P.-H. Optimization of the laccase-catalyzed synthesis of (±)-syringaresinol and study of its thermal and antiradical activities. 2016, ChemistrySelect;1, 5165–5171. http://dx.doi.org/10.1002/slct.201600543.

[10] Zardo I, Rodrigues NP, Sarkis JR, Marczak LD. Extraction and Identification by Mass Spectrometry of Phenolic Compounds from Canola Seed Cake. Journal of Science of Food and Agricultural. 2020;100:578–586. DOI: 10.1002/jsfa.10051

[11] Boscariol Rasera G, Hilkner MH, de Alencar SM, de Castro RJS. Biologically Active Compounds from White and Black Mustard Grains: An Optimization Study for Recovery and Identification of Phenolic Antioxidants. Industrial Crops and Products. 2019;135:294–300. DOI: 10.1016/j.indcrop.2019.04.059

[12] Flourat AL, Willig G, Teixeira ARS, Allais F. Eco-friendly extraction of sinapine from residues of mustard

production. Frontiers in Sustainable Food Systems 2009;3. DOI: 10.3389/fsufs.2019.00012

[13] Nandasiri R, Eskin NAM, Thiyam-Höllander U. Antioxidative Polyphenols of Canola Meal Extracted by High Pressure: Impact of Temperature and Solvents. Journal of Food Science. 2019;84: 3117–3128. DOI: 10.1111/1750-3841.14799

[14] Nawaz H, Shad MA, Rauf A. Optimization of Extraction Yield and Antioxidant Properties of Brassica *Oleracea Convar Capitata Var L.* Leaf Extracts. Food Chemistry. 2018;242: 182–187. DOI:10.1016/j.foodchem.2017.09.041

[15] Rauf A, Nawaz H, Shad M. Effect of Solvent Polarity and Extraction Time on in Vitro Antioxidant Properties of Brassica *Oleracea Convar Capitata Var L.* Seeds. Pakistan Journal of Pharmaceutical Science. 2018; 31(5): 1889–1897. DOI: 10.1016/j.foodchem.2017.09.041

[16] Oroian M, Leahu A, Dutuc A, Dabija A. Optimization of Total Monomeric Anthocyanin (TMA) and Total Phenolic Content (TPC) Extractions from Red Cabbage (*Brassica Oleracea Var. Capitata F. rubra*): Response Surface Methodology versus Artificial Neural Network. International Journal of Fodd engineering. 2017: 13. DOI:10.1515/ijfe-2016-0036

[17] Arnáiz E, Bernal J, Martín MT, Diego JC, Bernal JL, Recio LT. Optimisation of the Supercritical Fluid Extraction of Antioxidants from Broccoli Leaves. Food Analytical Methods. 2016;9:2174–2181. DOI: 10.1007/s12161-016-0399-4

[18] Yu X, Gouyo T, Grimi N, Bals O, Vorobiev E. Ultrasound enhanced aqueous extraction from rapeseed green biomass for polyphenol and protein valorization. Comptes Rendus Chimie. 2016;19:766–777. DOI:10.1016/j.crci.2016.03.007

[19] Dal Prá V, Bolssoni Dolwitsch C, Oliveira Lima F, Amaro de Carvalho C, Viana C, Cícero do Nascimento P, Barcellos da Rosa M. Ultrasound-Assisted Extraction and Biological Activities of Extracts of *Brassica Oleracea Var. Capitata*. Food Technology and Biotechnology. 2015;53: 102–109. DOI:10.17113/ftb.53.01.15.3533

[20] Szydłowska-Czerniak A, Tułodziecka A. Application of response surface methodology to optimize ultrasound-assisted extraction of total antioxidants from *Brassica napus* cultivars: Ultrasound effect on antioxidants extraction from rapeseed. European Journal of Lipid Science and Technology. 2015;117: 491–502. DOI: 10.1002/ejlt.201400310

[21] Teh SS, Niven BE, Bekhit AEDA, Carne A, Birch EJ. Microwave and pulsed electric field assisted extractions of polyphenols from defatted canola seed cake. International Journal of Food Science & Technology. 2015;50:1109–1115. DOI: 10.1111/ijfs.12749

[22] Wu H, Zhu J, Yang L, Wang R, Wang C. Ultrasonic-Assisted Enzymatic Extraction of Phenolics from Broccoli (*Brassica Oleracea L. Var. Italica*) Inflorescences and Evaluation of Antioxidant Activity inVitro. Food Sciences and Technology International. 2015; 21: 306–319. DOI:10.1177/1082013214536174

[23] Gonzales GB, Smagghe G, Raes K, Van Camp J. Combined Alkaline Hydrolysis and Ultrasound-Assisted Extraction for the Release of Nonextractable Phenolics from Cauliflower (*Brassica oleracea var. botrytis*) Waste. Journal of Agricultural and Food Chemistry. 2014;62:3371–3376. DOI:10.1021/jf500835q

[24] Teh SS, Birch EJ. Effect of ultrasonic treatment on the polyphenol content and antioxidant capacity of extract from defatted hemp, flax and canola seed

cakes. Ultrasonics Sonochemistry. 2014; 21:346–353. DOI: 10.1016/j.ultsonch.2013.08.002

[25] Campos D, Chirinos R, Barreto O, Noratto G, Pedreschi R. Optimized Methodology for the Simultaneous Extraction of Glucosinolates, Phenolic Compounds and Antioxidant Capacity from Maca (*Lepidium Meyenii*). Industrial Crops and Products. 2013;49: 747–754. DOI:10.1016/j.indcrop.2013.06.021

[26] Okić S, Cvjetko M, Božić Đ, Fabeh S, Toth N, Vorkapić-Furač J, Redovniković IR. Optimisation of Microwave-Assisted Extraction of Phenolic Compounds from Broccoli and Its Antioxidant Activity. International Journal of Food Science and Technology. 2012;47:2613–2619. DOI: 10.1111/j.1365-2621.2012.03143.x

[27] Li SL, Chen XM, Lin J. Optimization of Microwave-Assisted Extracting of Proanthocyanidins from Purple Cabbage and Evaluation of Antioxidant Activity In Vitro Available online: https://www.scientific.net/AMR.490-495.3500 (accessed on 15 January 2021).

[28] Kim HK, Lee GD, Kwon JH, Kim KH. Monitoring on Extraction Yields and Functional Properties of *Brassica Oleracea Var. Capita* Extracts. Food Science Biotechnology. 2005;14: 836–840. DOI: 10.1021/jf500835q

[29] FAOSTAT, 2020. Production/Yield Quantities of Mustard Seed [WWW Document], http://www.fao.org/faostat/en/#data/QV/visualize (accessed 05.06.19).

[30] Sehwag S, Das M,. A brief overview: present status on utilization of mustard oil and cake. Indian Journal of Traditional Knowledge. 2015;14:244–250.

[31] Sarwar G, Bell JM, Sharby TF, Jones JD. Nutritional evaluation of meals and meal fractions derived from rape and mustard seed. Canadian Journal of Animal Science. 1981;61: 719–733. DOI: 10.4141/cjas81-087.

[32] Newkirk R, Classen H, Tyler R. Nutritional evaluation of low glucosinolate mustard meals (*Brassica juncea*) in broiler diets. Poultry Science 1997;76: 1272–1277. DOI: 10.1093/ps/76.9.1272

[33] Thiyam-Holländer U, Aladedunye F, Logan A, Yang H, Diehl BWK. Identification and quantification of canolol and related sinapate precursors in Indian mustard oils and Canadian mustard products: Identification of canolol, sinapine and sinapic acid in mustard. European Journal of Lipid Science and Technology. 2014;116:1664–1674. DOI: 10.1002/ejlt.201400222

[34] Li, Y.; Li, J.; Su, Q.; Liu, Y. Sinapine reduces non-alcoholic fatty liver disease in mice by modulating the composition of the gut microbiota. Food Funct. 2019, 10, 3637–3649. DOI: 10.1039/c9fo00195f

[35] Dubie J, Stancik A, Morra M, Nindo C. Antioxidant Extraction from Mustard (*Brassica juncea*) Seed Meal Using High-Intensity Ultrasound. Journal of Food Science. 2013;78:E542–E548. DOI:10.1111/1750-3841.12085

[36] Galanakis CM, Goulas V, Tsakona S, Manganaris GA, Gekas V. A knowledge base for the recovery of natural phenols with different solvents. International Journal of Food Properties. 2013;16:382–396. DOI:10.1080/10942912.2010.522750.

[37] Chemat F, Abert-Vian M, Fabiano-Tixier AS, Strube J, Uhlenbrock L, Gunjevic V, Cravotto G. Green extraction of natural products. Origins, current status, and future challenges. Trends in Analytical Chemistry. 2019; 118: 248–263. DOI:10.1016/j.trac.2019.05.037

Chapter 12

Practicing Response Surface Designs in Textile Engineering: Yarn Breaking Strength Exercise

Nefise Gönül Şengöz

Abstract

Predicting properties of end product from known properties of raw material is an important part of quality control in manufacturing. Main concept in this research is to reach a specified property of end product from known properties of raw material by attaining response surface designs with feasible region. The Ne20–19.21 T/inch yarn breaking strength (response, desired value 450 cNs) is acquired from cotton fiber properties (variables). The relationship between response and variables are obtained in response surface drawings and contour plots. The area showing the desired value in contour plots are colored in lilac and are intersected to obtain the feasible regions. By reading backwards from the feasible region borders, the variable value ranges are reached which will give the desired value of the response is obtained. When this information to start the yarn production is ready, the cotton lots containing these fiber property value ranges will be bought or from raw material in hand we will be read which yarn breaking strength will occur at the end of production. It was concluded that response surface designs with feasible region are quick, practical, and effective tools, provide valuable results, contribute a lot to quality control, and are beneficial in textile quality control.

Keywords: design of experiments, response surface design, feasible region, central composite design (CCD), textile, yarn, fiber, breaking strength

1. Introduction

Textile engineering comprises manufacturing of apparel and home textiles used in daily life together with technical textiles. Nevertheless which kind of textile it is, from the point of view of textile production, quality control is very important in entire of them. The product has to possess properties in accordance to the areas where that textile product will be used and have to be welcomed by the people using them. In order to end up with a textile product with the properties preferred in the area of its usage, at the very beginning, the raw material chosen has to conform such properties to achieve the preferred properties at the end product. Quality control is performed in raw material, in every step of production downstream, and in the end product to guarantee the preferred properties of the end product keeping them in tolerances via various statistical methods. It should also be indicated here that quality control gained more importance in due time with the shortage of raw material, and lowering the costs because of high competitiveness in the market.

To succeed the end product to own the preferred properties, the properties of the raw material and the semi-products, production steps and machine setting are all very important and needs to be known in detail since they totally affect the properties of the end product. On the other hand, many research is practiced in textile engineering to predict the preferred properties of the end product from the known properties of the raw material and semi-products in the production downstream. The known properties are generally called variables and the predicted property is called a response(s) (**Figure 1**). Many distinct evaluative quality control tools in graphical, statistical, mathematical, and simulation methods are developed to investigate the relationships between variables and response(s). To name a few of these methods are histograms, data charts, analysis of variance, regression, correlation, control charts, artificial neural networks, discriminant analysis, principal component analysis, varimax, design of experiments, etc.

The statistical method Design of Experiments (DoE) evaluates the variable data which affects the response data by taking into consideration the variations in the variable data assuming to reflect the variation in the response data. Multiple variables can be worked and their effects on the response are obtained together with the important interactions between the variables. The purpose in using DoE is to determine the optimum variable values for the examined response. With the key concepts blocking, randomization and replication, DoE can reach the response in less time and is less costly in material and energy consumption, ensures quality in the product and reduces the need for inspection in quality control. DoE *are expressed by particular subgroups like* screening designs, and response surface designs, full factorial designs, fractional factorial designs, optimal designs, latin hypercube designs, quasi-random designs, mixture designs, discrete choice designs, space-filling designs, nonlinear designs, Taguchi designs, and augmented designs. Response surface design is applied in this research [1–10].

Response Surface Designs are a set of advanced DoE techniques in which the cause-and-effect relationships between known variables and response(s) are searched by a series of designed experiments to attain an optimized response(s) for robust manufacturing conditions. This modeling and analysis of problems method was introduced by George E. P. Box and K. B. Wilson in 1951 or around. Response surface designs are available for continuous factors, cover second-degree polynomial models, and consequently provide approximation, easy estimation and application even when variable data is small. There are two main types of response surface designs, Central Composite Designs (CCD) and Box–Behnken Designs. In this research, CCD is preferred because they can fit a full quadratic model, they are usually used when the plan of the design is appropriate for sequential experimentation, and they imply information from a factorial experiment. They have 5 levels/factor and they can include runs when all the variables are at their extreme settings [4, 5, 11–14].

If the response is "y" and variables are "x_1", "x_2", ..., "x_k" and deviation is "e", then the response is generally expressed by Eq. (1):

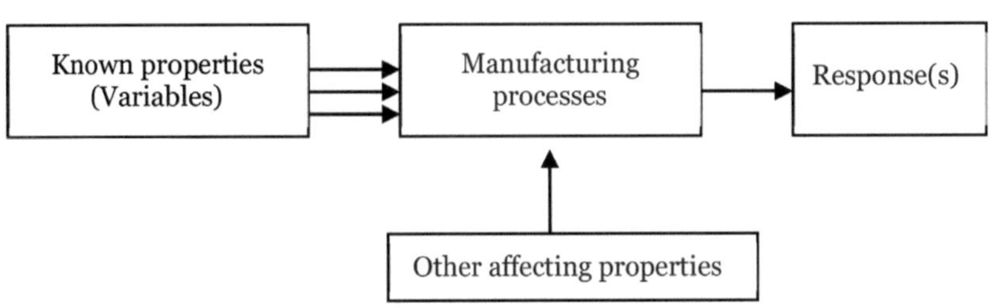

Figure 1.
Variables to response(s).

$$y = f\left(x_1,, x_2,,, \ldots x_k\right) + e \tag{1}$$

A linear model with two variables can be written as Eq. (2):

$$Y = \beta_0 + \beta_1 X_1 + \beta_2 X_2 + \beta_{12} X_1 X_2 + e \tag{2}$$

where; Y is the response for X_1 and X_2 variables, X_1X_2 term for interaction between X_1 and X_2, β's for regression coefficients, β_0 for the response of Y when both variables are zero; *e* is the experimental error; and for three variables (Eq. (3)):

$$\begin{aligned} Y = \beta_0 + \beta_1 X_1 + \beta_2 X_2 + \beta_3 X_3 + \beta_{12} X_1 X_2 + \beta_{13} X_1 X_3 \\ + \beta_{23} X_2 X_3 + \beta_{123} X_1 X_2 X_3 + e \end{aligned} \tag{3}$$

where; Y is the response for X_1, X_2, and X_3 variables, $k(k\text{-}1)/2 = 3^*2/2 = 3$ two-way interaction terms and 1 three-way interaction term, and β_0 for the response of Y when all variables are zero. When the data are analyzed, all the unknown β parameters are calculated and the coefficients of the X terms are tested to determine which ones are significantly different from zero; response surface designs are usually approximated by a second-order regression model because the higher-order effects are usually unimportant. A second-order regression model (full quadratic) for k number of variables for central composite design (CCD) can be written as Eq. (4):

$$\begin{aligned} Y = \beta_0 + \beta_1 X_1 + \ldots \beta_k X_k + \ldots \beta_{11}{}^2 X_1{}^2 + \ldots \beta_{kk}{}^2 X_k{}^2 \\ + \beta_{12} X_1 X_2 + \ldots \beta_{k-1,k} X_{k-1} X_k \end{aligned} \tag{4}$$

where; Y is the response for k number of variables, does not include three-way interaction terms but adds three more terms to the linear model, which are: $\beta_{11}X_{21} + \beta_{22}X_{22} + \beta_{33}X_{23}$ It should be noted that a full model contains many interaction terms of X^2's but in general these terms are not included and most DoE soft wares remove them out of the model; the quadratic terms model the curvatures in the response which is mapping a region of a response surface indicating how changes in variables affect a response variable; allows the levels of variables that optimize a response variable; and, determines the operating conditions to produce the product with preferred properties [3, 15–19].

The visual outputs of response surface designs are curved response surface drawings (**Figure 2**) and contour plots (**Figure 3**). They contribute to better interpretation of the behavior of the response and to clearly notice the conditions around the optimum response. Contour lines show the similar heights of x_1-x_2 couples at the same response value. In a response surface design study, both the response surface drawing and the contour lines support to understand the trend of a response in different values of variables effecting it.

A further approach to response surface designs is the feasible region.. If there are more than one response, then the different response surface drawings are put on top of each other and their intersection area is the feasible region which optimizes both response variable's values as seen in **Figure 4**. In this research, feasible regions are studied from another point of view where different contour plots of variable couples are superposed to provide the response value range covering them all. The variables' values are read backwards from the feasible region to reach the desired response. The variable values guarantee the desired response at the end of the production. So, the concept explained above to start with raw material possessing properties which

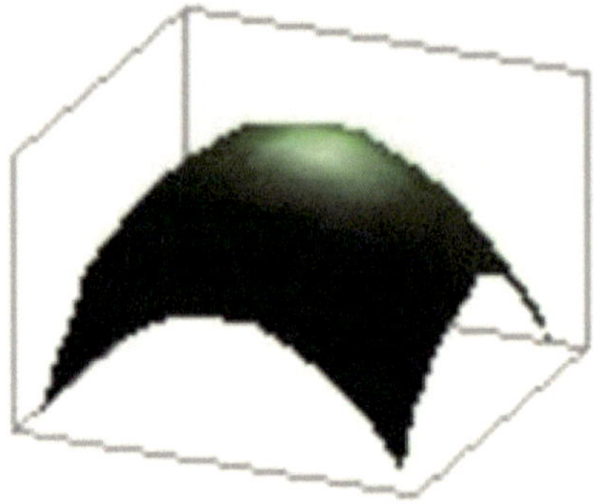

Figure 2.
Example of curved response surface drawing [13].

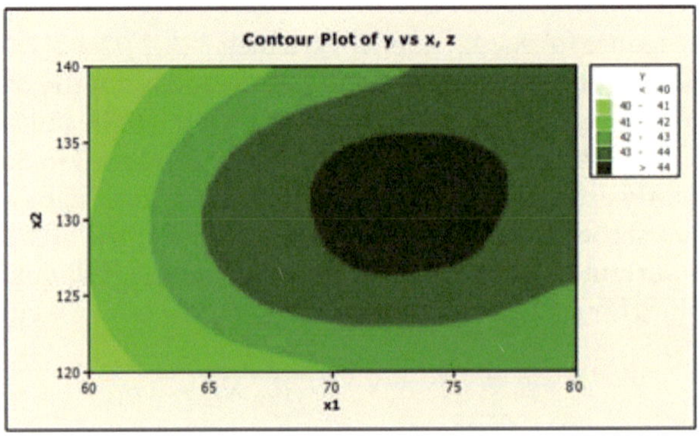

Figure 3.
Example of contour lines [25].

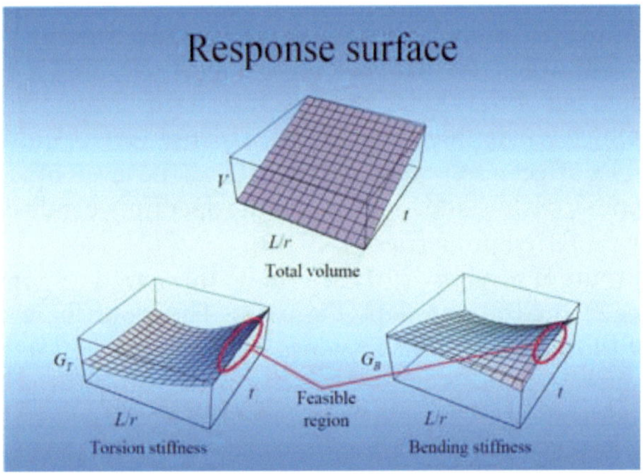

Figure 4.
Feasible region to optimize two response variable's values [25].

will end up with the desired property will be succeeded. This overall information demonstrates that response surface designs are quick, practical, and effective tools for both manufacturing and quality control. Response surface designs are used in many industries like mechanical, automotive, medical, chemical, electronics, etc., but very few in textile engineering, and none in textile quality control, except our previously published article, where response surface design with feasible region was applied to yarn irregularity property, now it is applied to yarn breaking strength property [5, 20–25].

2. Material and method

Textile production downstream mainly consists of fibers, yarns, fabrics, and confection; finishing may well be in every step. Fibers may be natural (cotton, wool, etc.) (**Figures 5-7**) or man-made (polyester, polyamide, etc.) (**Figure 8**) and have properties like length, fineness, breaking strength and breaking elongation, elasticity, maturity, regain, trash, oil content, color, static electricity, etc. Yarns may be spun from one kind of fiber, being 100% or blends of different fibers, may be spun in different techniques (ring, Open-End, etc.) (**Figure 9**) and have properties like count (in Ne, tex, etc. units) and twist (in T/inch or T/m unit) which are adjusted on the spinning machine, and breaking strength and breaking elongation, elasticity, abrasion resistance, moist content, oil content, irregularity, hairiness, static electricity, etc. which are outcomes relative to various variables. To provide a short information, some properties mentioned above are specified here: Yarn irregularity is a measure of deviation of fibers' orientation in the yarn from the orientation in the ideal yarn, which is expressed as the mean linear irregularity (U% unit) or coefficient of variation of the yarn mass (CVm% unit). Breaking strength for fibers and yarns is the maximum tensile force measured during the strength test which is expressed as BForce in cNs unit. Yarn

Figure 5.
Cotton boll [26].

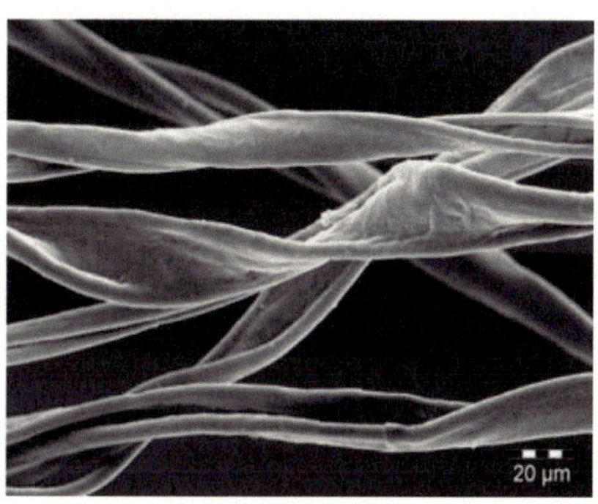

Figure 6.
Cotton fiber [27].

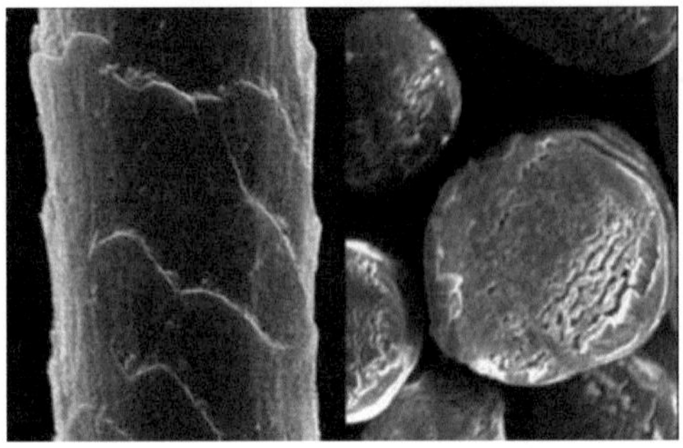

Figure 7.
Wool fiber [28].

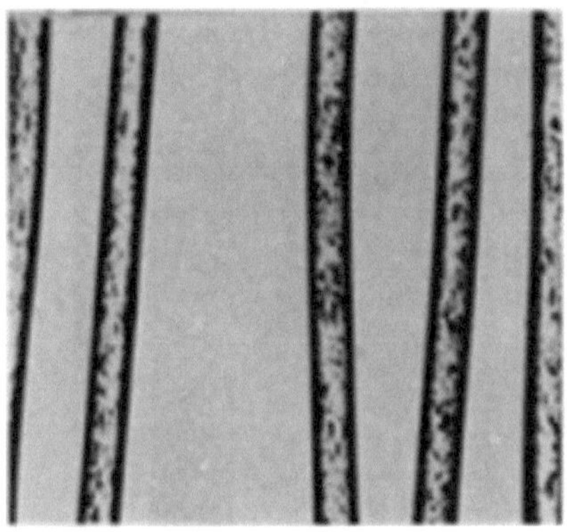

Figure 8.
Polyester fiber [29].

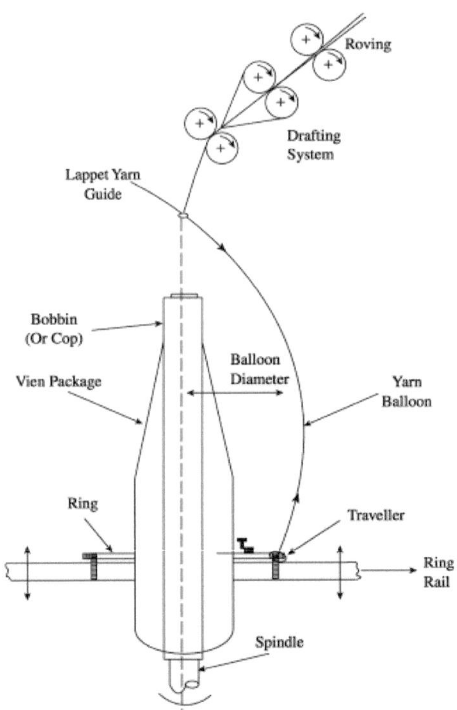

Figure 9.
Ring spinning technique of yarns [30].

breakages have many different causes like raw material, machine settings, climate conditions, human factors, etc., so it needs to be known before starting the production.

In this research, response surface designs with feasible region are applied to yarn breaking strength data which are considered as the response and fiber properties forming the same yarn are considered as the variables. The data for yarn breaking strength and fiber properties are real production data obtained from a company producing 100% cotton yarn in Uşak-Turkey. The fiber properties are fineness/micronaire index (Mic), maturity index (Mat), length (Len), fiber length uniformity index (Unf), short fiber index (SFI), fiber breaking strength (Str), fiber breaking elongation (Elg), moisture content (Moist), reflectance (Rd), yellowness (b), trash count (Tr_Cnt) and trash % area (Tr_Area), being twelve different ones. The lower the micronaire index (Mic) value the fibers are fine, the higher the Mic value the fibers are coarse. The higher the maturity index (Mat) value the cotton fibers constitute more cellulose layers in their cross-section. The higher the fiber length uniformity index (Unf) value the length distribution of the fibers in a lot are close to each other, in other words, the length difference is small. The lower the short fiber index (SFI) value the fiber lot contains less amount of short fibers. The higher the yellowness (b) value the cotton fibers are more mature. The trash count (Tr_Cnt) is the amount of trash counted on the measuring screen and is proportional with the lot, the trash % area (Tr_Area) is the area of the measuring screen the trash occupies and is proportional with the lot. In both Tr_Cnt and Tr_Area, and the rest length (Len), fiber breaking strength (Str), fiber breaking elongation (Elg), moisture content (Moist) and reflectance (Rd) the higher values indicate that property possesses high results in the tests.

The tests to obtain fiber property values were carried out in Uster®HVI Spectrum apparatus (**Figure 10**) in 20 ± 2 °C, 65 ± 2% relative humidity standard atmospheric conditions in the laboratory of the factory. The cotton fiber lots are from Adana region in Turkey. The production downstream is blowroom, carding, I.drawing, II.drawing, roving and ring spinning machines The yarn is 100% cotton, carded, ring spun, count Ne20, twist 19,21 turns per inch (T/inch), and in bobbin form. The yarn properties are irregularity (%U, %CVm), imperfections (Thin50, Thick50, Neps200), hairiness (H), yarn breaking strength (BForce), yarn breaking elongation (Elongation), tenasity (Rkm), work to break (Bwork), etc. The tests to obtain yarn breaking strength values were carried out in Uster Tensorapid 3 apparatus (**Figure 11**) which is used in the regular measurements of the factory. In this apparatus, breaking strength (BForce) in cNs unit, breaking strength (Tenacity) in cN/tex (Rkm) unit, breaking elongation (Elong) in %, and work to break (BWork) in cN.cm unit were also obtained but the yarn breaking strength (BForce) in cNs is processed with 114 repeats in this research. Yarn breaking strength testing is given in **Figure 12** and a yarn at break is seen in **Figure 13**.

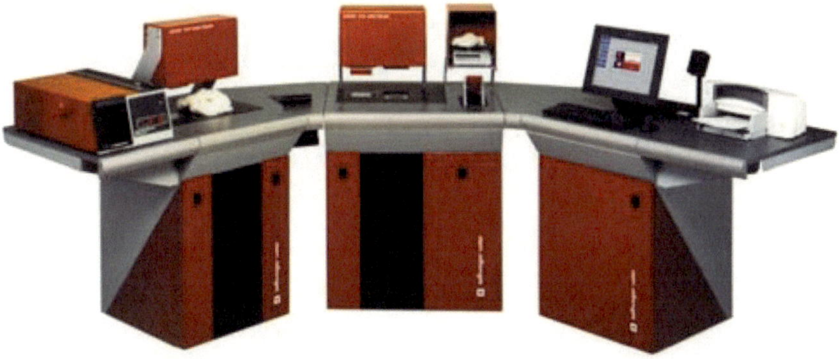

Figure 10.
Uster®HVI Spectrum apparatus [31].

Figure 11.
Uster Tensorapid 3 apparatus [32].

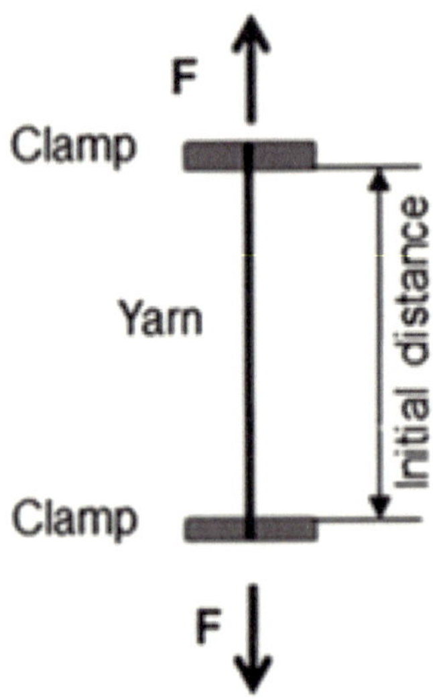

Figure 12.
Yarn breaking strength testing [33].

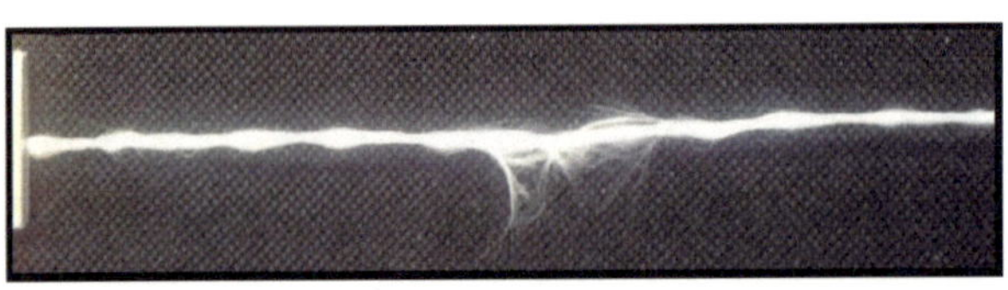

Figure 13.
A yarn at break [34].

Similar data of fibers and yarn was used in our previous research [25] where the yarn irregularity property in U% unit was studied via response surface designs with feasible region, but in this research yarn breaking strength property is studıed which is different from our former article.

	Mic	Mat	Len	Unf	SFI	Str	Elg	Moist	Rd	b	Tr_Cnt	Tr_Area
Mean	4,9781	0,9203	29,4522	85,5337	7,0765	30,3041	9,6827	7,9306	67,3531	8,2459	54,2449	1,5232
SD	0,1377	0,0130	0,4465	0,7832	0,4681	1,0551	0,289	0,8858	1,7809	0,4424	12,4504	0,3567
%CV	2,766	1,417	1,516	0,916	6,615	3,482	2,984	11,169	2,644	5,365	22,952	23,42

Table 1.
Means, SDs and CV% of fiber properties [25].

		%U	%CVm	H	Thin50	Thick50	Neps200	BForce	Elongation	Rkm	BWork
	Mean	11,28400	14,46330	6,01670	1,75000	216,91700	285,16700	442,74700	5,25630	14,99270	626,86300
Bobbin	SD	0,22695	0,34045	0,28067	1,87430	66,99100	150,88520	24,01450	0,46158	0,81246	79,31180
	%CV	2,011	2,354	4,665	107,103	30,883	52,911	5,424	8,781	5,419	12,652

Table 2.
Means, SDs and CV% of yarn properties [25].

The values for twelve fiber properties mentioned above consist of 98 different lots in bales with 5 repeats each. In this research, values in means are processed in order to get rid of the small differences between the lots, not to decline the power of response surfaces. The overall means, standard deviations and constant of variations are given in **Table 1**.

The values for yarn breaking strength consists of 30 lots in bobbin form with 10 repeats each. The values in means are also processed in order to get rid of the small differences between the lots, not to decline the power of response surfaces. The overall means, standard deviations and constant of variations are given in **Table 2**. Yarns given in **Table 2** are produced from the fibers given in **Table 1** in the factory from which the real production data are obtained.

The response (BForce of %100 cotton carded yarn) and variables (cotton fiber properties) are summerized in **Table 3**.

To elaborate the response and the variables considered for the experiments, they are implemented to **Figure 1** as seen in **Figure 14**. The main goal is to predict a property of the end-product (response) from the known properties of the raw material (variables), where in this research yarn breaking strength (BForce, marked bold) is the response and the fiber properties are the variables. This information will help the manufacturer to be aware of what will be reached at the end-product with the known properties of the raw material before starting production, so will conduct the factory more efficiently.

Response surface designs with feasible region which are regarded as satisfactory tools in textile quality control are applied in MINITAB program with central composite

Response	Variables									
BForce	Mic	Mat	Len	Unf	SFI	Elg	Rd. b	Tr_Cnt	Tr_Area	Str	Moist

Table 3.
Summary of the Response and the Variables.

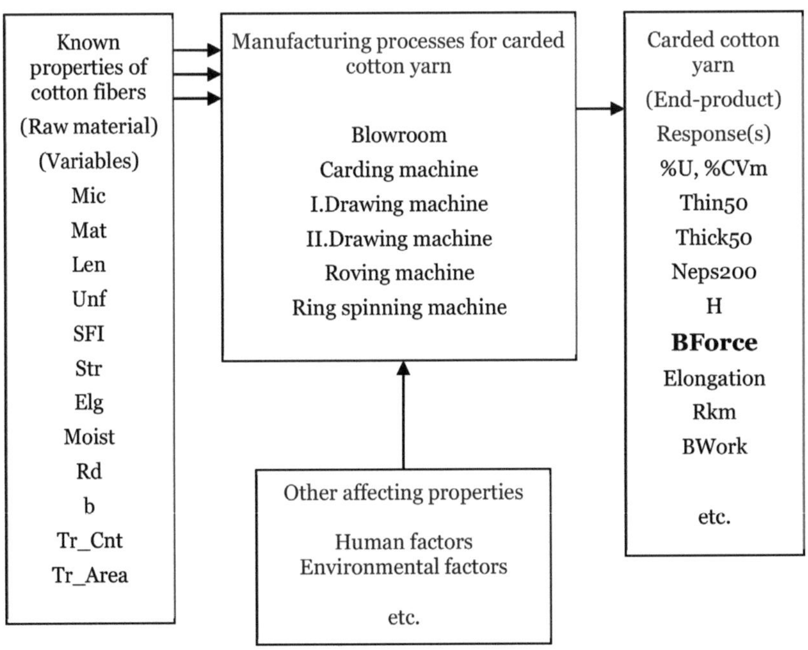

Figure 14.
Implemented variables to response(s).

Response Z axis	X axis variable versus Y axis variable being a different one every time			
BForce	Mic - Mat	Mic - Len	Mic - Unf	Mic - SFI
	Mic - Elg	Mic - Rd	Mic - b	(Seven combinations)
	Mic - Tr_Cnt	Mic - Tr_Area	Mic - Str	
	Mic - Moist	(Four combinations)		
	Totally eleven combinations			

Table 4.
Combinations for response surface drawings and contour plots.

design (CCD) method to obtain the relationship between the response (BForce) and the variables (fiber properties). The response is located in the Z axis, one variable (Mic) in the X axis and another variable in the Y axis, the variable in the Y axis being a different one every time. Eleven combinations of response surface drawings in 3D and contour plots are obtained and the combinations are listed in **Table 4**.

The area in the contour plots conforming greater than 450 cNs are colored in lilac and afterwards, the colored contour plots are put on top of each other by one

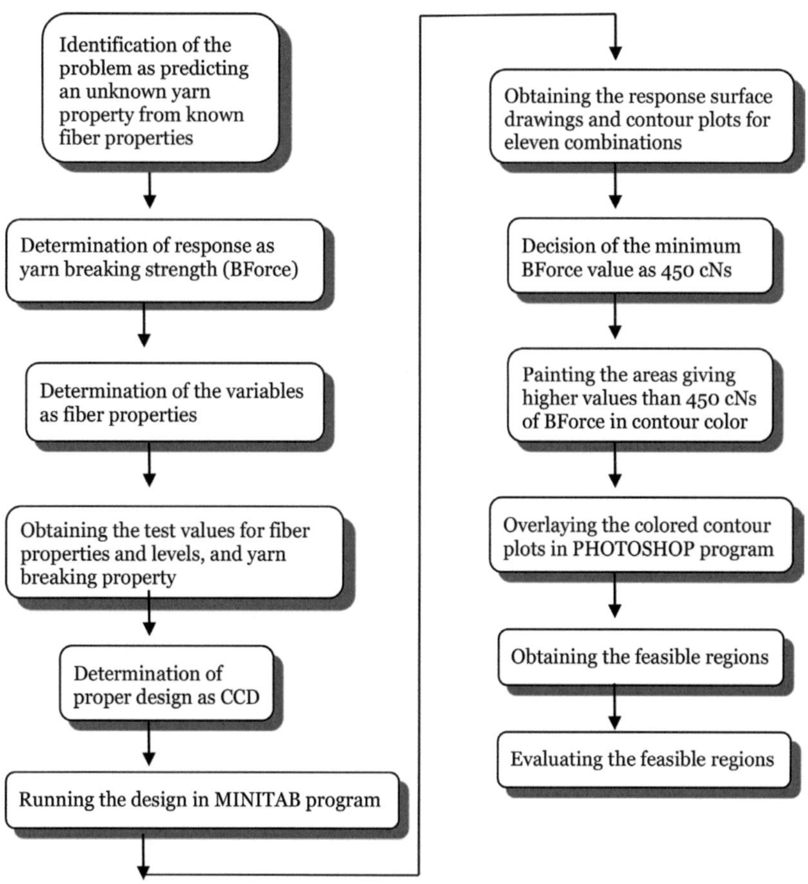

Figure 15.
The flow chart of the steps of the experimental study followed.

by in PHOTOSHOP program and the intersection of the lilac colored areas are noted as the intersection of the desired areas, where finally the borders of the feasible region is acquired and it is painted in red color. The feasible region is where the same greater than 450 cNs BForce (yarn breaking strength) values' areas are found for the most number of combinations because two feasible regions formed: One for the seven combinations in **Table 4** and the other for the four combinations in the same table. By reading backwards from the feasible regions to the BForce response in the response surface drawings and contour plots, the value range of the fiber properties which conform the yarn breaking strength higher than 450 cNs BForce values are determined adequately. The same study can be done for other yarn properties, say in a computer program for example, and all of their combination would give the fiber properties suitable for the desired yarn properties required in the end product.

A visual observation always provides better knowledge about a process [21, 35, 36]. To accentuate the concept of this research, the steps of the experimental study followed is summerized as a flow chart in **Figure 15**.

3. Results and discussion

The results of response surface designs, applied to yarn breaking strength and the properties of fibers forming the same yarns, which consists of response surface drawings, contour plots, lilac colored contour plots, and feasible regions are given in this section and are discussed.

Yarn breaking strength (BForce) is taken as the response and response surface design with feasible region is applied as variables of cotton fiber properties which are fineness/micronaire index (Mic), maturity index (Mat), length (Len), fiber length uniformity index (Unf), short fiber index (SFI), fiber breaking strength (Str), fiber breaking elongation (Elg), moisture content (Moist), reflectance (Rd), yellowness (b), trash count (Tr_Cnt) and trash % area (Tr_Area). Eleven combinations formed which are Mic-Mat, Mic-Len, Mic-Unf, Mic-SFI, Mic-Str, Mic-Elg, Mic-Moist, Mic-Rd, Mic-b, Mic-Tr_Cnt, and Mic-Tr_Area. Response surface drawings (**Figures 16** and **17** I. Colomns) and contour plots (**Figures 16** and **17** II. Colomns) are obtained for all eleven combinations.

In the contour plots, the different shades between contour lines implicate different value ranges for yarn breaking strength (**Figures 16** and **17** II. Colomns) and it is noticed that all the eleven combinations have an area of contour line conforming yarn breaking strength higher than 450 cNs except combinations for Mic-Mat and Mic-Elg. being 448 cNs. Since the difference is small and the t-test gave $p = 0{,}741$ meaning the difference is insignificant, the desired yarn breaking strength value (response BForce) is taken as greater than 450 cNs for the response surface design with feasible region in this research. The condition of being 450 cNs is also tested with the mean value for BForce being 442,747 cNs in **Table 2** which creates a normal distribution curve. $p = 0{,}155$ is obtained which is statistically insignificant and also proves that the data used is distributing normal.

In **Figure 16**, the response surface drawings in the I. Colomn, the contour plots in the II. Colomn, and the lilac colored contour plots in the III. Colomn are given for the seven combinations of eleven ones which are Mat, Len, Unf, SFI, Elg, Rd., and b versus Mic. In **Figure 17**, the response surface drawings in the I. Colomn, the contour plots in the II. Colomn, and the lilac colored contour plots in the III. Colomn are given for the four combinations of eleven ones which are Tr_Cnt, Tr_Area, Str, and Moist versus Mic. In **Figure 18**, the feasible region acquired by superposing the lilac colored contour plots in **Figure 16** III. Colomn is seen in red color, and, in **Figure 19**, the

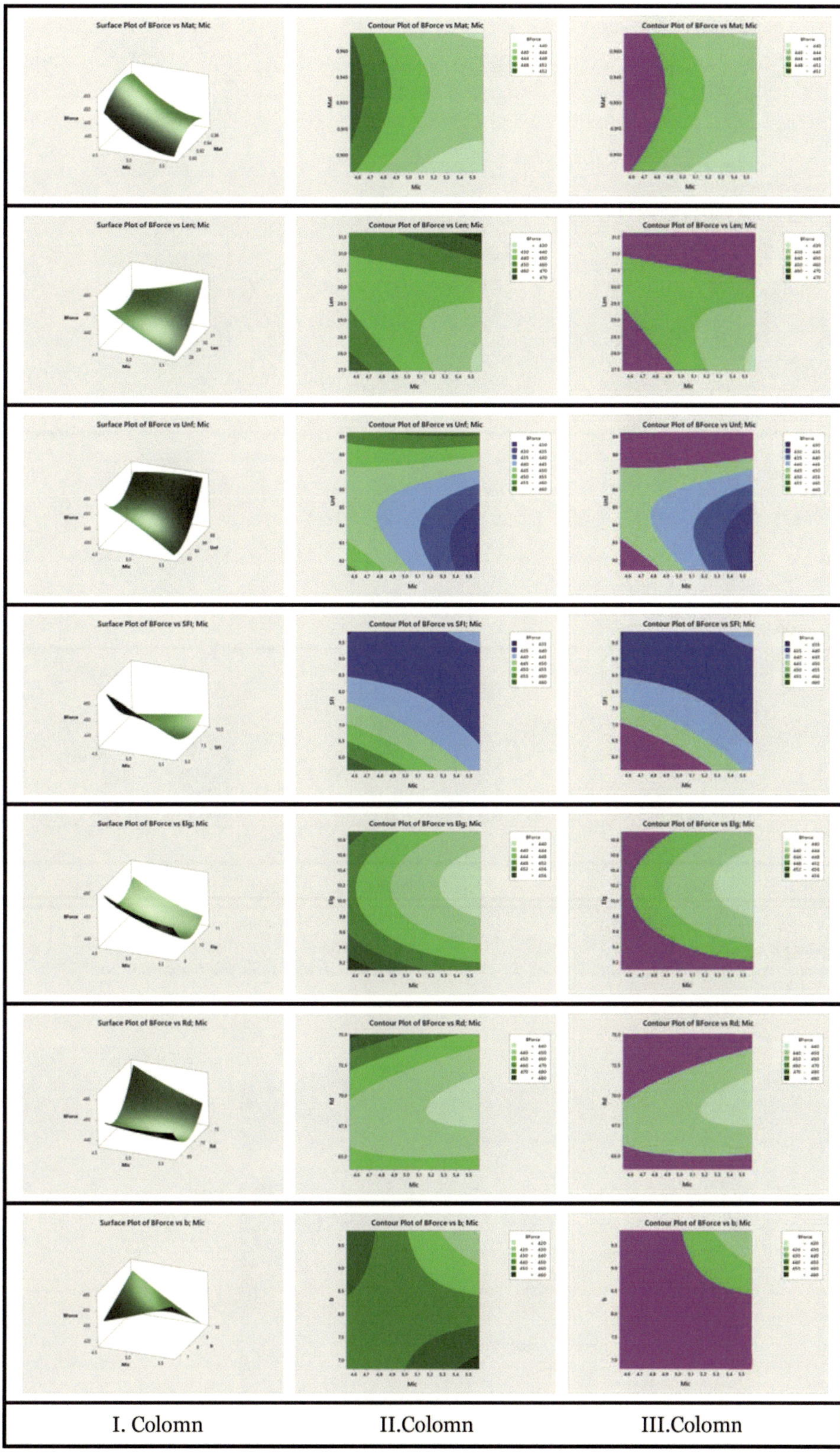

Figure 16.
BForce as response and Mic, Mat, Len, Unf, SFI, Elg, Rd., and b as variables, seven combinations.

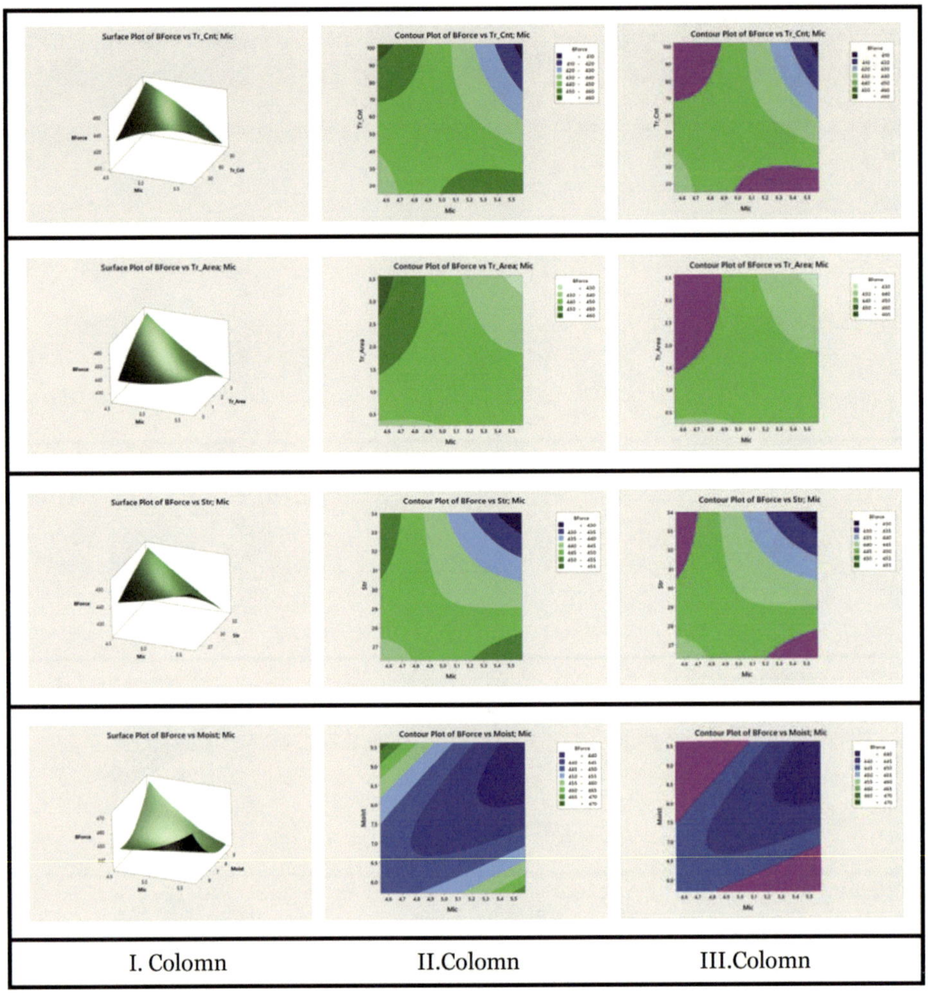

Figure 17.
BForce as response and Mic, Mat, Tr_Cnt, Tr_Area, Str, and Moist as variables, four combinations.

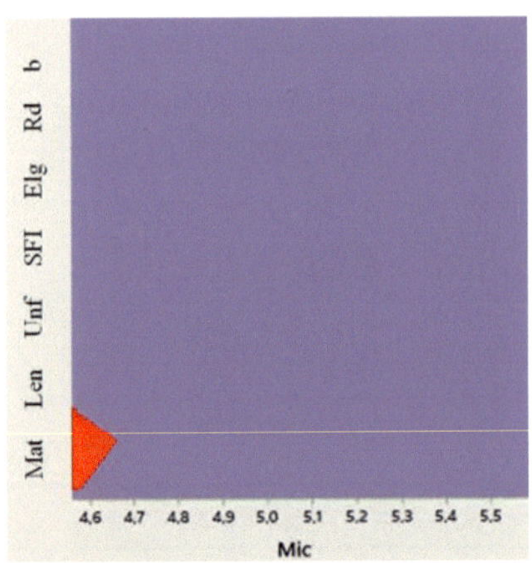

Figure 18.
Feasible region for BForce as response of seven combinations.

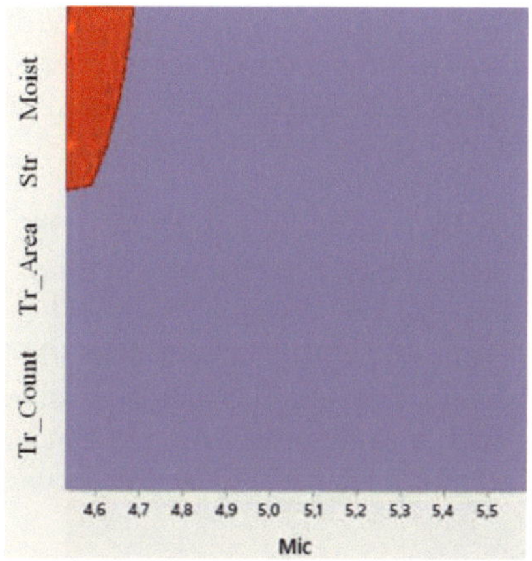

Figure 19.
Feasible region for BForce as response of four combinations.

feasible region acquired by superposing the lilac colored contour plots in **Figure 17** III. Colomn is seen in red color, both conforming yarn breaking strength higher than 450 cNs. Two feasible regions formed because when all the colored contour plots were put on top of each other there occurred to be two regions. To get a better understanding of the feasible regions, they are separated into two figures.

Examining the I. Colomns of **Figures 16** and **17**, it can be discussed that the behavior of each eleven combination is different from each other. This means each fiber property has a different influence on the yarn breaking strength, while one has an increasing impact at its high values the other possesses a decreasing one or increasing impact at both high and low values the other a decreasing one. General speaking; the curvature of response surface looks downwards in five combinations which are Mic-Mat, Mic-b, Mic-Tr_Cnt, Mic-Tr_Area, and Mic-Str, while upward looking curvature of response surfaces are in six combinations which are Mic-Len, Mic-Unf, Mic-SFI, Mic-Elg, Mic-Rd, and Mic-Moist. The downwards looking curvatures indicate high values of yarn breaking strength at their top sections, the upwards looking curvatures indicate high values of yarn breaking strength at their side sections. The relationship between the fiber properties outgiving downwards and upwards curvatures needs to be discussed. The downwards curvatures come from fiber properties fineness, maturity index, yellowness, fiber breaking strength, trash count, and trash % area. The first four properties are dependent on agricultural factors so it is reasonable to behave similar in their curvatures. The last two properties are dependent on harvesting factors and they also behave the same as agricultural factors. The relationship between the agricultural factors and the harvesting factors is another aspect of investigation aroused in this research. The upwards curvatures come from fiber properties fineness, length, fiber length uniformity index, short fiber index, fiber breaking elongation, reflectance, and moisture content. All properties except moisture content are dependent on agricultural factors so it is reasonable to behave similar in their curvatures. However, the opposite curvature behavior than the downwards curvature needs to be investigated. Moisture content property is dependent on storing and transporting factors. The relationship between the opposite observated agricultural factors and the storing and transporting factors is also another aspect of investigation. Another discussion of the response surface drawings is the attitude of the edges of the curvatures. The shape, length, height, inclination, and deflection of the curvatures seem to be important and should be evaluated. These arguments aroused in this research and will be analyzed in imminent investigations.

Examining the II. Colomns of **Figures 16** and **17**, it can be discussed that these impacts can apparently be seen in different shades of color in which different yarn breaking strength values are specified. Considering both the response surface drawings and the contour plots, it is attained that yarn breaking strength is higher than 450 cNs when;

- Cotton fiber maturity (Mat) is inbetween 0,908–0,967 and cotton fiber fineness (Mic) is less than 4,63; meaning that cotton fibers are mature and fine;

- Cotton fiber length (Len) starts to get longer than 30.02 mm and cotton fiber fineness is higher than 4,5, and, length starts to get shorter than 29,38 mm and fineness is less than 4,94; meaning that long and fine to coarse cotton fibers, and, short and fine cotton fibers have the similar impact on both sides;

- Cotton fiber lot uniformity (Unf) starts to increase at 87,2 and cotton fiber fineness is higher than 4,5, and, uniformity is less than 83,1 and fineness is less than 48,4; meaning that uniformity other than 83,1–87,2 range with fine to coarse cotton fibers have the similar impact on both sides;

- Cotton fiber lot short fiber index (SFI) is less than 7,0 and cotton fiber fineness is less than 5,25; meaning that cotton fiber lots should contain less short fibers and be fine;

- Cotton fiber elongation (Elg) is inbetween 9,1–10,9 and cotton fiber fineness is higher than 4,5; meaning that cotton fibers should elongate and be fine unless coarse as in the middle of the contour plot;

- Cotton fiber reflectance (Rd) starts to increase at 69,5 and less than 65,8 and cotton fiber fineness is higher than 4,5; meaning that cotton fibers should possess reflectance other than 65,8–69,5 range and be fine unless coarse as in the middle;

- Cotton fiber yellowness (b) is other yellownesses that start to increase at 8,48 and cotton fiber fineness is higher than 5,02; meaning that cotton fibers should be slightly yellow and fine to coarse;

- Cotton fiber lot trash count (Tr_Cnt) starts to increase at 68 and cotton fiber fineness is less than 4,87, and, trash count starts to decrease at 28 and fineness is higher than 4,98; meaning that trash count other than 28–68 range with fine and coarse cotton fibers, respectively, have the similar impact on both sides;

- Cotton fiber lot trash % area (Tr_Area) is higher than 1,4 and cotton fiber fineness is less than 4,89; meaning that cotton fiber lots should contain rather high trash % area and be fine;

- Cotton fiber breaking strength (Str) starts to increase at 30.5 cNs and cotton fiber fineness is less than 4,69, and, fiber breaking strength starts to decrease at 27,8 cNs and fineness is higher than 5,21; meaning that strong and fine, and, less strong and coarse cotton fibers have the similar impact on both sides;

- Cotton fiber moisture content (Moist) starts to increase at 7,5 and cotton fiber fineness is less than 4,92, and, fiber moisture content starts to decrease at 6,76 and fineness is higher than 4,8; meaning that normal to high moisture and fine cotton fibers, and, less moisture and coarse cotton fibers have the similar impact on both sides.

The information the feasible regions provide is achieved by reading backwards from the feasible regions in **Figures 18** and **19**. For each variable (fiber properties), the feasible region borders are corresponded with that variable's scale in the contour plots one by one and the value ranges of the variables are determined. The results are given in **Table 5**. Cotton fiber lots retaining fiber properties within these ranges will reach to yarn Ne20–19.21 T/inch having yarn breaking strength higher than 450 cNs.

The results obtained is discussed in two different points of view:

- In a factory where cotton Ne20–19.21 T/inch ring spun yarn is produced, cotton lots have to be supplied and then the fiber properties have to be tested and results obtained. Then, the new fiber property results have to be crosschecked with the fiber properties given in **Table 5** and see if they fall in between these ranges. If they do, this means the Ne20–19.21 T/inch yarn which will be produced from those lots will possess a breaking strength higher than 450 cNs. If some of them do not fall in between these limits, then we have to read backwards from **Figure 18** or **Figure 19** to figure out what the yarn breaking strength (BForce) will be in the new produced yarn;

- In a factory where cotton Ne20–19.21 T/inch ring spun yarn is produced, cotton lots will be ordered with the property value ranges in **Table 5**, and cotton bales possessing these properties will be bought. In order to do so, there has to be a system in the country where every cotton bale will have its fiber property test results stuck on them in a central warehouse, so the sales of the bales will be according to the values in the factory's order, the bales possessing these features will be chosen and sold to the factory. Furthermore, lots with much better fiber properties can be ordered.

In this research, one yarn property (breaking strength-BForce) is studied while yarn irregularity property (U%) was studied in our previous work [24, 37] however the same work can be done for different yarn properties such as tenacity (cN/tex; Rkm), breaking elongation (%), yarn irregularity (CVm%), hairiness (H Index), product of yarn count and tenacity multiplication (CSP), constant of variation of count (CV_C%), constant of variation of twist (CV_{Twist}%) constant of variation of tenacity ($CV_{Tenacity}$%), and imperfections (thin places, thick places, neps (piece/

	Variable	Range
From **Figure 16**	Mic	4,58 – 4,69
	Mat	0,890 – 0,908
	Len	27,5 – 28,25
	Unf	81,5 – 82,9
	SFI	5,5 – 6,41
	Elg	8,83 – 9,42
	Rd	64,17 – 66,2
	b	6,88 – 7,4
From **Figure 17**	Tr_Cnt	68,9–100
	Tr_Area	2,25 – 3,00
	Str	31,22 – 34,00
	Moist	8,18 – 9,5

Table 5.
Range of fiber properties for reaching yarn Ne20–19.21 T/inch breaking strength higher than 450 cNs.

km)), comparisons of these properties, ring and rotor yarns, fabric properties, different machine settings, etc. The variables can be chosen from the point of view of the important properties and it can be worked with less or more effecting variables to perform feasible regions. When choosing the effecting variables, the opposite can be done, so that the yarn properties can be the effecting variables, a fiber property may be the response variable, and the feasible regions can be accomplished vice versa, the concept being knowing what will be reached at the end before starting the production, therefore the advantages of this work is obvious. The originality of this concept is that if all this work will be incorporated into a computer program for statistical quality control, then the same research will be done for every different yarn property and for every fiber lot arriving the factory. New feasible regions will occur in due time, when the factory will make different settings in machines or renew machines in production downstream they will compare the changes in the feasible regions and conclude if the new ones help for better or worse. The developed computer program can also be used in many different industrial applications which will yield to more developed evaluative comments for the feasible regions in due time.

4. Conclusions

In industry, it is important to predict a specific property of an end product from the known properties of raw material. There are many statistical methods for prediction in literature but response surface designs with feasible region is not benefitted much, even it is an effective and versatile tool in prediction. In this research, the concept of using response surface designs with feasible region is accomplished in textile engineering quality control and is concluded that response surface designs with feasible region is important from the point that it is a quick, practical, and comprehensive tool for predicting properties of an end product from the raw material.

Yarn breaking strength property is studied by response surface designs with feasible region and this property is predicted from the fiber properties, fibers the raw material forming the yarn. The data used in this research is obtained from a company producing 100% cotton yarn in Uşak-Turkey, the company bought the cotton lots from Adana-Turkey, and the company carries out these measurements regularly during their daily production, so the important point here is that real production data are used and is guaranteed that the results achieved from this research will be suitable for production.

MINITAB program is used obtain response surface drawings and contour plots where response is the 100% cotton yarn breaking strength (BForce) and the variables are cotton fiber properties, fineness/micronaire index (Mic), maturity index (Mat), length (Len), fiber length uniformity index (Unf), short fiber index (SFI), fiber breaking strength (Str), fiber breaking elongation (Elg), moisture content (Moist), reflectance (Rd), yellowness (b), trash count (Tr_Cnt) and trash % area (Tr_Area), revealing the relationship between the yarn breaking strength and fiber properties, yarn in bobbin form. The desired value of yarn breaking strength is decided higher than 450 cNs. After obtaining the response surface drawings and contour plots, the areas in contour plots conforming desired values higher than 450 cNs were marked in lilac color, the colored contour plots were put on top on each other in PHOTOSHOP program, the intersecting area of gave the two feasible regions which are marked in red. Then, the borders of the feasible region is read backwards by corresponding the feasible region with that variable's scale in the contour plot for each different cotton fiber property. This procedure concludes the determination of the range of values of fiber properties which will reach to yarn Ne20–19.21 T/inch having the desired value higher than 450 cNs for the response yarn breaking strength.

The behavior of the response surface drawings is different in each eleven combination. It is concluded that each fiber property has a different influence on the yarn breaking strength, such as, whilst one has an increasing impact at its high values the other possesses a decreasing one, or, increasing impact at both high and low values the other decreasing, naming them downwards and upwards looking curavutures. Analyzing the downwards and upwards looking curvatures generally, it is concluded that fineness, maturity index, yellowness, fiber breaking strength, trash count, and trash % area have curvatures looking downwards, the first four being dependent on agricultural factors and the last two dependent on harvesting factors. Both agricultural and harvesting factors having similar effect on response, yarn breaking strength. On the other hand, fineness, length, fiber length uniformity index, short fiber index, fiber breaking elongation, reflectance, and moisture content have curvatures looking upwards, all except moisture content are dependent on agricultural factors, moisture content depends on storing and transporting factors. In this case both agricultural, and, storing and transporting factors having similar effect on response, yarn breaking strength. The reason why some conclude in downwards look and some conclude upwards look needs further research. Besides, the behavior of the edges of the curvatures such as their shape, length, height, inclination, and deflection are divergent and demand research in detail. Future research will be continued on the relationships between the agricultural, harvesting, storing, and transporting factors, the opposite manner of the curvatures, attitude of the edges of the curvatures, which all aroused in this research.

The information provided by the feasible region is evaluated in two different ways; first, the fiber property results of a new arriving cotton fiber lot to the factory can be compared with the value ranges of fiber properties achieved for yarn Ne20–19.21 T/inch yarn breaking strength higher than 450 cNs in this research and conclude if the desired value for yarn breaking strength will be achieved at the end of yarn production, conversely if the value ranges do not match, the new yarn breaking strength can also be determined by reading backwards from the feasible region; and the second, when ordering cotton fiber lots to the factory, to give the limits of cotton fiber properties and buy the cotton bales encompassing those values from a central warehouse, which needs a new system of cotton fiber trade in the country.

Response surface designs with feasible region serve to the main goal of predicting a property of the end-product (response) from the known properties of the raw material (variables), where in this research 100% cotton yarn breaking strength (BForce) is exercised the response and the cotton fiber properties forming the yarn are the variables, in our previous research yarn irregularity property (U%) was exercised. The information acquired by response surface designs with feasible region will help the manufacturer to be aware of what yarn breaking strength will be reached at the end-product with the known properties of the raw material cotton fibers while they are still in the bales before starting yarn production, so will run the factory more efficiently, will yeild less waste, will achieve less costs and higher profit, have better relationships with customers, will have positive effects on the economy of the country and the world. The same work can be done for all different properties of textiles where the variables can be chosen from the point of what is aimed at the end products, being responses. Further work will be done to achieve these goals, also by chosing different desired values. On the other hand, working response surface designs in textile engineering quality control is an original concept, and will be incorporated into a computer program for statistical quality control, and this will give the opportunity to get response surface drawings and contour plots and feasible regions for every different yarn property and for every fiber lot arriving the factory, as well as in different fields of textiles, leading to standardization at the far end, also to be convenient to be utilized in many different industry branches.

Acknowledgements

The author gratefully acknowledges the data provided by Kaynak İplik A.Ş.-Uşak-Turkey.

Author details

Nefise Gönül Şengöz
Usak University Engineering Faculty Textile Engineering Department, Bir Eylül Kampüsü, Uşak, Turkey

*Address all correspondence to: nefisegonul.sengoz@usak.edu.tr

References

[1] Antony J. Design of Experiments for Engineers and Scientists. 1st ed. Great Britain: Elsevier Science & Technology Books; 2003. 190p.

[2] JMP 8 Design of Experiments Guide. 2nd ed. Cary, North Carolina: Sas Publishing; 2009. 273p.

[3] Box G E P, Hunter J S, Hunter W G. *Statistics for experimenters: design, discovery and innovation*. Hoboken, New Jersey: Wiley; 2005. 633p.

[4] Dean A, Morris M, Stufken J, Bingham D. Handbook of Design and Analysis of Experiments. Boca Raton: CRC Pres; 2015. 924p.

[5] Montgomery D C. Introduction to Statistical Quality Control. 4th ed. New York: Wiley; 2001. 796p.

[6] Design of Experiments. [Internet]. Available from http://www.mathworks.com/design-of-experiments. [Accessed: 2020-11-03]

[7] What is DOE? Design of Experiments Basics for Beginners. [Internet]. Available from https://www.sartorius.com/en/knowledge/science-snippets/what-is-doe-design-of-experiments-basics-for-beginners-507170. [Accessed: 2020-11-03]

[8] What is Design of Experiments (DOE)? [Internet]. Available from https://asq.org/quality-resources/design-of-experiments. [Accessed: 2020-11-05]

[9] Design of experiments. [Internet]. Available from https://en.wikipedia.org/wiki/Design_of_experiments. [Accessed: 2020-11-06]

[10] Design of Experiments – A Primer. [Internet]. Available from https://www.isixsigma.com/tools-templates/design-of-experiments-doe/design-experiments-%E2%90%93-primer/. [Accessed: 2020-11-06]

[11] Response surface methodology. [Internet]. Available from https://en.wikipedia.org/wiki/Response_surface_methodology. [Accessed: 2020-11-18]

[12] Box-Behnken Response Surface Methodology. [Internet]. Available from https://www.theopeneducator.com/doe/Response-Surface-Methodology/Box-Behnken-Response-Surface-Methodology. [Accessed: 2020-12-02]

[13] Response Surface Methodology. [Internet]. Available from https://support.minitab.com/en-us/minitab/18/help-and-how-to/modeling-statistics/doe/supporting-topics/response-surface-designs/response-surface-central-composite-and-box-behnken-designs/. [Accessed: 2020-12-02]

[14] Lesson 11: Response Surface Methods and Designs. [Internet]. Available from https://online. stat.psu.edu/stat503/book/export/html/681. [Accessed: 2020-12-02]

[15] Bradley N. The Response Surface Methodology [thesis]. South Bend, USA: Indiana University; 2007.

[16] Response surface designs. [Internet]. Available from https://www.itl.nist. gov/div898/handbook/pri/section3/pri336.htm. [Accessed: 2020-11-19]

[17] What is experimental design? [Internet]. Available from https://www.itl. nist.gov/div898/handbook/pri/section1/pri11.htm. [Accessed: 2020-12-13]

[18] DOE Contour Plot. [Internet]. Available from https://www.itl.nist.gov/div898/handbook/eda/section3/eda33a1.htm. [Accessed: 2020-12-18]

[19] Design. [Internet]. Available from https://www.theopeneducator.com/doe/Response-Surface-Methodology/

Design-Response-Surface-Methodology. [Accessed: 2020-12-18]

[20] Koç B, Kaymak-Ertekin F. Response Surface Methodology and Food Processing Applications. Ege University Food Faculty Articles. 2013;32(8):1-8.

[21] Aydar A Y. Utilization of Response Surface Methodology in Optimization of Extraction of Plant Materials. IntechOpen. 2018. http://dx.doi.org/10.5772/intechopen.73690

[22] Bozacı E. Investigation of Biosurfactant Usage in Raw Wool Scouring by Response Surface Methodology. Tekstil ve Konfeksiyon. 2017;27(4):382-392.

[23] Tetteh E K, Rathilal S. Application of Magnetized Nanomaterial for Textile Effluent Remediation Using Response Surface Methodology. Materials Today Proceedings. 2020. https://doi.org/10.1016/j.matpr.2020.03.827

[24] Kusuma H S, Mahfud M. Response Surface Methodology for Optimization Studies of Microwave-assisted Extraction of Sandalwood Oil. J Mater Environ Sci. 2016;7(6):1958-1971.

[25] Şengöz N G, Arslan P. Response Surface Designs in Quality Control: Yarn Irregularity Exercise. Tekstil ve Konfeksiyon. 2017;27(3):289-299.

[26] Textile Study Center Online Library for Textile Engineering. [Internet]. Available from https://textilestudycenter.com/properties-of-cotton-fibers/. [Accessed: 2021-01-02]

[27] Asian Textile Studies. [Internet]. Available from http://www.asiantextilestudies.com/cotton.html. [Accessed: 2021-01-02]

[28] Wool Fiber. [Internet]. Available from https://www.google.com/search?q=wool+fiber+longitudinal+view&tbm=isch&ved=2ahUKEwjSn5LGsZHuAhXVIRoKHSJiBNkQ2-cCegQIABAA&oq=wool+fiber&gws_lcp=CgNpbWcQARgCMgIIADICCAAyAggAMgIIADIECAAQHjIECAAQHjIECAAQHjIECAAQHjIECAAQHjIECAAQHjoECAAQQ1DXo0ZYnL1GYPfORmgAcAB4AIABlAGIAc8FkgEDMi40mAEAoAEBqgELZ3dzLXdpei1pbWfAAQE&sclient=img&ei=x_X6X5KgA9XDaKLEkcgN&bih=1297&biw=1621&rlz=1C1AOHY_trTR709TR710#imgrc=gRwApDApRfr1aM. [Accessed: 2021-01-06]

[29] Khan A, Abir N, Rakib M A N, Bhuiyan E, Howlader MR. A Review Paper on Textile Fiber Identification. IOSR Journal of Polymer and Textile Engineering. 2017;04:14-20. DOU:10.9790/019X-04021420

[30] Ring spinning technique of yarns. [Internet]. Available from https://www.google.com/search?q=breakage+of+yarn&rlz=1C1AOHY_trTR709TR710&tbm=isch&source=iu&ictx=1&fir=B0GZ5iqOyrR_nM%252C-NYiJqzV6b6emM%252C_&vet=1&usg=AI4_-kR2Mz0ipMl_iOhEAPBrLgXYsV_m0w&sa=X&ved=2ahUKEwj_qqPA65HuAhVPA2MBHa_3BZEQ9QF6BAgOEAE#imgrc=B0GZ5iqOyrR_nM&imgdii=fdPJ05-sMmsYxM. [Accessed: 2021-01-06]

[31] USTER. USTER®HVI 1000 The fiber classification and analysis system [Internet]. 2013. Available from https://www.uster.com/fileadmin/user_upload/customer/customer/Instruments-use_folder_PRODUCTS/Broschueren/en_USTER_HVI_1000_web_brochure.pdf [Accessed: 2021-01-07]

[32] USTER. USTER Tensorapid 3 Apparatus. [Internet]. Available from https://www.google.com/search?q=uster+tensorapid+3+manual&rlz=1C1AOHY_trTR709TR710&source=lnms&tbm=isch&sa=X&ved=2ahUKEwj_s8PNxpHuAhWutIsKHc-AAQQQ_AUoAXoECBUQAw&biw=1621&bih=1297. [Accessed: 2021-01-07]

[33] Yarn breaking strength testing. [Internet]. Available from https://www.google.com/search?q=yarn+breaking+strength+clamps&tbm=isch&ved=2ahUKEwi7jYu0vZHuAhUHM-wKHeGzB5IQ2-cCegQIABAA&oq=yarn+breaking+strength+clamps&gs_lcp=CgNpbWcQA1DTyQRYmOUEYKHqBGgBcAB4AIABdogBwgaSAQMyLjaYAQCgAQGqAQtnd3Mtd2l6LWltZ8ABAQ&sclient=img&ei=NgL7X7uaGYfmsAfh556QCQ&bih=1297&biw=1621&rlz=1C1AOHY_trTR709TR710. [Accessed: 2021-01-07]

[34] A yarn at break. [Internet]. Available from https://www.google.com/search?q=yarn+breaking&tbm=isch&ved=2ahUKEwju9YmS7JHuAhXEgqQKHSwKAC8Q2-cCegQIABAA&oq=yarn+break&gs_lcp=CgNpbWcQARgB MgQIABATMgQIABATMgQIABATMgQIABATUKyyAVjruQFg-98BaABwAHgAgAF1iAHZApIBAzAuM5gBAKABAaoBC2d3cy13aXotaW1nwAEB&sclient=img&ei=NzP7X67-IcSFkgWslID4Ag&rlz=1C1AOHY_trTR709TR710. [Accessed: 2021-01-08]

[35] Güvercin S, Yıldız A. Optimization of Cutting Parameters Using The Response Surface Method. YTU Sigma J Eng & Nat Sci.2018; 36 (1):113-121.

[36] Sağbaş A, Yılmaz F B, Altınışık F. RSM Tekniği Uygulanarak Derlin Malzemesinin Optimum Aşınma Değerinin Tahmin Edilmesi. In: Proceedings of the TİMAK – Tasarım İmalat Analiz Kongresi; 26-28 April 2006; Balıkesir. İstanbul: TİMAK; 2006. p. 170-176

[37] Arslan P. Interpretational Analysis for Yarn Quality Control [thesis]. Uşak: Uşak University; 2011. Supervisor: Asist.Prof.Dr.N.Gönül Şengöz.

Chapter 13

Application of Response Surface Methodology in Food Process Modeling and Optimization

Solomon Worku Kidane

Abstract

Modeling and optimization is an important task in food manufacturing. It enables one to understand and describe processes which in turn help establish quantified relationship between input and output variables. Modeling and optimization help to make informed decision on a process with the objective of improving efficiency and minimizing cost while maintaining quality. Response surface Methodology (RSM) has been employed in modeling and optimizing several food processing operations including baking, cooking, roasting, drying, extrusion, fermentation and many others. Moreover, RSM has been extensively used in product formulation and ingredient optimization. This chapter describes the application of RSM in food process modeling and optimization. The steps to be followed, the experimental designs that can be used and the interpretation of response surfaces developed are described. Moreover, selected application of RSM in food process modeling and optimization are reviewed and presented.

Keywords: Food process modeling, optimization, response surface

1. Introduction

Improving system performance without increasing cost of production and process time while maintaining the required quality attributes is the main objective of food processing and manufacturing. Finding the optimum processing condition and recipe (formulation) for food products of high quality and high marketability is paramount importance for successful product. The method used for coming up optimal processing condition and combination of ingredients with the best output is called optimization [1, 2]. Modeling precedes optimization and helps establish a quantitative relationship between independent and dependent/response variables. In the food industry, models are used for exactly the same purpose as in the scientific world. They help practitioners and scientists to think about processes that are too complicated to understand in every detail [3].

Modeling and optimization of processes including food processes has been done through focussing on the effect of changes in one parameter on a response keeping all other factors constant. This is called one-variable-at-a-time technique [4–6]. The major limitation of this method is that the interactive effects among the variables are not accounted for and there is a lack of explanation of the complete effect of the factors on the response or an overview of the variables' behavior within the entire experimental space. Moreover, one-variable-at-a-time method increases the number of experimental runs required to conduct the research, which eventually leads to increased cost and time to do the research [4, 5, 7]. In order to address this limitation, optimization studies should be conducted by applying procedures like response surface methodology (RSM) where multiple factors are considered at a time. RSM has been found to be an effective method for food product modeling and optimization [2, 5, 8].

RSM is a collection of statistical and mathematical techniques useful for developing, improving, and optimizing processes. It also has important applications in the design, development, and formulation of new products as well as in the improvement of existing product designs [2, 5, 8, 9]. The most extensive applications of RSM are in the industrial world, including the food industry, particularly in situations where multiple input variables potentially influence the quality characteristics of the product or the process. RSM has been extensively used in modeling and optimizing food processing operations and formulation of products. Major food process operations like drying, extrusion, fermentation, baking and cooking operations have been modeled and optimized using RSM [1–5]. Moreover, food product formulations and product design and development has been carried out using RSM [1, 2, 5]. Several experimental designs including factorial designs, central composite design with its variants, D-optimal design, and mixture designs have been used with RSM [5, 7].

Besides analyzing the effects of the independent variables, RSM generates an empirical model which describes the process under study. The term *Response Surface Methodology* was derived from the graphical view created after fitting the mathematical model [1, 2, 5, 7]. The objectives of this chapter were to present a brief historical and theoretical overview of RSM, describe its application in food process modeling and optimization and product formulation, highlight interpretation of response surfaces and graphical optimization techniques (overlay plots) and review previous works where RSM has been used. Moreover, in this chapter the steps to model and optimize food processes and formulations using RSM are presented and different experimental designs used in RSM are also described.

2. Theory and steps in carrying out RSM

Though RSM was developed in the 1950s, its application in food process operations began in 1960s [2, 7]. The RSM's major advantage is generating large amount of information from a reduced number of experimental runs that are required to evaluate multiple parameters and their interactions [6]. The relationship between the independent variables and the response can be represented by Eq. (1) [1]

$$y = f\left(x_1, x_2, x_3 \cdots x_n\right) + \varepsilon \tag{1}$$

where y is the response, f is the unknown function of response, x_1, x_2, ... x_n denote the independent variables and n is the number of independent variables, ε is error that represents other sources of variability which is not explained by the mathematical relationship (by the function, f).

The modeling and optimization procedure using RSM is normally carried out in stages. Though different steps or stages are reported in literature, all the steps outlined in different literature have similarities or commonalities. The steps in general include (1) identification of independent variables and their levels, (2) selection of the experimental design, (3) selection of a regression model and prediction and verification of model equation, (4) graphical presentation of the model equation and (5) prediction and determination of optimal operating conditions [1, 2, 6–8].

2.1 Selection/identification of key process or independent variables and their levels

Many factors often affect food manufacturing process, both recipe/ingredient related and process parameter related. The independent variables to be studied are selected based on experience, research results obtained from literature or preliminary experiments [5]. If there are too many variables involved, as is the case in most new food product development, screening procedure should be used to identify those that critically influence the responses of interest. Screening designs allow the researcher to look at the effects of several variables each of which takes on two levels with less number of runs [5, 7]. Those significant variables are then selected for further optimization. Screening designs like Plackett-Burman and Saturated fractional factorial designs are commonly used in food processing and formulation [2, 10]. Some specifically designed preliminary experiments are conducted using screening designs and they enable the food researcher to estimate the effect of each factor and to select the most significant and critical variables from the potential variables with minimum experimental efforts.

2.2 Selection of the experimental design

An important aspect in applying RSM is the design of experiments. After selection of the food quality attributes of interest (response) and identifying the significant independent variables, the next step of statistical food product design and development is to design an appropriate experiment. Some computer packages offer optimal designs based on the special criteria and input from the user. These designs differ from one an other with respect to their selection of experimental points, number of runs and blocks. The 3^n factorial, the central composite design (CCD), the Box–Behnken Design (BBD), the D-optimal designs (constrained designs) and mixture designs are commonly used in RSM [2, 5, 11, 12]. The following sections introduce the designs briefly.

2.2.1 Full factorial (3^n factorial design)

A 3^n factorial design is suitable for supporting the building of a quadratic model, if there are less than four significant variables ($n \leq 4$) selected for modeling in the food systems and chemical processes [6]. A 3^n experimental design supplies 3^n degrees of freedom, in which one is fixed for determining the total average value β_0 (constant term) in the model. The remaining (3^n - 1) degrees of freedom then allow estimation and calculation of the effects of each factor, the interactions between and among factors, and the curvature in the system. A 3^n factorial design is constructed by the combination of all the possible test levels of each variable. It can be divided into four subgroups: a 2^n factorial plan with 2^n trials, 2n central points of all the surfaces, border middle points and one central point (in practice, this should be repeatedly performed) (**Table 1** and **Figure 1**) [2, 9].

2.2.2 Central composite designs (CCD)

The Central Composite Designs (CCD) is the foundation of the RSM and is popularly used to estimate parameters of a full second-degree model in all scientific research areas [2, 5, 9]. One of the advantages of CCD is its efficiency with respect to the smaller number of runs required with each factor having 3 or 5 levels (**Table 1** and **Figure 2**) [13]. The other advantage of CCD is that it can be constructed in

Run no.	3^3 factorial design			Central composite design			Box–Behnken design		
	X_1	X_2	X_3	X_1	X_2	X_3	X_1	X_2	X_3
1	0	−1	0	−1	−1	1	0	1	−1
2	1	1	0	−1	−1	−1	0	−1	1
3	1	0	0	−1	1	−1	−1	0	1
4	1	−1	1	1	1	1	0	1	1
5	−1	1	0	1	−1	−1	1	0	−1
6	0	1	0	1	1	−1	−1	−1	0
7	1	1	1	1	−1	1	1	1	0
8	−1	−1	−1	−1	1	1	−1	0	−1
9	1	0	−1	0	0	$+\alpha$	0	−1	−1
10	0	0	−1	0	$-\alpha$	0	1	−1	0
11	0	0	1	$-\alpha$	0	0	−1	1	0
12	1	−1	0	$+\alpha$	0	0	1	0	1
13	−1	1	−1	0	$+\alpha$	0	0	0	0
14	1	1	−1	0	0	$-\alpha$			
15	0	1	−1	0	0	0			
16	0	1	1						
17	−1	−1	1						
18	1	−1	−1						
19	−1	0	−1						
20	0	−1	1						
21	−1	0	1						
22	−1	1	1						
23	−1	0	0						
24	−1	−1	0						
25	1	0	1						
26	0	−1	−1						
27	0	0	0						

Table 1.
Number of runs for designs with three factors used in RSM.

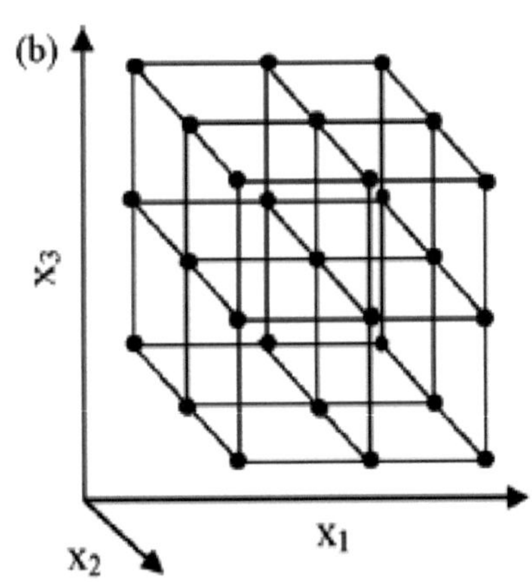

Figure 1.
Graphical presentation of 3^n factorial design (the dots are the design/experimental points) [6].

a sequential program of experimentation by building onto information gathered previously from a 2^n factorial design. If a linear model based on a 2^n factorial design turns out to be insignificant, then some extra trials can be designed, according to the principles of a CCD, to repair the model. All these data will be used to build a quadratic model. This is also known as the build-up principle of the CCD. Normally, a quadratic model would meet the needs for accuracy in practical product development and process modeling [6, 8, 9].

2.2.3 Box-Behnken design

The Box–Behnken Design (BBD) comprises a specific subset of the factorial combinations from the 3^n factorial design. These designs are formed by combining 2^n factorials with incomplete block designs [2]. The resulting designs are usually very efficient in terms of the number of required runs, and they are either rotatable or nearly rotatable [13]. In addition, in a BBD, the experimental points are situated on a hypersphere equally distant from the central point (**Figure 3**) (**Table 1**). Applying this design is popular in food processes due to its economical design. BBD is appropriate to evaluate interaction between factors and especially to study processes without extreme points (where high levels of factors involved in the process is difficult to implement) such as high temperature and pressure next to each other [2, 9]. Several studies employed BBD to optimize food process operations.

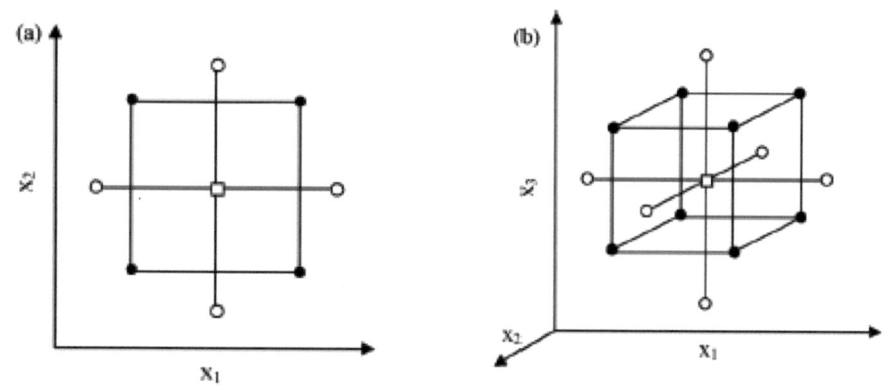

Figure 2.
Graphical presentation of central composite design for two factors (a) and three factors (b) [6] ((•)points of factorial design; (◦) axial points; (□) centre points).

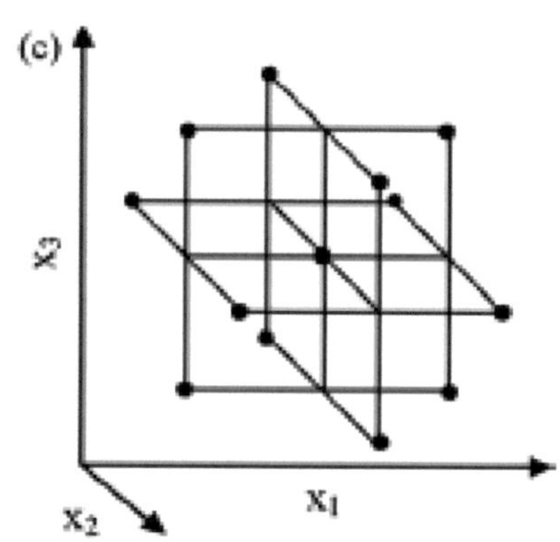

Figure 3.
Graphical presentation of the Box–Behnken design for three factors [6].

2.2.4 D-Optimal design (constrained designs)

The factorial designs are not always applicable for some food processes because of functional or technical restriction. At times there are combinations of some factor levels that are not practically possible to conduct the experiment. Examples of such are combinations of high levels of all the factors or low levels of all the factors in a given experiment. For example, in a roasting operation combining the highest temperature and the longest time may result in a product that is over roasted which is not fit for sensory evaluation whereas the high temperature can be combined with other shorter roasting times. Every trial under factorial design must be performed and the trial number increases rapidly beyond affordable limit when the number of factors increase. On the other hand, though CCD offers a smaller number of trials, it requires the exact setting of the test levels at the defined values and cannot be changed or is not flexible to handle constraints. D-optimal design was developed to overcome these shortcomings or exclude practically unsound scenarios [6, 13]. In D-optimal design, the test level of each variable can be selected flexibly and a variable can be tested at as many levels as the researcher wants. The number of levels of the different factors can be different or same. D-optimal designs are computer-generated. "D-optimal" means that these designs are selected from the list of valid candidate runs that provide as much orthogonality between the columns of the design matrix as possible. D-optimal designs have been used in optimizing food ingredients (D-optimal mixture designs) and process conditions [12, 14–16].

2.3 Selection of a regression model, prediction and model verification

Building a model is one of the most important steps in food process and product design. After the experiments have been conducted and the data collected, the intended model is fitted to the data by using regression analysis least square minimization technique. The two important criteria for selecting a usable and precise model from the alternative equations are: the model with the highest precision for accurate application and the model with the simplest form for easy application [5]. Polynomials have been used extensively in empirical modeling of chemical, biological, and food research systems. They provide a simple curvilinear relationship between a number of variables, possess a clearly defined optimum, and use simple computational algorithms by using the least square minimization method for estimation of the model coefficients in the model. Low-degree polynomials, such as a first-degree polynomial with interaction terms or a quadratic polynomial, are the most appropriate models to adequately describe food processes [1, 2, 5, 8]. The second order model can be written as follows [2, 5]:

$$Y = \beta_0 + \sum_{j=1}^{n} \beta_j X_j + \sum_{j<k=2}^{n} \beta_{jk} X_j X_k + \sum_{j=1}^{n} \beta_{jj} X_j^2 \quad (2)$$

where β_0, β_j, β_{jj} and β_{jk} are regression coefficients for intercept, linear, quadratic and interaction terms respectively and X_j, and X_k, are coded independent variables.

Following the estimation of regression coefficients, the estimated response could be estimated/predicted using the model equation. Moreover, one must check whether the model adequately describes the relationship between the dependent and the response variables, i.e., fits well to the experimental data. Several techniques could be used to check the adequacy of the developed model. These techniques include residual analysis, prediction error sum of squares (PRESS) residuals, and testing of the lack of fit. The overall predictive capability of the model is commonly explained by the coefficient of determination (R^2). However, R^2 alone is not a measure of the model's accuracy. R^2 indicates the percentage of variability in the response explained by the changes in the independent variables [9].

2.4 Graphical presentation of the model equation

Model building is not the only and ultimate objective of food process and product design. The interest of food product and process designers focus on the effect of different factors on the quality attributes. The question usually is which variables and in which ranges have significant effects on specific quality attributes or response variables. The other question could be under what conditions should food be processed to get a pre-defined quality attribute. Only a significant and precise model can supply reliable and essential information for the food researcher. Generally, two approaches are used to extract this information from the model: the graphical and numerical method. The predictive models are used to generate contours and response surfaces within the experimental range [13]. The response surface plot is the theoretical three-dimensional plot (3D surface) showing the relationship between the response and the independent variables (**Figures 4a** and **5a**). The two-dimensional display of the surface plot is called contour plot (**Figures 4b** and **5b**). In the contour plot, lines of constant response are drawn in the plane of the independent variables. It is a two-dimensional screen of the surface plot, in which, ranges of constant dependent variables is drawn in the plane of the independent variables. Indeed, the contour plots improve the researcher's understanding of the shape of a responses surface [1, 2, 9].

Proper interpretation of contour plots is an important part of the optimization exercise. When the contour plot displays ellipses or circles, the centre of the system refers to a point of maximum (**Figure 4**) or minimum (**Figure 5**). response. Sometimes, contour plot may display hyperbolic or parabolic system of the contours. In this case, the stationary point is called a saddle point and it is neither a maximum nor a minimum point (**Figure 6**). These plots give useful information about the model fitted but they may not represent the true behavior of the system. It is important to keep in mind that the contours or the 3D surfaces represent contours or surfaces of estimated response and the general nature of the system that arises as a result of a fitted model, not the true structure [1, 5, 8].

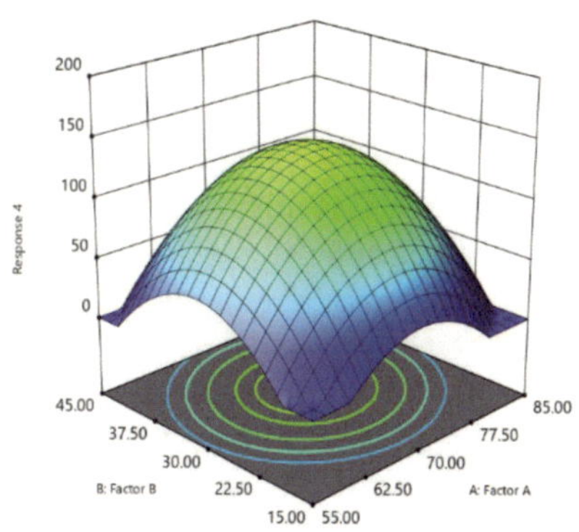

(a)

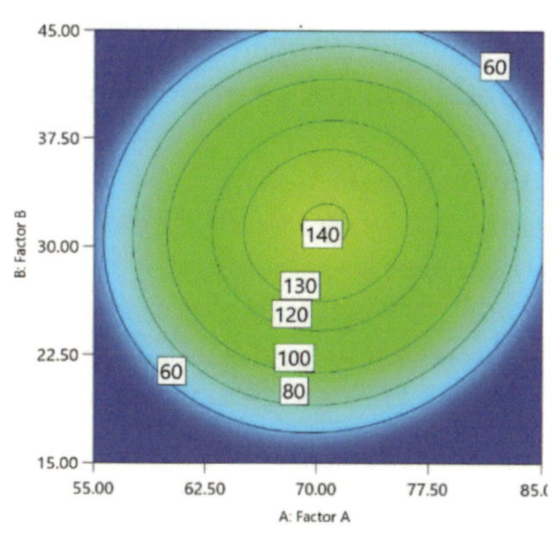

(b)

Figure 4.
Graphical presentation of 3D surface (a) and contour plot (b) where there is a maximum response.

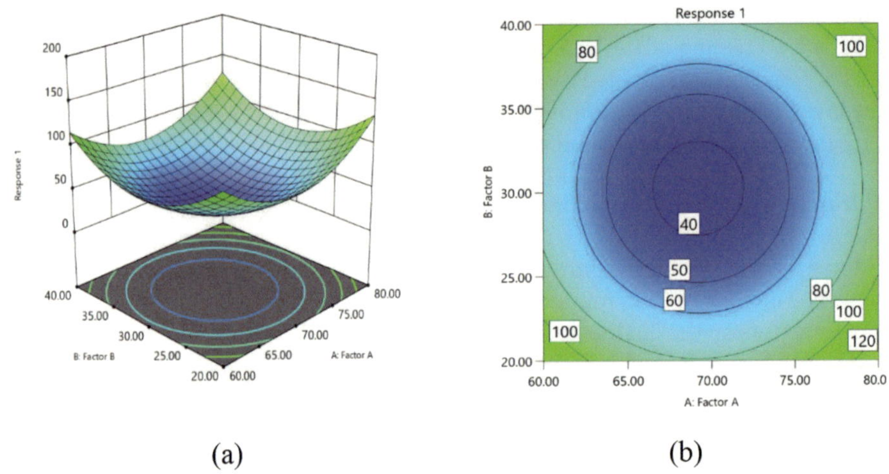

Figure 5.
Graphical presentation of 3D surface (a) and contour plot (b) where there is a minimum response.

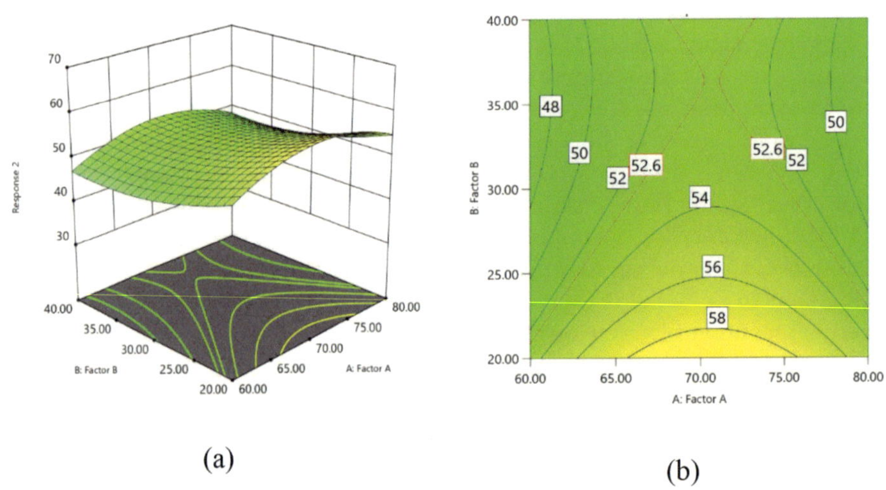

Figure 6.
Graphical presentation of 3D surface (a) and contour plot (b) where there is a saddle point point (no maximum and no minimum).

2.5 Prediction and determination of optimal operating conditions

Prediction of food quality attributes enables the researcher to estimate the response variable given the independent variables in the experimental region where no trials have been conducted. Prediction also helps in calculating the possible independent variables for a given response value. Apart from prediction, researchers are also interested in optimization which is an important step in statistical food process and product design. Optimization gives more detailed information about the level combinations of the independent variables that will result in optimum food quality attributes. This information from the optimization is reliable only if the model built is significant and adequately describes the relationship between the independent and the response variables.

In food and beverages, the researcher must often deal with multiple quality attributes (physicochemical properties and sensory attributes) as desirable responses. There are several aspects that complicate the process of choosing a best alternative when considering multiple attributes to the decision-making. There are almost no perfect practical decisions where it is possible to get the optimal result for each response or criterion in a single choice. Therefore, for most situations, it is necessary to make trade-offs between the different objectives among the quality attributes. As a result, optimizing based on multiple objectives should provide mechanisms for incorporating the experimenter's priorities and preferences [9]. An optimum product may be achieved with different combinations of levels of the variables. The optimal levels of the independent variables that give the 'best' product can be determined using numerical and graphical techniques [5, 13].

2.5.1 Numerical optimization

The numerical method is most universal optimization approach. Though it cannot show overall (visual) information about the system, it performs complicated mathematical optimizations and gives specific combinations of levels of the independent variables that gives the best result. The minimum or maximum point of a second order equation is the point where the first derivative of the function is equated to zero [1, 2]:

$$\text{If } y = f(x_1, x_2) = \beta_0 + \beta_1 x_1 + \beta_2 x_2 + \beta_{11} x_1^2 + \beta_{22} x_2^2 + \beta_{12} x_1 x_2 \quad (3)$$

The maximum or minimum point is found by equating the partial derivative of the polynomial equation (Eq. (3)) with respect to the independent variables as shown in Eqs. (4) and (5) [1, 2]:

$$\frac{\partial y}{\partial x_1} = \beta_1 + 2\beta_{11} x_1 + \beta_{12} x_2 = 0 \quad (4)$$

$$\frac{\partial y}{\partial x_2} = \beta_2 + 2\beta_{22} x_2 + \beta_{12} x_1 = 0 \quad (5)$$

The partial derivatives equated to zero are solved to find the values of x_1 and x_2. The values of x_1 and x_2 determined are the coded values of the independent variables that give the maximum or minimum value of the response.

Food processors and developers usually are interested in optimization of multiple responses simultaneously. A common approach to optimize multiple responses is to use simultaneous optimization technique which makes use of desirability function. The desirability function approach is one of the most widely used methods in industry for the optimization of multiple-response processes. The general approach is first to convert each response y_i into desirability function d_i that varies over the range 0 to 1 [13]. If the objective or target T for the response Y is a maximum value, the individual desirability functions are structured as (Eq. (6)) [9, 13]:

$$d = \begin{cases} 0 & Y < L \\ \left(\frac{Y - L}{T - L}\right)^r & L \le Y \le U \\ 1 & Y > T \end{cases} \quad (6)$$

If the target T for response y is minimum the individual desirability functions are structured as (Eq. (7)):

$$d = \begin{cases} 1 & Y < T \\ \left(\dfrac{U-Y}{U-T}\right)^r & T \le Y \le U \\ 0 & Y > U \end{cases} \tag{7}$$

where r is weight, L is lower value and U is upper value. Then, for *m* responses, the design variables are chosen to maximize the overall desirability D (Eq. (8)) [9, 13]

$$D = \left(d_1 \times d_2 \times d_3 \cdots d_m\right)^{\frac{1}{m}} \tag{8}$$

2.5.2 Graphical optimization

Graphical optimization is preferred when the process variables are few. In graphical optimization the contour plots for each response are superimposed (overlayed) to obtain an overlay plot [9, 13]. **Figure 7** shows an overlay plot for the two responses plots (contour plots). These are contours for which desired values for response 3 ranges from 65 to 68 units and the desired values for response 4 ranges from 98 to 110 units. These ranges of the two responses were judged to be acceptable. If these ranges represent important attributes that must be met by the process, the shaded portion of the overlay plot (**Figure 7**) indicates that there are a number of combinations of factor A and factor B that result in a satisfactory process and a food product that meets the targeted objectives. The experimenter has the opportunity to visually examine the overlay plot to determine appropriate operating conditions, and select a region that is most desirable given other practical considerations are feasible. According to the overlay plot (**Figure 7**) ranges of factor A and B that give best results are from about 61.5 to 69 and from 22 to 25 units, respectively.

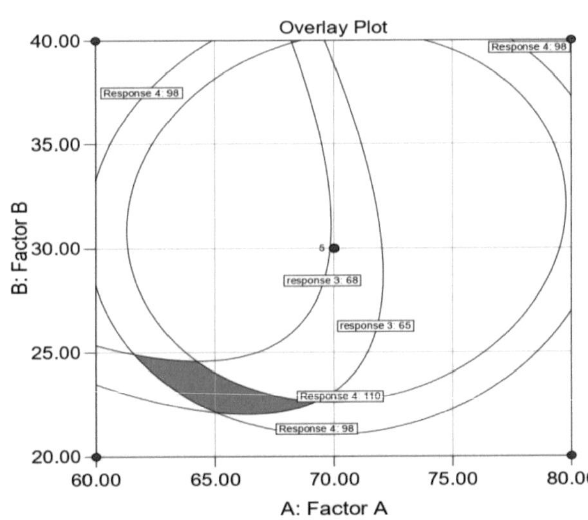

Figure 7.
Overaly plot of two responses (response 3 and 4).

When there are more than three independent variables, overlaying (superimposing) contour plots becomes difficult because the contour plot is two dimensional, and n - 2 of the independent variables have to be held constant to construct the overlay plot. Often a lot of trial and error is required to determine which factors to hold constant and what levels to select to obtain the best view of the surface [9].

3. Case study of roasting process

3.1 The process and variables

In coffee roasting operation roasting time and temperature are critical parameters in terms of affecting the quality of roasted beans and the quality of the brewed coffee. In this case study roasting time ranging from 20 min to 40 min and roasting temperature ranging from 160–200°C were used. An experiment was designed using Design-expert (Version 13). Central Composite Design was used with the levels (low, middle, upper and star levels) as indicated in the table below (**Table 2**) and a total of 13 runs were generated. The response variables studied are acceptability tests in terms of color, aroma, flavor, taste and overall acceptability measured using a 9-point hedonic scale ranging from "*1 = Dislike extremely*" and "*9 = like extremely*".

3.2 Data analysis and interpretation of response surface

Polynomial equations were fitted to the data and response surfaces were generated for each response variable as presented in the **Figure 8a–e**. All the sensory attributes increase with increase in roasting time and temperature followed by a decrease in the sensory attributes of the brewed coffee as roasting time and temperature increased further. Such responses are naturally expected because short roasting time and low roasting temperature result in under-roasted beans which influence the acceptability scores negatively. Similarly, longer roasting time combined with high roasting temperature may result in over-roasted beans which definitely affect the acceptability score of brewed coffee adversely [17].

3.3 Graphical and numerical optimization

Graphical optimization gives an overview or range of operating conditions which results in all the response variables to be within the desired value. To determine a region for optimal roasting time and temperature aimed at obtaining an acceptable product in terms of color, flavor, aroma, taste, and overall acceptability, the contour plots of the five responses are superimposed to come up with an overlay plot (**Figure 8f**). This optimum region provides the coordinates of possible optimal levels of roasting time and temperature. The criteria for the optimal region were test score between 7 (*like moderately*) to 9 (*like extremely*) for each attribute. Thus, roasting temperature ranging from 169–188°C and roasting time of 27 min to 36 min could be used to obtain an acceptable brewed coffee. Using the criteria

Factors	Coded levels				
	−1.414(−α)	**−1**	**0**	**1**	**+1.414(+α)**
Temperature (°C)	151.7	160	180	200	208.28
Time (min)	15.87	20	30	40	44.14

Table 2.
Actual and coded values of factor levels.

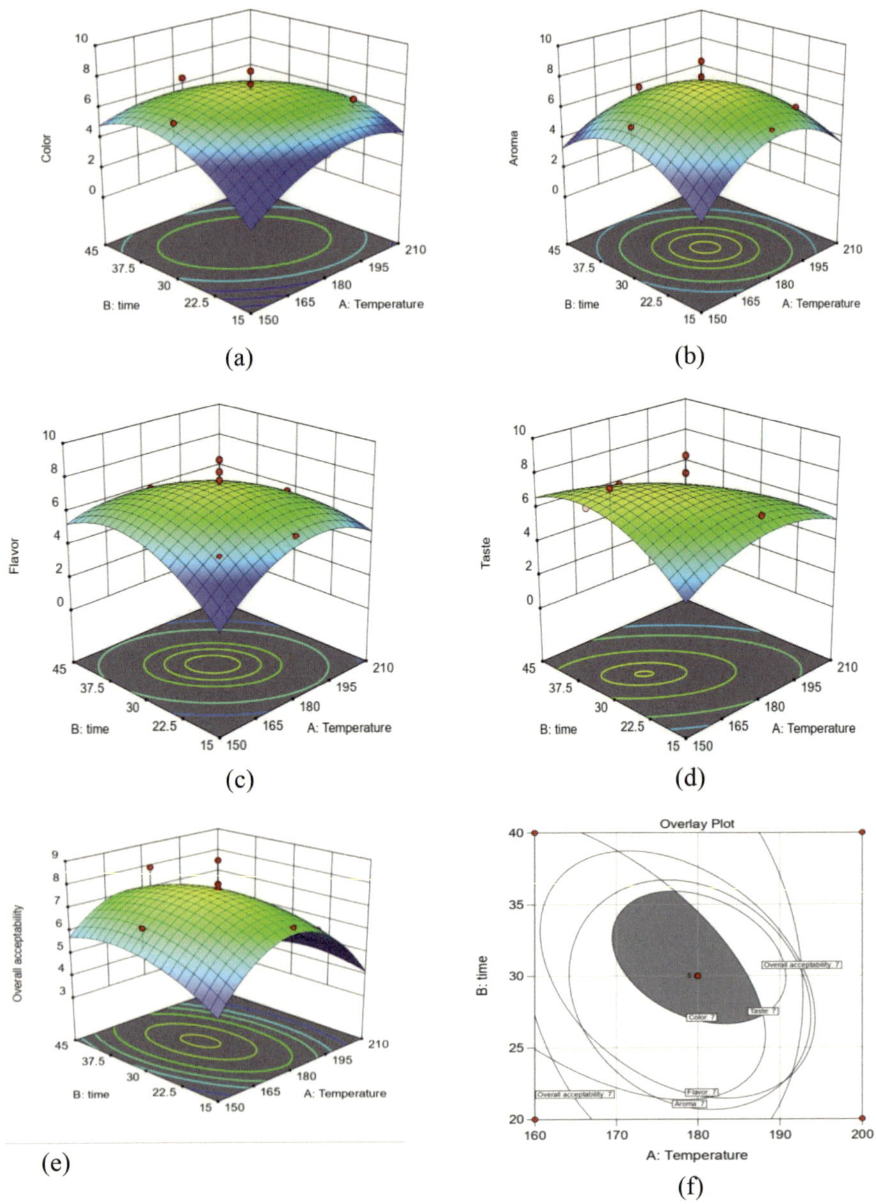

Figure 8.
Response surface of the different quality atributes (a–e) of brewed coffee as a function of coffee bean roasting time and temperature and (f) overlay plot.

of maximizing all the sensory attributes, best results were obtained using roasting time of 30 min and roasting temperature of 177°C with a desirability value of 0.694.

4. Selected applications of RSM in the food process modeling and optimization

4.1 Application RSM in optimizing baked products

Baking is a method of preparing food that uses dry heat, typically in an oven. Several studies have been conducted to optimize the baking conditions including baking temperature and time and the ingredients used to come up an acceptable product. Some of the studies include the effect of inulin on textural and sensory characteristics of sorghum based high fiber biscuits using response surface methodology [18], effect of different ingredients on the mixing and fermentation times required [19], the effect of the interaction of red rice flour and the microbial transglutaminase enzyme in the production of prebiotic gluten-free breads [20], the effect of hydroxypropylmethyl cellulose, yeast β-glucan, and whey protein levels [21], the effect of whole oat flour, maltodextrin and isomalt on textural and sensory characteristics of biscuits, optimization of composite flour biscuits [22], gluten-free bread fortified with soy flour and dry milk [23], enzymatic treatment using RSM on the quality of bread flour [24] and many others have been reported in literature. These and other studies on baking process used RSM to model and optimize baking processes. Full factorial designs, CCD, BBD were among the experimental designs used.

4.2 Application of RSM in optimization of cooking and roasting parameters

Thermal processes like cooking and roasting are commonly used food processing unit operations. Critically important in these operations is finding the optimal combinations of the operating conditions. RSM has been extensively and successfully used to optimize the process parameters. Some of such studies are optimization of cooking protocol for rice bean to improve the efficiency of conventional process [25], of high-pressure processing of black tiger shrimp (Penaeus monodon) [26], process optimization for high-pressure processing of black tiger shrimp (Penaeus monodon) [27], development of sensory acceptable, low-salt, shelf-stable frankfurters [28], optimization of the effect of frying temperature, and frying time on some physicochemical, textural, and sensory properties of wheat chips [29], optimization of initial water content, saturated steam pressure and processing time for roasted chick pea [30], optimization of microwave roasting of peanuts [31], optimization of leavened dough sunflower oil frying process conditions [32], optimization of roasting time and temperature of coffee beans [17] have been studied are reported. RSM experimental designs and numerical and graphical optimization were used in the studies.

4.3 Application of RSM in formulation of products

One important area in in food processing is development of new formulation for products using various ingredients. Deciding on the relative proportion of the ingredients calls for application of scientific procedure or experimental design. RSM has been effectively used to optimize ingredients or raw materials. Such studies include wheat dextrin yoghurt formulation [33], fat-reduced ice cream formulation employing inulin as fat replacer [34], the effect of Homogenized Infant Foods [35], optimization of honey, vinegar and tomato powder to make sweet and sour chicken meat spread [36], optimization of inulin, cocoa powder, and sucrose to develop a dessert made with soymilk [37], optimization of a stable palatable oil-in-water emulsion made with soy protein and red guava juice [38] and optimisation of soy protein and pink guava juice to develop soy-based desserts [39].

Special designs for formulation studies called mixture designs have been widely used together with RSM to deal with food formulation related problems. Some studies reported are optimization of pasteurized milk with soymilk powder and mulberry leaf tea using user defined mixture design [40], optimization of diverse chloride salts on the growth parameters of *L. pentosus* using D-optimal mixture design [12], optimization of natural fermentative medium for selenium-enriched yeast by D-optimal mixture design [14], optimization of wheat, sprouted mung bean and sorghum composite flour bread using D-optimal mixture design [16], optimization of blending ration of three different fruits in jam making using augmented simplex mixture design [15], optimization of sugar, peanut and chocolate using constrained mixture design to develop chocolate peanut spread [41] and formulation of yoghurt using augmented simplex-centroid mixture design [42], were reported in literature.

4.4 Application in RSM in drying, extrusion and fermentation processes

Some common operations in the food industry like extrusion, drying, fermentation etc. have been successfully modeled and optimized. Development of functional yoghurt via soluble fiber fortification [43], optimization of ripening temperature, ripening time, the level of rennet on the quality of cheese [44], bleaching condition on soyabean oil [45] has been performed using RSM. The effect of stevia and inulin on physicochemical and rheological properties of mango nectar [46], optimization of high-pressure process to extend shelf life of apple juice [47], Optimizing the thermal assisted high-pressure process parameters for a sugarcane based mixed beverage [48], optimization of ultrasonication parameters on chemical and microbiological properties of sour cherry juice [49] has be done using RSM,

Some examples of optimization fermentation process using RSM are the effect of fermentation conditions on the phytochemical composition, sensory qualities and antioxidant activity of green tea infusion [50], the effect of steaming time (20–50 min) and fermentation time (12–48 hr) [51]. Drying operations have also been modeled and optimized using RSM which included optimization of drying conditions on the quality of fruit cubes [52], the effect of spray drying condition on the quality of apricot juice powder [53] the effect of hot air and microwave drying condition [54].

Extrusion is a versatile operation in the production of wide range of extruded products and extraction. Extrusion conditions like temperature, feed moisture content and screw speed are the dominate parameters and are extensively studies. Studies in extrusion using RSM included optimization extrusion conditions of soybean flour and achda [55], antinutritional factors and protein and starch digestibility of lentil splits [56] optimization of carrot pomace powder [57], and physical and functional properties of extruded snack foods [58]. The effect of banana peel and rice bran extrusion for value addition has also reported [59]. Optimization of extraction conditions for extraction of olive oil [60], flavonoids from shallot skin [61], and phenolic compounds from fruits [62] using RSM have also been reported.

5. Conclusion

Response surface methodology has been extensively and effectively used to model and optimize food processes. It is important to follow the steps and use proper experimental designs in order to obtain valid results. Modeling and optimization of both processing conditions and of ingredients in food formulations has been done widely by applying RSM. The advancement in statistical packages to design experiments and analyze data has contributed immensely in statistical and computer-aided food product design.

Author details

Solomon Worku Kidane
Department of Food and Nutrition Sciences, Faculty of Consumer Sciences, University of Eswatini, Eswatini

*Address all correspondence to: wsolomon@uniswa.sz; solowkj@yahoo.com

References

[1] Baş D, Ismail H. Boyacýı İH. Modeling and optimization I: Usability of response surface methodology. Journal of Food Engineering. 2007: 78; 836-845

[2] Yolmeh M, Jafari SM, Applications of response surface methodology in the food Industry processes. Food Bioprocess Technology. 2017; 10: 413-433.

[3] Hofmeester JJM. The use of models in process development the case of fermentation processes. In: Tijskens LMM, Hertog MLATM and Nicolaï BM, editors. Food process modelling: Woodhead Publishing Limited; 2001. p275-287.

[4] Czitrom V. One-factor-at-a-time versus designed experiments. The American Statistician. 1999; 53 (2): 126-131.

[5] Ruguo HU. Food product design: A computer-aided statistical approach. Netsle R&D center, Inc.;1999. 225p.

[6] Bezerraa MA, Santelli ER, Oliveiraa EP, Villar LS, Escaleiraa, LA. Response surface methodology (RSM) as a tool for optimization in analytical chemistry. 2008; 76: 965-977.

[7] Nugent A. Response surface methodology and consumer-driven product Optimization. In. Beckley H, Herzog LJ, Foley MM, editors, 2nd ed. Accelerating new food product design and development. Wiley, 2017. p323-364

[8] Nwabueze TU. Basic steps in adapting response surface methodology as mathematical modelling for bioprocess optimisation in the food systems. International Journal of Food Science and Technology. 2010; 45: 1768-1776.

[9] Myers RH, Montgomery DC, Anderson-Cook CM. Response surface Methodology: Process and product optimization using designed experiments. Wiley; 2016; 825p

[10] Modia VK, Prakash M. Quick and reliable screening of compatible ingredients for the formulation of extended meat cubes using Plackett–Burman design. Lwt. 2008; 41:878-882.

[11] Dingstad G, Egelandsdal B, Næs T. Modelling methods for crossed mixture experiments—a case study from sausage production. Chemometrics and Intelligent Laboratory Systems. 2003; 66: 175– 190.

[12] Arroyo-López FN, Bautista-Gallego J, Chiesa A, Durán-Quintana MC, Garrido-Fernández A. Use of a D-optimal mixture design to estimate the effects of diverse chloride salts on the growth parameters of Lactobacillus pentosus. Food Microbiology. 2009; 26: 396-403.

[13] Montgomery, DC. Design and Analysis of Experiments. 9th ed. Wiley; 2017. 734p

[14] Yin H, Chen Z, Zhenxin Gu Z, Han Y. Optimization of natural fermentative medium for selenium-enriched yeast by D-optimal mixture design. LWT - Food Science and Technology. 2009; 42: 327-331.

[15] Dlamini N, Solomon WK.. Optimization of blending ratios of jam from Swazi indigenous fruits; tincozi (Syzygium cordatum), tineyi (Phyllogeiton zeyheri) and umfomfo (Cephalanthus natalensis oliv.) using mixture design. Cogent Food Agric. 2019;5(1):1-16.

[16] Manyatsi N, Solomon WK, Shelembe JS. Optimization of blending ratios of wheat-maize-sprouted mungbean (Vigna radiata L.) composite

flour bread using D-optimal mixture design. Cogent Food Agric. 2020;6(1):1-19

[17] Anisa A, Solomon WK, Solomon A. Optimization of roasting time and temperature for brewed hararghe coffee (*Coffea Arabica* L.) using central composite design. International Food Research Journal. 2017; 24(6): 2285-2294.

[18] Banerjee C, Singh R, Jha A, Mitra J. Effect of inulin on textural and sensory characteristics of sorghum based high fibre biscuits using response surface methodology. Journal of Food Science and Technology, 2014; 51(10): 2762-2768.

[19] Calderón-Domínguez G, Farrera-Rebollo R, Arana-Errasquín R, Mora-Escobedo R. The effect of varying the mixing formula on the quality of a yeast sweet bread and also on the process conditions,as studied by surface response methodology. International Journal of Food Science and Technology. 2005; 40: 157-164.

[20] Gusmă T A, Gusmão RP, Moura HV, Silva HA, Cavalcanti-Mata ME, Duarte ME. Production of prebiotic gluten-free bread with red rice flour and different microbial transglutaminase concentrations: modeling, sensory and multivariate data analysis. Journal of Food Science and Technology. 2019; 56(6): 2949-2958.

[21] Kittisuban P, Ritthiruangdej P, Suphantharika P. Optimization of hydroxypropylmethylcellulose, yeast b-glucan, and whey protein levels based on physical properties of gluten-free rice bread using response surface methodology. LWT - Food Science and Technology. 2014;57: 738-748.

[22] Okpala LC, Okoli EC. Optimization of composite flour biscuits by mixture response surface methodology. Food Science and Technology International. 2012; 19(4): 343-350.

[23] Sanchez HD, Osella CA, Torre MA. Use of Response surface methodology to optimize gluten-free bread bortified with soy flour and dry milk. Food Science amd Technology International. 2004; 10(1): 5-9.

[24] Shafisoltani M, Salehifar M, Hashemi M. Effects of enzymatic treatment using Response Surface Methodology on the quality of bread flour. Food Chemistry. 2014; 148: 176-183.

[25] Bepary RH, Wadikar DD. Optimization of rice bean cooking parameters for the production of instant/convenience foods using response surface methodology. Journal of Food Processing and Preservation. 2018; 1-11.

[26] Kaur BP, Rao PS. Process optimization for high-pressure processing of black tiger shrimp (Penaeus monodon) using response surface methodology. Food Science and Technology International. 2016; 23(3): 197-208.

[27] O'Neilla CM, Cruz-Romeroa MC, Duffyb G, Kerrya JP. The application of response surface methodology for the development of sensory accepted low-salt cooked ham using high pressure processing and a mix of organic acids. Innovative Food Science and Emerging Technologies. 2018; 45: 401-411.

[28] O'Neill CM, Cruz-Romero MC, DuffyG, Kerry JP. The application of response surface methodology for development of sensory acceptable, low-salt, shelf-stable frankfurters using highpressure processing and a mix of organic acids. European Food Research and Technology. 2019; 245: 1277-1291.

[29] Kayacier A, Yüksel F, Karaman S. Response surface methodology study for optimization of effects of fiber level, frying temperature, and frying time on some physicochemical, textural, and

sensory properties of wheat chips enriched with apple fiber. Food Bioprocess Technol, 2014; 7:133-147.

[30] Mrad R, Assy P, Maroun RG, Louka N. Multiple optimization of polyphenols content, texture and color of roasted chickpea pre-treated by IVDV using response surface methodology. LWT - Food Science and Technology.2015; 62: 532-540

[31] Raigar RK, Upadhyay R, Mishra HN. Optimization of microwave roasting of peanuts and evaluation of its physicochemical and sensory attributes. Journal of Food Science and Technology. 2017; 54(7):2145-2155.

[32] Turan S, Keskin S, Solak R. Determination of the changes in sunflower oil during frying of leavened doughs using response surface methodology. Journal of Food Science and Technology.2021; https://doi.org/10.1007/s13197-021-04980-2

[33] Peerkhan N, Nai, S. Optimization of wheat dextrin yogurt formulation using response surface methodology. Journal of Food Science and Technology. 2020; doi:doi.org/10.1007/s13197-020-04683-0

[34] Pintor A, Severiano-Pérez P, Totosaus A. Optimization of fat-reduced ice cream formulation employing inulin as fat replacer via response surface methodology. Food Science and Technology International.2013; 20(7): 489-500.

[35] Martinez B, Rincon F, Abellan P. Improving the nutritive value of homogenized infant foods using response surface methodology. Journal of Food Science. 2004; 69(1): SNQ38-SNQ43.

[36] Arya A, Mendiratta SK, Singh TP, Agarwal R, Bharti SK. Development of sweet and sour chicken meat spread based on sensory attributes: process optimization using response surface methodology. Journal of Food Science Technology. 2017; 54(13):4220-4228.

[37] Chattopadhyay S, Raychaudhuri U, Chakraborty U. Optimization of soy dessert on sensory, color, and rheological parameters using response surface methodology. Food Science and Biotechnology. 2013; 22(1): 47-54.

[38] Granato D, Ribeiro JCB, Castro IA, Masson ML. Sensory evaluation and physicochemical optimisation of soy-based desserts using response surface methodology. Food Chemistry. 2010; 121: 899-906

[39] Granato D, Castro IA, Ellendersen LSV, Luciamasson M. Physical Stability Assessment and sensory optimization of a dairy-free emulsion using response surface methodology. Journal of Food Science, 2010; 75(2): S149-S155.

[40] Sangsopha J, Moongngarm A, Johns N P, Grigg PN . Optimization of pasteurized milk with soymilk powder and mulberry leaf tea based on melatonin, bioactive compounds and antioxidant activity using response surface methodology. Heliyon. 2019; 5: e02939.

[41] Chu, CA, Resurreccion AVA. Sensory profiling and characterization of chocolate peanut spread using response surface methodology. Journal of Sensory Studies. 2005; 20: 243-274.

[42] Karnopp AR, Oliveira KG, Andrade EF, Postingher BM, Granato D. Optimization of an organic yogurt based on sensorial, nutritional, and functional perspectives. Food Chemistry, 2017; 233: 401-411.

[43] Mudgil, D., Barak, S., & Khatkar, B. S. Development of functional yoghurt via soluble fiber fortification utilizing enzymatically hydrolyzed guar gum. Food Bioscience. 2016; 14: 28-33.

[44] Alizadeh M, Hamedi M, Khosroshahi A. Optimizing sensorial suality of Iranian white brine cheese using response surface methodology. Journal of Food Science. 2005; 70(4): S299-S303.

[45] Ortega-García J, Medina-Juárez LA, Gámez-Meza N, Noriega-Rodriguez J A. Optimisation of bleaching conditions for soybean oil using. Food Science and Technology International. 2005; 11(6): 443-449.

[46] Alizadeh M, Hamedi M, Khosroshahi A. Optimizing sensorial suality of Iranian white brine cheese using response surface methodology. Journal of Food Science. 2005; 70(4): S299-S303.

[47] Juarez-Enriquez E, Salmeron-Ochoa I, Gutierrez-Mendez N, Ramaswamy HS, Ortega-Rivas E. Shelf life studies on apple juice pasteurised by ultrahigh hydrostatic pressure. LWT - Food Science and Technology. 2015; 62: 915-919

[48] Raj AS, Chakraborty S, Rao PS. Optimizing the thermal assisted high-pressure process parameters for a sugarcane based mixed beverage using response surface methodology. Journal of Food Process Engineering. DOI:10.1111/jfpe.13274

[49] Tűrken T, Erge HS. Effect of ultrasound on some chemical and microbiological properties of sour cherry juice by response surface methodology. Food Science and Technology International. 2017; 23(6): 540-549.

[50] Liu Y, Luo L, Liao C, Chen L, Wang J, Zeng L. Effects of brewing conditions on the phytochemical composition, sensory qualities and antioxidant activity of green tea infusion: A study using response surface methodology. Food Chemistry, 2018;, 269: 24-34.

[51] Pradhananga M. Effect of processing and soybean cultivar on natto quality using response surface methodology. Food Science and Nutrition. 2019; 7: 173-182.

[52] Diamante L M, Yamaguchi YResponse surface methodology optimization of dried apple-blackcurrant cubes. Journal of Food Processing and Preservation. 2013; 37: 1084-1093.

[53] Kim JH, Kim JH, Eun JB. Optimization of spray drying process of Japanese apricot (Prunusmume Sieb. et Zucc.) juice powder using nondigestiblemaltodextrin by response surface methodology (RSM). Journal of Food Science and Technology. 2021; doi:doi.org/10.1007/s13197-020-04917-1

[54] Tian Y, Zhang Y, Zeng S, Zheng Y, Chen F, Guo, Z., . . . Zheng B. Optimization of microwave vacuum drying of lotus (*Nelumbo nucifera* Gaertn.) seeds by response surface methodology. Food Science and Technology International. 2021; 18(5): 477-488.

[55] Anuonye J C, Inyang CU, Chinma CE, James SY. Sensory Properties of Extruded Blends of 'Acha' and Soybean Flour –A Response Surface Analysis. Nigerian Food Journal. 2012; 30(1): 101 – 108.

[56] Rathod RP, Annapure US. Effect of extrusion process on antinutritional factors and protein and starch digestibility of lentil splits. LWT - Food Science and Technology. 2016; 66: 114-123

[57] Upadhyay A, Sharma HK, Sarkar BC.. Optimization of carrot pomace powder incorporation on extruded product quality by response surface methodology. Journal of Food Quality. 2010; 33: 350-369.

[58] Yağcı S, Göğűs F. Response surface methodology for evaluation of physical

and functional properties of extruded snack foods developed from food-by-products. Journal of Food Engineering. 2008; 86: 122-132.

[59] Jan K, Riar CS, Saxena DC. Value addition to food industry by-products and wastes (deoiled rice bran and banana peel) by optimizing pellets' formulation using response surface methodology: characterisation and classification by PCA approach. Journal of Food Processing and Preservation. 2017; 41: 1-13.

[60] Meziane S. Optimization of oil extraction from olive pomace using response surface methodology. Food Science and Technology International. 2012; 19(4): 315-322.

[61] Thu HN, Minh KN, Thi TL, Van PN. Optimization of extraction of flavonoids from shallot skin using response surface methodology based on multiple linear regression and artificial neural network and evaluation of its xanthine oxidase inhibitory activity. Journal of Food Measurement and Characterization. 2021; doi:doi.org/10.1007/s11694-021-00811-2

[62] Rohilla S, Mahanta CL. Optimization of extraction conditions for ultrasound-assisted extraction of phenolic compounds from tamarillo fruit (*Solanum betaceum*) using response surface methodology. Journal of Food Measurement and Characterization. 2021; doi:doi.org/10.1007/s11694-020-00751-3.